Bayesian Modeling of Spatio-Temporal Data with R

CHAPMAN & HALL/CRC Interdisciplinary Statistics Series

Series editors: B.J.T. Morgan, C.K. Wikle, P.G.M. van der Heijden

Recently Published Titles

Modern Directional Statistics
C. Ley and T. Verdebout

Survival Analysis with Interval-Censored Data: A Practical Approach with Examples in R, Sas, and Bugs
K. Bogaerts, A. Komarek, and E. Lesaffre

Statistical Methods in Psychiatry and Related Field: Longitudinal, Clustered and Other Repeat Measures Data
R. Gueorguieva

Bayesian Disease Mapping: Hierarchical Modeling in Spatial Epidemiology, Third Edition
A.B. Lawson

Flexbile Imputation of Missing Data, Second Edition
S. van Buuren

Compositional Data Analysis in Practice
M. Greenacre

Applied Directional Statistics: Modern Methods and Case Studies
C. Ley and T. Verdebout

Design of Experiments for Generalized Linear Models
K.G. Russell

Model-Based Geostatistics for Global Public Health: Methods and Applications
P.J. Diggle and E. Giorgi

Statistical and Econometric Methods for Transportation Data Analysis, Third Edition
S. Washington, M.G. Karlaftis, F. Mannering and P. Anastasopoulos

Parameter Redundancy and Identifiability
D. Cole

Mendelian Randomization: Methods for Causal Inference Using Genetic Variants, Second Edition
S. Burgess, S. G. Thompson

Bayesian Modeling of Spatio-Temporal Data with R
Sujit K. Sahu

For more information about this series, please visit: https://www.crcpress.com/Chapman--HallCRC-Interdisciplinary-Statistics/book-series/CHINTSTASER

Bayesian Modeling of Spatio-Temporal Data with R

Sujit K. Sahu

CRC Press
Taylor & Francis Group
Boca Raton London New York

CRC Press is an imprint of the
Taylor & Francis Group, an **informa** business

A CHAPMAN & HALL BOOK

First edition published 2022
by CRC Press
6000 Broken Sound Parkway NW, Suite 300, Boca Raton, FL 33487-2742

and by CRC Press
4 Park Square, Milton Park, Abingdon, Oxon, OX14 4RN

ISBN: 978-0-367-27798-7 (hbk)
ISBN: 978-1-032-20957-9 (pbk)
ISBN: 978-0-429-31844-3 (ebk)

DOI: 10.1201/9780429318443

Publisher's note: This book has been prepared from camera-ready copy provided by the authors

*To my family, parents, teachers and
all those who lost their lives during the 2020 pandemic.*

Contents

Introduction

My motivation for writing this textbook comes from my own practical experience of delivering numerous courses on Bayesian and spatio-temporal data modeling in the University of Southampton and in many overseas institutions in Australia, Italy, and Spain. During these courses, I have met a large number of students and researchers in applied sciences who are interested in practical and sound Bayesian modeling of their data but are prevented from doing so due to their lack of background knowledge of the statistical models, complicated notations, and complexity of the computing methods to fit those models. Having seen this repeatedly in those courses, I am highly delighted to present this book which, I hope, will act as a bridge between the current state of the art in statistical modeling methods for spatial and spatio-temporal data and their myriad of applications in applied sciences.

Preface

Introduction

Scientists in many applied fields such as atmospheric and environmental sciences, biology, climate, ecology, economics, and environmental health are increasingly creating large data sets that are often both geographically and temporally referenced. Such data are called spatio-temporal data, which vary both in space and time. Such data arise in many contexts, e.g. air pollution monitoring, disease mapping, and monitoring of economic indicators such as house prices, poverty levels, and so on. There are many other important science areas where it is routine to observe spatio-temporal data sets. Examples include hydrology, geology, social sciences, many areas of medicine, such as brain imaging, wildlife population monitoring and tracking, and machine vision.

Spatio-temporal data sets are often multivariate, with many important predictors and response variables, and may exhibit both spatial and temporal correlation. For example, there may be decreasing temporal trends in air pollution in a study region or a clear spatial north-south divide in precipitation or economic growth. For such data sets, interest often lies in detecting and analyzing spatial patterns and temporal trends, predicting the response variable in both space and time. These are usually hard inferential tasks which are the key subject matter topics of this text. In addition, making broad sense of the data set at hand, in the first place, is also a very important topic of interest—one that will also be discussed in this book.

Armed with both spatial and temporal variations, spatio-temporal data sets come in various spatial and temporal resolutions. Considering spatial resolutions, there are three broad types of data: point referenced data, areal data, and point pattern data. Point referenced data are referenced typically by a single point in space, e.g. a latitude-longitude pair. Examples of such data include air pollution monitoring data and data from a fixed number of precipitation gauges. This book only models point referenced data which are continuous, or deemed to be continuous and can be assumed to be Gaussian. Analysis of point referenced count data has not been considered here.

Areal data sets include aggregated cancer rates for a local authority area or a health surveillance district, index of multiple deprivations for different post-code regions, percentage of green space for each locality in a town, etc.

The third data type, point pattern data, arises when an event of interest, e.g. outbreak of a disease, occurs at random locations. The pattern of the random locations over time and space is studied to discover the disease epidemic of interest. There are many excellent articles and textbooks on analyzing point pattern data. This data type will not be considered any further in this text.

Jargon abound in spatial and temporal statistics: stationarity, variogram, isotropy, Kriging, autocorrelation, Gaussian processes, nugget effect, sill, range, spatial smoother, internal and external standardization, direct and indirect standardization, CAR models, and so on. Even more cumbersome are the terms which are mere variations, e.g. semi-variogram, partial sill, ordinary Kriging, universal Kriging, and co-Kriging. No doubt that all these have their respective places in the literature, but this important jargon can be studied outside the realm of statistical modeling. Indeed, this is encouraged so that we fully grasp and appreciate these concepts without the cloud of statistical modeling. During the modeling stage, this jargon takes its own respective place in enhancing the descriptive and predictive abilities of the adopted models. For example, as will be seen later on, Bayesian prediction using a Gaussian process-based model will automatically render Kriging. More about modeling, especially Bayesian modeling, follows.

Modeling is an integral part of statistical data analysis and uncertainty quantification. Modeling is essentially a method to assign rules for the observed data. Each and every individual data point, preferably not their summaries, is given a probability distribution to mimic the force of nature, which gave rise to the data set. Statistical models, expressed through the distributions for the data points, must respect the known or hypothesized relationships between the data points. For example, it is to be expected that there is high spatial and temporal correlation among the observed values. Different statistical models then will specify different spatial, temporal, or spatio-temporal structures for those correlations. Of course, unlike in a simulation study, we will not know the "correct" dependence model which may have generated the data. Hence, statistical modeling comes with a plethora of methods to perform model comparison, validation, and selection. Once validated and selected, statistical models enable us to make inferences on any aspect of the data-generating process or any aspects of the future data points that would be generated by the underlying data-generating mechanisms. A caveat in this philosophy is that a wrong model may be selected since the oft-quoted remark goes, "all models are wrong but some are useful." However, this also makes modeling stronger since by explicit modeling, we understand more about the characteristics of the data and by rejecting the failed models, we narrow down the choices in pursuit of the correct model. Further strength in modeling comes from the fact that by assuming a statistical model, we make all the assumptions explicit so that by examining the model fit in detail, we can scrutinize the assumptions too. Procedure-based data analysis methods usually do not lend themselves to such levels of explicit scrutiny.

Why Bayesian modeling? Tremendous growth in computing power and the philosophical simplicity of the Bayesian paradigm are the main reasons for adopting and championing Bayesian modeling. The celebrated Moore's law (Moore, 1975) states that the number of transistors in an integrated circuit would double every two years. Bayesian modeling, performed through Bayesian computation, exploits the growth in computing power in so much as doubly more complex models are fitted and analyzed in similar regular time intervals. Seemingly complex models are easily and rapidly solved by harnessing the ever-growing computing power.

Philosophical simplicity of the Bayesian methods comes from the fact that probability is the only sensible measure of uncertainty. Hence, in any situation appropriately defined probabilities are given as answers to questions regarding uncertainty quantification. This probability-based formalization is appealing to all students and practitioners of statistical modeling. In the Bayesian paradigm all the "unknowns", e.g. parameters and missing and future observations, are given their respective distributions and by conditioning on the observed data the Bayes theorem is used to obtain the refined probability distributions for all those "unknowns" in the model. These refined probability distributions are used naturally to make inferences and evaluate uncertainties. This appealing and universal procedure is applied to all statistical data analysis and modeling problems—regardless of the sub-topic of the data set of interests.

The complexities in probability calculations in the Bayesian paradigm are, as can be expected, handled by modern computing methods. The community-developed R software packages are at the forefront in this field of computation. R is not a *statistical* package that only allows implementation of complex statistical procedures developed by salaried individuals in closed door offices. Rather, R provides an open-source computing platform for developing model fitting and data analysis algorithms for all types of practical problems. Writing R code for data analysis and model fitting is surprisingly easily done through community-provided R extension packages, examples, and illustrations. As yet untested and unfitted models provide challenging open problems to researchers worldwide and solutions appear as research papers and user contributed R packages that benefit both the developers and the users. Currently, there are more than 10,000 user contributed R packages that have enabled the users to perform extensive model fitting tasks that were not possible before.

Who should read this book?

This book is primarily intended for the army of post-graduate students who are embarking on a future research carrier in any branch of applied sciences including statistics. Early career researchers trained in a discipline other than statistics will also find the book useful when they decide to turn to use Bayesian statistical methods for their modeling and data analysis purposes. The examples presented in the book are meant to be appealing to modelers, in the broad field of applied sciences both natural and social. The primary aim here is to reach post-graduate researchers, PhD students, and early career researchers who face the prospect of analyzing spatio-temporal data.

The book can be used for teaching purposes too. The materials presented are accessible to a typical third-year mathematics major undergraduate student, or a taught post-graduate student in statistics or a related area with some background in statistical modeling, who is looking to do a research project on practical statistics. Such a student is able to learn the methods in Chapters 4 and 5 and then subsequently apply those for practical modeling presented in the later chapters.

The methods Chapters 4 and 5 do not discuss, and neither require any knowledge of spatial and spatio-temporal modeling. Hence these two introductory chapters can be studied by students and researchers interested in learning Bayesian inference and computation methods. These two chapters also contain a number of exercises and solutions that will be appealing to beginner students in Bayesian statistics. Introductory Bayesian modeling methods of Chapters 6 and 10 together with the methods of Chapters 4 and 5 may be used as a textbook for a one-semester course for final year undergraduate or master's students.

Why read this book?

At the moment there is no dedicated graduate-level textbook in the analysis of spatio-temporal data sets using Bayesian modeling. This book takes an inter-disciplinary approach that is designed to attract a large group of researchers and scientists who do not necessarily have a strong background in mathematical statistics. Hence a primary feature of this book is to provide a gentle introduction to the theory and current practice so that the readers can quickly get into Bayesian modeling without having to fully master the deep statistical theories underpinned by rigorous calculus-based mathematics.

There are many competing textbooks in this research area. Here are the most relevant titles with our commentary:

1. *Spatial and Spatio-temporal Bayesian Models with R – INLA* by Blangiardo and Cameletti (2015). This is a very practically oriented textbook that describes Bayesian methodology for spatial and spatio-temporal statistics.

2. *Spatio-Temporal Statistics with R* by Wikle et al. (2019). This is a very authoritative textbook discussing the main statistics relevant

for analyzing spatio-temporal data. It also discusses hierarchical modeling and model selection methods using `INLA` and other software packages. This is perhaps the closest to the current book although the main emphasis of the current book is to present Bayesian modeling and validation of spatio-temporal data.

3. *Hierarchical Modeling and Analysis for Spatial Data* by Banerjee et al. (2015). This book highlights hierarchical Bayesian modeling at an excellent theoretical and practical level. However, it again requires a very good background in mathematical statistics to fully appreciate the methodologies presented in the book.

4. *Analysis and Modelling of Spatial Environmental Data* by Kanevski and Maignan (2004). This book uses a user-friendly software, GSO Geostat Office, much like Microsoft Office under MS Windows operating system, to present tools and methods for analyzing spatial environmental data. It claims to present complete coverage of geostatistics and machine learning algorithms and illustrate those with environmental and pollution data. However, it does not discuss methods for analyzing spatio-temporal data.

5. *Spatio-Temporal Methods in Environmental Epidemiology* by Shaddick and Zidek (2015). This book is more geared toward epidemiology and does not do Bayesian modeling for point referenced data using Gaussian processes.

What is this book all about?

The main subject matter of this book is modern spatio-temporal modeling and data analysis in the Bayesian inference framework. The book aims to be an accessible text describing the theory and also the practical implementation of the methods in this research area. The primary aim of the book is to gradually develop spatio-temporal modeling methods and analysis starting from a simple example of estimating the mean of the normal distribution assuming a known variance. The book aims to describe the methods, both theory and practice, using an accessible language. As much as is possible, the book provides the required theory so that the interested reader can fully grasp the key results in spatio-temporal model fitting and predictions based on them. However, the theory can also be side-stepped so that a reader with less mathematical background can still appreciate the main results. Indeed, this is facilitated throughout by numerical illustrations using the purposefully developed R package `bmstdr`. Using the package, the interested readers can quickly check the theory to enhance their understanding of the key concepts in model fitting and validation. The book also aims to provide a plain, simple English language explanation of the methods and equations presented so that the book will also appeal to those researchers whose first degree is not necessarily in mathematics or a related subject.

How to read this book?

The answer to this question depends on the level of background knowledge in statistics, and more particularly in Bayesian modeling, of the reader. A beginner in statistical modeling may need to go through the first six chapters as presented sequentially. The chapter-specific R commands are designed to bring the theory within easy reach of the reader so that a deeper understanding can take place instantaneously. However, the theoretical developments and proofs can be skipped, and the reader may proceed straight to the numerical results and the associated plain English language commentary. An expert in Bayesian modeling can skip the Bayesian methods and computing of Chapters 4 and 5 and jump to the point referenced spatial data modeling of Chapter 6 or aerial data modeling of Chapter 10.

All the main practical results in the book, i.e., the figures and tables, are fully reproducible using the code and data provided online. These are downloadable from the publicly available github repository *https://github.com/ sujit-sahu/bookbmstdr.git* and *https://www.sujitsahu.com/bookbmstdr/*. Readers are able to download those and reproduce the results in order to have a deeper understanding of the concepts and methods.

This book does not provide tutorials on learning the R language at all. Instead, it provides commentary on the algorithm used and R commands adopted to obtain the tables and figures. Novice users of R can first reproduce the reported results without understanding the full code. Then they can change data and other parameters to fully grasp the functionality of the presented R code. The `ggplot2` library has been used to draw all the maps and summary graphs. A glossary of `ggplot2` commands is provided to help the reader learn more about those online.

Acknowledgments

I would especially like to thank all my undergraduate and graduate students who have contributed to research in Bayesian modeling and computation. I would also like to thank all my co-authors and mentors for their collective wisdom, which has enriched my learning throughout my academic career.

1

Examples of spatio-temporal data

1.1 Introduction

This chapter introduces several motivating data sets that have been actually used to perform and illustrate various aspects of spatio-temporal modeling. The main purpose here is to familiarize the reader with the data sets so that they can grasp the modeling concepts as and when those are introduced. A typical reader, who is primarily interested in applied modeling but not in theoretical developments, is able to quickly scan all the data sets presented here with a view to choosing one that is closer to their line of research for following subsequent modeling developments for that particular data sets.

The chapter is organized in three main sections. Section 1.2 discusses the three broad spatio-temporal data types. Section 1.3 provides six examples of what are known as point referenced data, and Section 1.4 provides five examples of areal unit data sets. Two particular data sets, one on air pollution monitoring in the state of New York from Section 1.3 and the other on number of Covid-19 deaths in England during the 2020 global pandemic from Section 1.4, are used as running examples in different chapters of this book.

The ideas behind the spatio-temporal modeling can be broadly cross-classified according to: (a) their motivation, (b) their underlying objectives, and (c) the scale of data. Under (a) the motivations for models can be classified into four classes: (i) extensions of time series methods to space, (ii) extension of random field and imaging techniques to time, (iii) interaction of time and space methods, and (iv) physical models. Under (b) the main objectives can be viewed as either data reduction or prediction. Finally, under (c) the available data might be sparse or dense in time or space respectively, and the modeling approach often takes this scale of data into account. In addition, the data can be either continuously indexed or discretely indexed in space and/or time. Based on these considerations, especially (i)–(iii), statistical model building and their implementation take place.

DOI: 10.1201/9780429318443-1

1

1.2   Spatio-temporal data types

Data are called spatio-temporal as long as each one of them carries a location and a time stamp. This leaves open the possibility of a huge number of spatio-temporal data types even if we exclude the two extreme degenerate possibilities where data are observed either at a single point of time or at a single location in space. Analysts often opt for one of the two degenerate possibilities when they are not interested in the variation due to either space or time. This may simply be achieved by suitably aggregating the variation in the ignored category. For example, one may report the annual average air pollution levels at different monitoring sites in a network when daily data are available. Aggregating over time or space reduces variability in the data to be analyzed and will also limit the extent of inferences that can be made. For example, it is not possible to detect or apportion monthly trends just by analyzing annual aggregates. This text book will assume that there is spatio-temporal variation in the data, although it will discuss important concepts, as required, for studying spatial or temporal only data.

One of the very first tasks in analyzing spatio-temporal data is to choose the spatial and temporal resolutions at which to model the data. The main issues to consider are the inferential objectives of the study. Here one needs to decide the highest possible spatial and temporal resolutions at which inferences must be made. Such considerations will largely determine whether we are required to work with daily, monthly or annual data, for example. There are other issues at stake here as well. For example, relationships between variables may be understood better at a finer resolution than a coarser resolution, e.g. hourly air pollution level may have a stronger relationship with hourly wind speed than what the annual average air pollution level will have with the annual average wind speed. Thus, aggregation may lead to correlation degradation. However, too fine a resolution, either temporal and/or spatial, may pose a huge challenge in data processing, modeling and analysis without adding much extra information. Thus, a balance needs to be stuck when deciding on the spatio-temporal resolution of the data to be analyzed and modeled. These decisions must be taken at the data pre-processing stage before formal modeling and analysis can begin.

Suppose that the temporal resolution of the data has been decided, and we use the symbol t to denote each time point and we suppose that there are T regularly spaced time points in total. This is a restrictive assumption as often there are cases of irregularly spaced temporal data. For example, a patient may be followed up at different time intervals; there may be missed appointments during a schedule of regular check-ups. In an air pollution monitoring example, there may be more monitoring performed during episodes of high air pollution. For example, it may be necessary to decide whether to model air pollution at an hourly or daily time scales. In such situations, depending on the main purposes of the study, there are two main strategies for modeling. The first one is to model at the highest possible regular temporal resolution, in which case there would be many missing observations for all the unobserved time points. Modeling will help estimate those missing observations and their uncertainties. For example, modeling air pollution at the hourly resolution will lead to missing observations for all the unmonitored hours. The second strategy avoiding the many missing observations of the first strategy is to model at an aggregated temporal resolution where all the available observations within a regular time window are suitably averaged to arrive at the observation that will correspond to that aggregated time. For example, all the observed hourly air pollution recordings are averaged to arrive at the daily average air pollution value. Although this strategy is successful in avoiding many missing observations, there may still be challenges in modeling since the aggregated response values may show un-equal variances since those have been obtained from different numbers of observations during the aggregated time windows. In summary, careful considerations are required to select the temporal resolution for modeling.

In order to model spatio-temporal data, there is an obvious need to keep track of the spatial location, denoted by $\mathbf{s}$ in a region $\mathbb{D}$ say, and the time point, $t \in \mathbb{R}$ of an observation, say $y(\mathbf{s}, t)$ of the random variable $Y(\mathbf{s}, t)$. There may be additional covariate information available which we denote by $\mathbf{x}(\mathbf{s}, t)$. Different data types arise by the ways in which the points $\mathbf{s}$ are observed in $\mathbb{D}$.

Attention is now turned to exploring the spatial aspects of the data sets when time is fixed. In spatial analysis there are three broadly different data types:

(i) point referenced data,

(ii) areal unit data,

(iii) point pattern data.

Typical *point referenced data* arise when each data point is described by a single non-random point in a map, e.g. a latitude-longitude pair. For example, a response, e.g. a value of air pollution level, is observed at a particular monitoring site within a network. For point referenced data $\mathbf{s}$ varies continuously over a fixed study region $\mathbb{D}$. For example, we may have observed the response $y(\mathbf{s}, t)$ and the covariates at a set of n locations denoted by $\mathbf{s}_i, i = 1, \ldots, n$ and at T time points so that $t = 1, \ldots, T$. The set of spatial locations can either be fixed monitoring stations, like in an air pollution example, or can vary with time, for example, data obtained from a research ship measuring ocean characteristics as it moves about in the ocean.

Areal unit data are observed in typical spatially aggregated domains, e.g. administrative geographies like postcodes, counties or districts. For such data the spatial reference is an area in a map. Examples of such data types abound in disease mapping by public health authorities, such as Covid-19 death rates. Section 1.4 provides several examples.

The third data type, *a point pattern*, arises when an event of interest, e.g. outbreak of a disease, occurs at random locations. Thus, the randomness of the observation locations is what separates a point pattern data from a point referenced data, although the locations in a point pattern are referenced by points, each of which may be described by a latitude-longitude pair. The usual concept of a response variable is typically absent in a point pattern data set, although there may be several explanatory variables for explaining the pattern of points. As has been noted before, this textbook will not discuss or analyze point pattern data any further. The interested reader is referred to many excellent textbooks such as Baddeley et al. (2015) and Diggle (2014).

1.3 Point referenced data sets used in the book

1.3.1 New York air pollution data set

We use a real-life data set, previously analyzed by Sahu and Bakar (2012a), on daily maximum 8-hour average ground-level ozone concentration for the months of July and August in 2006, observed at 28 monitoring sites in the state of New York. We consider three important covariates: maximum temperature (xmaxtemp in degree Celsius), wind speed (xwdsp in nautical miles), and percentage average relative humidity (xrh) for building a spatio-temporal model for ozone concentration. Further details regarding the covariate values and their spatial interpolations for prediction purposes are provided in Bakar (2012). This data set is available as the data frame nysptime in the bmstdr package. The data set has 12 columns and 1736 rows. The R help command ?nysptime provides detailed description for the columns.

In this book we will also use a temporally aggregated spatial version of this data set. Available as the data set nyspatial in the bmstdr package, it has data for 28 sites in 28 rows and 9 columns. The R help command ?nyspatial provides detailed descriptions for the columns. Values of the response and the covariates in this data set are simple averages (after removing the missing observations) of the corresponding columns in nysptime. Figure 1.1 represents a map of the state of New York together with the 28 monitoring locations.

```
library(bmstdr)
library(ggsn)
nymap <- map_data(database="state",regions="new york")
p <- ggplot() +
  geom_polygon(data=nymap, aes(x=long, y=lat, group=group),color="black
      ", size = 0.6, fill=NA) +
  geom_point(data =nyspatial, aes(x=Longitude,y=Latitude)) +
  labs( title= "28 air pollution monitoring sites in New York", x="
      Longitude", y = "Latitude") +
```

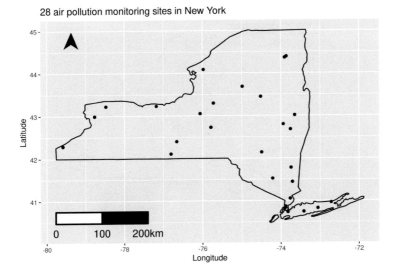

FIGURE 1.1: The 28 air pollution monitoring sites in New York. The R code used to draw this figure is included below and the code lines are explained in Code Notes 1.1.

```
scalebar(data =nymap, dist = 100, location = "bottomleft", transform
    =T, dist_unit = "km",
            st.dist = .05, st.size = 5, height = .06, st.bottom=T,
                model="WGS84") +
north(data=nymap, location="topleft", symbol=12)
p
```

> ♣ **R Code Notes 1.1. Figure 1.1** The map_data command extracts the boundary polygons. The ggplot command draws the figure where (i) geom_polygon draws the map boundary, (ii) geom_point adds the points, (iii) scalebar draws the scale of the map and (iv) north puts the north arrow. In the polygon drawing the option fill=NA allows the map to be drawn with a blank background without any color filling. The functions scalebar and north are provided by the ggsn library which must be linked before calling the plotting command.

The two data sets nyspatial and nysptime will be used as running examples for all the point referenced data models in Chapters 6 and 7. The spatio-temporal data set nysptime is also used to illustrate forecasting in Chapter 9. Chapter 3 provides some preliminary exploratory analysis of these two data sets.

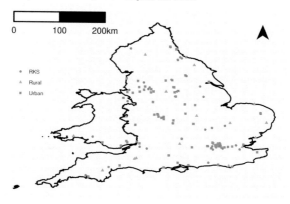

FIGURE 1.2: A map of England and Wales showing the locations of 144 AURN air pollution monitoring sites.

1.3.2 Air pollution data from England and Wales

This example illustrates the modeling of daily mean concentrations of nitrogen dioxide (NO_2), ozone (O_3) and particles less than $10\mu m$ (PM_{10}) and $2.5\mu m$ ($PM_{2.5}$) which were obtained from $n = 144$ locations from the Automatic Urban and Rural Network (AURN, *http://uk-air.defra.gov.uk/networks*) in England and Wales. The 144 locations are categorized into three site types: rural (16), urban (81) and RKS (Road amd Kerb Side) (47). The locations of these 144 sites are shown in Figure 1.2.

Mukhopadhyay and Sahu (2018) analyze this data set comprehensively and obtain aggregated annual predictions at each of the local and unitary authorities in England and Wales. It takes an enormous amount of computing effort and data processing to reproduce the work done by Mukhopadhyay and Sahu (2018) and as a result, we do not attempt this here. Instead, we illustrate spatio-temporal modeling for NO_2 for 365 days in the year 2011. We also obtain an annual average prediction map based on modeling of the daily data in Section 8.2.

> ♣ **R Code Notes 1.2. Figure 1.2** The basic map data for this figure is the 2011 map layer data set `infuse_ctry_2011`. The `readOGR` command in the `rgdal` package reads the shape files. The `tidy` command from the `broom` package converts the map polygons into a data frame that can draw a map using the `ggplot`. The monitoring locations are plotted using the `geom_point` function. The `ggsn` library functions `scalebar` and `north` are used respectively to insert the map scale and the north arrow as in the previous figure.

1.3.3   Air pollution in the eastern US

This example is taken from Sahu and Bakar (2012b), where we consider modeling the daily maximum 8-hour average ozone concentration data obtained from 691 monitoring sites in the eastern US, as shown in Figure 1.3. These pollution monitoring sites are made up of 646 urban and suburban monitoring sites known as the National Air Monitoring Stations/State and Local Air Monitoring Stations (NAMS/SLAMS) and 45 rural sites monitored by the Clean Air Status and Trends Network (CASTNET).

We analyze daily data for $T = 153$ days in every year from May to September since this is the high ozone season in the US. We consider these data for the 10 year period from 1997 to 2006 that allows us to study trend in ozone concentration levels. Thus, we have a total of $1,057,230$ observations and among them approximately 10.44% are missing, which we assume to be at random, although there are some annual variation in this percentage of missingness.

The main purpose of the modeling exercise here is to assess compliance with respect to the primary ozone standard which states that the 3-year rolling average of the annual *4th* highest daily maximum 8-hour average ozone concentration levels should not exceed 85 ppb, see e.g., Sahu et al. (2007). Figure 1.4 plots the *4th* highest maximum and their 3-year rolling averages with a superimposed horizontal line at 85. As expected, the plot of the rolling averages is smoother than the plot of the annual *4th* highest maximum values. The plots show that many sites are compliant with respect to the standard, but many others are not. In addition, the plot of the 3-year rolling averages shows a very slow downward trend. Both the plots show the presence of a few outlier sites which are perhaps due to site-specific issues in air pollution, for example, due to natural disasters such as forest fires. This data set is analyzed in Section 8.3.

> ♣ **R Code Notes 1.3. Figure 1.3** The basic map data for this figure has been hand picked by selecting the specific states and regions whose boundaries have been drawn. Those state names are available as the `eaststates` object in the data files for this example. The `map_data` command in the `ggplot2` package extracts the required boundaries. The horizontal lines are drawn using the `geom_abline` function.

1.3.4   Hubbard Brook precipitation data

Measuring total precipitation volume in aggregated space and time is important for many environmental and ecological reasons such as air and water quality, the spatio-temporal trends in risk of flood and drought, forestry management and town planning decisions.

691 sites in the Eastern US

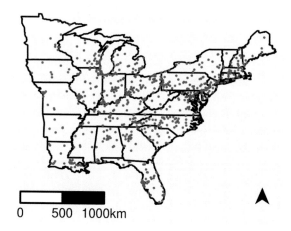

FIGURE 1.3: A map showing 691 ozone monitoring sites in the eastern US.

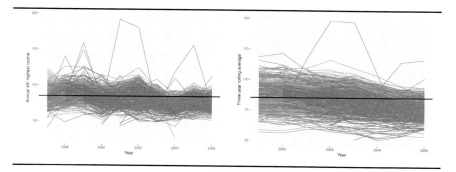

FIGURE 1.4: Time series plots of the ozone concentration summaries from 691 sites. Left panel: annual 4*th* highest maximum and right panel: 3-year rolling average of the annual 4*th* highest maximum. The solid black line at the value 85 of the *y*-axis was the government-regulated ozone standard at the time.

The Hubbard Brook Ecosystem Study (HBES), located in New Hampshire, USA and established in 1955, continuously observes many environmental outcome variables such as temperature, precipitation volume, nutrient volumes in water streams. HBES is based on the 8,000-acre Hubbard Brook Experimental Forest (see e.g. https://hubbardbrook.org/) and is a valuable source of scientific information for policy makers, members of the public, students and scientists. Of-interest here is a spatio-temporal data set on weekly precipitation volumes collected from 22 rain-gauges from 1997 to 2015.

Figure 1.5 shows the locations of the 23 rain gauges (without RG22) divided in two sub-areas called south and north facing catchments. Gauges RG1-RG11 are located in the south facing catchments in the north of the map and the remaining gauges are part of the north facing catchments seen in the south of the map. The south facing catchment consists of six watersheds, labeled W1-W6, and the north facing catchment consists of the remaining three watersheds, W7-W9. The main modeling objective here is to study spatio-temporal trend in precipitation volumes in the watersheds. We do not have the required data for reproducing this map. This data set is analyzed in Section 8.4.

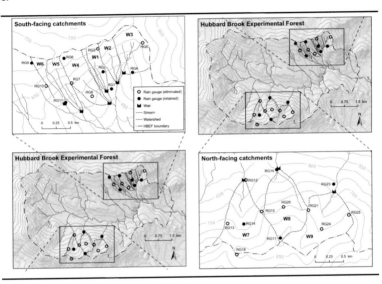

FIGURE 1.5: Maps of precipitation gauges and watersheds.

1.3.5 Ocean chlorophyll data

Taken from Hammond et al. (2017), this example studies long-term trends in chlorophyll (chl) levels in the ocean, which is a proxy measure for phytoplankton (marine algae). Phytoplankton is at the bottom of food chain and provides the foundation of all marine ecosystem. The abundance of phytoplankton affects the supply of nutrients and light exposure. Global warming can potentially affect the phytoplankton distribution and abundance, and hence it is of much scientific interest to study long-term trends in chl which influences the abundance of phytoplankton.

Figure 1.6 shows a map of the 23 ocean regions of interest where we have observed satellite-based measurements. The main modeling objective here is to study long-term trends in chl levels in these 23 oceanic regions. Section 8.5 assesses these trend values.

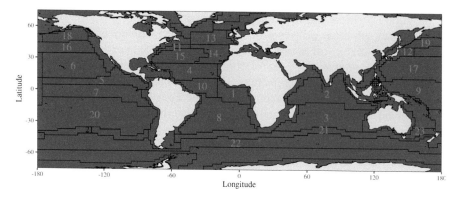

FIGURE 1.6: Map showing 23 regions of interest in the world's oceans.

♣ **R Code Notes 1.4. Figure 1.6** The map data for this figure has been obtained from the internet as the Longhurst world map version 4 published in 2010 and the world map `ne_110m_land` which plots the boundaries of the 110 countries. The `readOGR` function from the `rgdal` library has been used to read the map shape files. The `tidy` function from the `broom` library converts the read map data sets to data frames that can plotted using the `ggplot` function. The `fill=4` option in `geom_polygon` has color filled the oceans. The `ggplot` function `geom_text` has been used to place the region numbers in the map.

1.3.6   Atlantic ocean temperature and salinity data set

This example is taken from Sahu and Challenor (2008) on modeling deep ocean temperature data from roaming Argo floats. The Argo float program, see for example, http://www.argo.ucsd.edu, is designed to measure the temperature and salinity of the upper two kilometers of the ocean globally. These floats record the actual measurements which are in contrast to satellite data, such as the ones used in the ocean chlorophyll example in Section 1.3.5, which provide less accurate observations with many missing observations. Each Argo float is programmed to sink to a depth of one kilometer, drifting at that depth for about 10 days. After this period the float sinks a further kilometer to a depth of two kilometers and adjusting its buoyancy rises to the surface, measuring temperature and conductivity (from which salinity measurements are derived) on the way. Once at the surface, the data and the position of the float are transmitted via a satellite. This gives scientists access to near real-time data. After transmitting the data the float sinks back to its 'resting' depth of one kilometer and drifts for another ten days before measuring another temperature and salinity profile at a different location. Argo data are freely available via the international Argo project office, see the above-mentioned website.

We consider the data observed in the North Atlantic ocean between the latitudes 20° and 60° north and longitudes 10° and 50° west. Figure 1.7 shows the locations of the Argo floats in the deep ocean. The figure shows the moving nature of Argo floats in each of the 12 months. The primary modeling objective here is to construct an annual map of temperature at the deep ocean along with its uncertainty. The time points at which the data are observed are not equi-lagged, and we do not assume this in our modeling endeavor. Modeling required to produce an annual temperature map of the North Atlantic ocean is performed in Section 8.6.

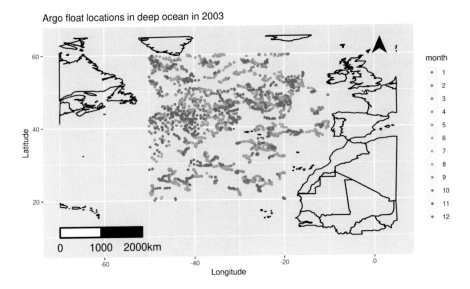

FIGURE 1.7: Locations of moving Argo floats in the deep ocean in 2003.

♣ **R Code Notes 1.5. Figure 1.7** The basic map data for this figure is obtained from the R command

```
map_data("world", xlim=c(-70, 10), ylim=c(15, 65))
```

The bmstdr data set `argo_floats_atlantic_2003` contains the other information plotted in this map.

1.4 Areal unit data sets used in the book

1.4.1 Covid-19 mortality data from England

This data set presents the number of deaths due to Covid-19 during the peak from March 13 to July 31, 2020 in the 313 Local Authority Districts, Counties and Unitary Authorities (LADCUA) in England; see Figure 1.8. There are 49,292 weekly recorded deaths during this period of 20 weeks. Figure 1.9 shows a map of the number of deaths and the death rate per 100,000 people in each of the 313 LADCUAs. Contrasting the two plots, it is clear that much spatial variation is seen in the right panel of the death rates per 100,000 people. The boxplot of the weekly death rates shown in Figure 1.10 shows the first peak during weeks 15 and 16 (April 10th to 23rd) and a very slow decline of the death numbers after the peak. The main purpose here is to model the spatio-temporal variation in the death rates. This data set will be used as a running example for all the areal unit data models in Chapter 10. Chapter 3 provides some further preliminary exploratory analysis of this data set.

FIGURE 1.8: A map of the local authorities and nine administrative regions in England. Air pollution monitoring sites are shown as blue dots in the map.

♣ **R Code Notes 1.6. Figure 1.8** The raw map boundary data for this figure has been obtained from the May 2020 publication of the Office for National Statistics in the UK. The tidied map data sets used in the plotting are provided as the data files and map shape files. The `ggplot` commands `annotate("text")` and `annotate("segment")` are used to draw the line and text annotations for the cities. The command `geom_path` has been used to draw the polygons for the local authorities. Other details regarding the plots are provided in the published exact code which the user may use to reproduce the figure.

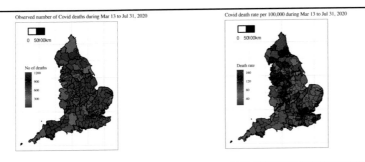

FIGURE 1.9: Raw number of Covid deaths (left) and the number of Covid deaths per 100,000 people (right).

♣ **R Code Notes 1.7. Figure 1.9** See the notes for Figure 1.8. The map data frame identifies the polygons by a column called `"id"`. The data to be plotted as colors in the map has a column called `"mapid"`. This id key is used to merge the map boundary data frame and the Covid-19 death data frame, e.g. `engtotals` in the `bmstdr` package. The merged data frame is then used obtain the plots in the figure using the `ggplot` function. The color scale bar accompanying each plot has been obtained using the `scale_fill_gradientn` function in `ggplot`. Other details regarding the plots are provided in the published exact code which the user may use to reproduce the figure.

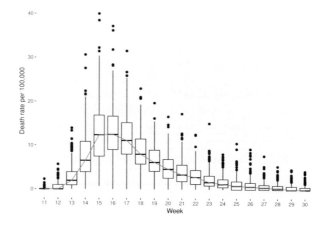

FIGURE 1.10: Boxplots of weekly death rates per 100,000 population.

♣ **R Code Notes 1.8. Figure 1.10** The function `geom_boxplot` plots this graph. The medians are joined by having the option

```
stat_summary(fun=median, geom="line", aes(group=1, col="red"))
```

in the `ggplot` function.

1.4.2 Childhood vaccination coverage in Kenya

The Demographic and Health Surveys (DHS) program[1] routinely collects several data sets for monitoring health at a global level. This example is based on a 2014 vaccination coverage data set for the country Kenya in East Africa. The data set contains the number of children aged 12-23 months who had received the first dose of measles-containing vaccine (MCV1) at any time before the survey in 2014. Figure 1.11 plots the observed vaccination proportions in 2014. A substantial analysis of this and several related data sets has been conducted by Utazi et al. (2021). Modeled in Section 11.2, this example aims to assess vaccination coverage rates in the different counties in Kenya.

---

[1]https://dhsprogram.com/

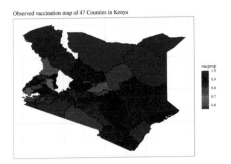

FIGURE 1.11: Observed vaccination proportions in 2014.

♣ **R Code Notes 1.9. Figure 1.11** We provide all the necessary code required to draw this figure. Assume that the data file Kenya_vaccine.csv is in the sub-directory `datafiles` and the map files are in the sub-directory `mapfiles`.

```
library(rgdal); library(broom); library(ggplot2);
mpath <- "mapfiles/"; dpath <- "datafiles/"
Kmap <- readOGR(dsn=path.expand(mpath), layer="
    sdr_subnational_boundaries2")
kdat <- read.csv(file=paste0(dpath, "Kenya_vaccine.csv"))
kdat$vacprop <- kdat$yvac/kdat$n
adf <- tidy(Kmap)
adf$id <- as.numeric(adf$id)
kdat$id <- kdat$id-1
udf <- merge(kdat, adf)
head(udf)

a <- range(udf$vacprop)
vmap <- ggplot(data=udf, aes(x=long, y=lat, group = group, fill
    =vacprop)) +
  scale_fill_gradientn(colours=colpalette, na.value="black",
      limits=a) +
  geom_polygon(colour="black",size=0.25) +
  theme_bw()+theme(text=element_text(family="Times")) +
  labs(title= "Observed vaccination map of 47 Counties in Kenya",
      x="", y = "", size=2.5) +
  theme(axis.text.x = element_blank(), axis.text.y =
      element_blank(),axis.ticks = element_blank())
plot(vmap)
```

1.4.3   Cancer rates in the United States

The Centers for Disease Control and Prevention in the United States provides downloadable cancer rate data at various geographical levels, e.g. the 50 states. Such a data set can be downloaded along with various information e.g. gender and ethnicity and types of cancer. However, due to the data identifiability and data protection reasons, some of the smaller rate counts (which arises due to finer classification by the factors) rates are not made public. Hence, for the purposes of illustration of this book, we aim to model aggregated annual data at the state level. The full data set provides state-wise annual rates of cancer from all causes during from 2003 to 2017. Figure 1.12 provides a map of the aggregated cancer rates per 100,000 people from all causes during from 2003 to 2017 for the 48 contiguous states. This is an example of a *choropleth map* that uses shades of color or gray scale to classify values into a few broad classes, like a histogram. The figure shows higher total incidence rates in the northeast compared to south-west. Florida also shows a higher rate which may be attributed to a larger share of the retired elderly residents in the state. The full spatio-temporal data set will be analyzed in Section 11.3.

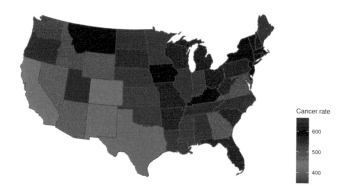

FIGURE 1.12: A choropleth map of cancer rate per 100,000 people from all causes from 2003–2017.

♣ **R Code Notes 1.10. Figure 1.12** The main function for plotting
the map of the USA is the `plot_usmap` in the package `usmap`.

```
plot_usmap(data = us48cancertot, values = "totrate", color = "
    red", exclude=c("AK", "HI")) +
scale_fill_gradientn(colours=colpalette,na.value="black",
    limits=range(us48cancertot$totrate), name = "Cancer rate")
    +
theme(legend.position = "right")
```

The observed standardized mortality rates, see discussion in Section 11.3
on how to obtain those, for ten selected states are shown in Figure 1.13.
These states are hand-picked to represent the full range of the SMR values.
The research question that is of interest here is, "is there an upward trend
in these rates after accounting for spatio-temporal correlation and any other
important fixed effects covariates?" This is investigated in Section 11.3.

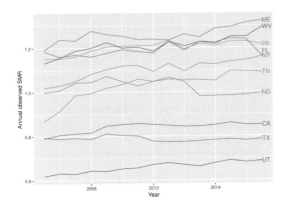

FIGURE 1.13: Standardized mortality rates of cancer deaths in 10 selected
states in the US.

♣ **R Code Notes 1.11. Figure 1.13** The function `geom_dl` in
the `directlabels` library has been used to annotate the names of the
states. The plotted states have been hand picked to see the overall
range of the mortality rates.

1.4.4 Hospitalization data from England

Monthly counts of the numbers of hospitalizations due to respiratory diseases from the 323 Local and Unitary Authorities (LUA) in England for the 60 months in the 5-year period 2007 to 2011 are available from the study published by Lee et al. (2017). These counts depend on the size and demographics of the population at risk, which are adjusted for by computing the expected number of hospital admissions E_{it} using what is known as *indirect standardization*, see Section 2.12, from national age and sex-specific hospitalization rates in England.

In this example, the study region is England, UK, partitioned into $i = 1, \ldots, n = 323$ Local and Unitary Authorities (LUA), and data are available for $t = 1, \ldots, T = 60$ months between 2007 and 2011. Counts of the numbers of respiratory hospitalizations for LUA i and month t are denoted by Y_{it}, for $i = 1, \ldots, 323$ and $t = 1, \ldots, 60$, which have a median value of 111 and a range from 6 to 2485. The monthly time scale matches the study by Greven et al. (2011), whereas the majority of studies such as Lee et al. (2009) utilize yearly data. An advantage of the monthly scale is that it requires less aggregation of the data away from the individual level, but it does mean that Y_{it} could include admissions driven by both chronic and acute pollution exposure.

The spatial (left panel) and temporal (bottom panel) patterns in the Standardized Morbidity Ratio, $\text{SMR}_{it} = Y_{it}/E_{it}$ are displayed in Figure 1.14, where a value of 1.2 corresponds to a 20% increased risk compared to E_{it}. The figure shows the highest risks are in cities in the center and north of England, such as Birmingham, Leeds and Manchester, while the temporal pattern is strongly seasonal, with higher risks of admission in the winter due to factors such as influenza epidemics and cold temperature. This data set is used as an example in Section 11.4 of this book.

> ♣ **R Code Notes 1.12. Figure 1.14** This example uses a previous version of the map of England containing 323 local authorities. The data used for this map have been provided in the data files for this example.

1.4.5 Child poverty in London

This data set concerns monitoring of annual child poverty levels in 32 boroughs and the City of London, UK. The child poverty measure we consider is a broad proxy for relative low-income child poverty as set out in the UK Child Poverty Act 2010. The purpose here is to perform an analysis at a local level. The measure shows the proportion of children living in families in receipt of out-of-work (means-tested) benefits or in receipt of tax credits where their reported income is less than 60 percent of UK median income. This example

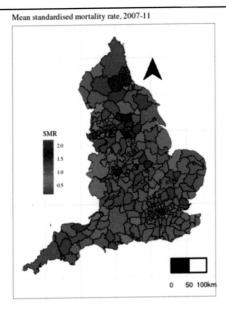

Mean standardised mortality rate, 2007-11

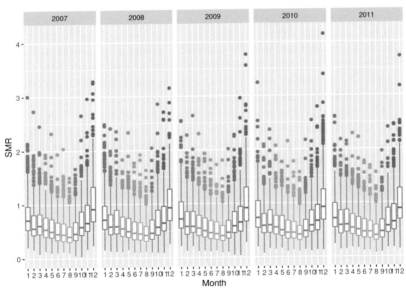

FIGURE 1.14: Top panel: A map of average SMR. Bottom panel: Boxplot of monthly SMR values over 5 years.

will analyze child poverty data for 10 years from 2006 to 2015. The average
levels are plotted in Figure 1.15, and Figure 1.16 shows a declining trend in
poverty levels. Spatio-temporal modeling will be performed for this data set
in Section 11.5.

> ♣ **R Code Notes 1.13. Figure 1.15** The `gCentroid` function in the
> library `rgeos` finds the centroids of the 32 boroughs and the City of Lon-
> don. The `geom_text` function in `ggplot` allows us to put the borough
> names in the map at the centroid locations. The color categories have
> been chosen by looking at the distribution of the percentages. The gg-
> plot function `scale_fill_manual` has been used to plot the legend color
> guide bar where the option `guide_legend(reverse=TRUE)` reversed the
> ordering of the colors in the legend.

1.5 Conclusion

This chapter outlines several data examples that will be used to illustrate
spatial and spatio-temporal model fitting and validation for both points referenced and areal unit data. The objective in each example is to introduce the
reader to these data sets and the associated modeling problems so that they
can be motivated to study the methodology introduced in the later chapters.
A practically motivated reader is also encouraged to read only the examples
which resemble their own practical modeling problems.

This chapter provides only commentary, except for the New York and
Kenya example, on how to draw the maps. It does not give the actual code
to reproduce the figures for brevity. Such code lines are provided online from
github[2], so that the reader can experiment with code directly in R rather than
read those as cumbersome chunks of the text here. The commentary provided,
however, touches upon the main steps and tricks needed to reproduce the
figures. Such tricks will be useful elsewhere too.

1.6 Exercises

1. Reproduce all the figures presented in this chapter using the code
   and data provided online from github.

---

[2]https://github.com/sujit-sahu/bookbmstdr.git

Percentage of children living in poverty, 2006-15

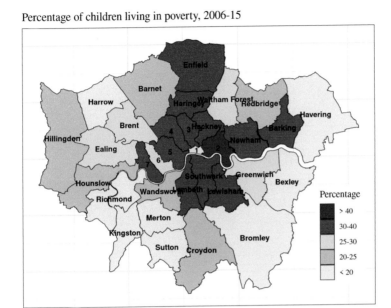

FIGURE 1.15: A map of the City of London and the 32 boroughs in London. 1 = City of London, 2 = Tower Hamlets, 3 = Islington, 4 = Camden, 5 = Westminster, 6 = Kensington and Chelsea, 7 = Hammersmith and Fulham.

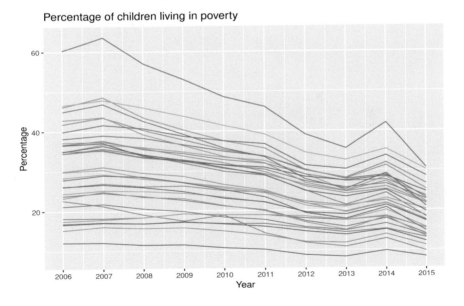

FIGURE 1.16: Percentage of children living in poverty in London.

2. Collect shape files for your own study region of interest. Draw and annotate your map with interesting information that presents important aspects of the data set to be modeled.

2

Jargon of spatial and spatio-temporal modeling

2.1 Introduction

One of the greatest difficulties in understanding any new technique or, more metaphorically, a spoken language lies in learning the keywords or the vocabulary. These keywords, called jargon here, are quite naturally key to understanding and hence correct use of the techniques – spatio-temporal modeling in the context of this book. The main aim of this chapter is to define and build this vocabulary quite independently in a stand-alone manner without going into the complex world of notation-based modeling. This vocabulary is also required when one intends to perform software-based procedural spatial, temporal, or spatio-temporal data analysis without explicitly admitting to doing modeling. Chapter 6 provides more discussion regarding procedure-based data analysis and explicit notation-based statistical modeling.

The jargon (both singular and plural) in spatio-temporal modeling obviously come from two different modeling worlds: spatial modeling and temporal modeling. Spatio-temporal modeling benefits from the rich interaction of the concepts and terms which come from the two constituents modeling worlds. The jargon we will define are related to stochastic processes, stationarity, variogram, isotropy, Matèrn covariance function, Gaussian Processes, space-time covariance function, Kriging, auto-correlation, Moran's I and Geary's C, Internal and external standardization, spatial smoothers, CAR models and point processes. Remarkably, some but not all of these terms are used in both of these worlds – a fact we bear in mind in this chapter.

2.2 Stochastic processes

We are familiar with the term random variable to denote measurements of outcomes of random experiments. For example, a random variable, say Y, could denote the height of a randomly chosen individual in a population of interest, such as a class of seven years old pupils in a primary school. If we

DOI: 10.1201/9780429318443-2

measure the height of n randomly chosen children, then we denote these heights by notations $y_1, \ldots, y_n$ which are the recorded numerical values. We use the corresponding upper case letters $Y_1, \ldots, Y_n$ to denote the physical random variable: heights of n randomly selected children.

The concept of random variables is sound and adequate enough when we intend to model and analyze data which are not associated with a continuous domain. For example, the population of seven years old is countable and finite – so the associated domain is not continuous. Further examples include hourly air pollution values recorded at the top of the hour at a particular monitoring site within a city. However, this concept of random variables is not adequate enough when we allow the possibility of having an uncountably infinite collection of random variables associated with a continuous domain such as space or time, or both. For example, suppose that we are interested in modeling an air pollution surface over a city. Because the spatial domain is continuous here, having an uncountably infinite number of locations, we shall require a richer concept of uncountably infinite number of random variables. Hence, we welcome the arrival of the concept of stochastic processes.

A *stochastic process* is an uncountably infinite collection of random variables defined on a continuous domain such as space, time, or both. Hence, the discrete notation, $Y_i, i = 1, 2, \ldots$ for random variables is changed to either $Y(\mathbf{s})$, $Y(t)$ or $Y(\mathbf{s}, t)$ where $\mathbf{s}$ denotes a spatial location, described by a finite number of dimensions such as latitude, longitude, altitude, etc. and t denotes a continuously measured time point depending on the data collection situation: only spatial, only temporal, and spatio-temporal, respectively. In the spatial only case, we shall use $Y(\mathbf{s})$ to denote a spatial stochastic process or simply a spatial process defined over a domain $\mathbb{D}$, say. In the temporal only case, $Y(t)$ is used to denote the temporally varying stochastic process over a continuous time period, $0 \leq t \leq T$. When time is considered to be discrete, e.g. hourly, then it is notationally convenient to use the notation Y_t instead of the more general $Y(t)$. In this case, Y_t is better referred to as a time series.

In this book we shall not consider time in the continuous domain at all since such analyses are much more theoretically rich, requiring deeper theoretical understanding but practically not so common in the subject area of the examples described previously in Chapter 1. Henceforth, we will use the notation t to denote time in a discrete sense and domain. With t being discrete, which of the two notations: $Y(\mathbf{s}, t)$ and $Y_t(\mathbf{s})$, should be adopted to denote our spatio-temporal data? Both the notations make sense, and it will be perfectly fine to use either. In this book we adopt the first, slightly more elaborate, notation $Y(\mathbf{s}, t)$ throughout, although the subscript t will be used to denote vector-valued random variables as necessary. Hence, the notation $y(\mathbf{s}, t)$ will denote a realization of a spatial stochastic process at location $\mathbf{s}$ and at a discrete-time point t.

Different properties and characteristics of a stochastic process give rise to different jargon in the literature. For example, a stochastic process may have a mean and a variance process, may be very haphazard or very stable over time and space and so on. In the sections below we describe the key jargon.

2.3 Stationarity

An object is stationary if it does not move from a fixed position. To be stationary, a stochastic process must possess certain stable behavior. A stochastic process, constituting of an infinite collection of random variables, cannot be a constant everywhere since otherwise, it will not be stochastic at all. Hence it makes sense to define stationarity of particular properties, e.g. mean and variance. The type of stationarity depends on the stationarity of the particular property of the stochastic process. In the discussion below, and throughout, we shall assume that the stochastic process under consideration has finite mean and variance, respectively denoted by $\mu(\mathbf{s})$ and $V(\mathbf{s})$ for all values of $\mathbf{s}$ in $\mathbb{D}$.

A stochastic process, $Y(\mathbf{s})$, is said to be *mean stationary* if its mean is constant over the whole domain $\mathbb{D}$. For a mean stationary process $Y(\mathbf{s})$, $\mu(\mathbf{s}) = E(Y(\mathbf{s}))$ is a constant function of $\mathbf{s}$. Thus, the mean surface of a mean stationary stochastic process will imply a one-color map depicting the mean over the domain $\mathbb{D}$. Such a map will not exhibit any spatial trend in any direction. Note that this does not mean that a particular realization of the stochastic process, $Y(\mathbf{s})$ at n locations $\mathbf{s}_1, \mathbf{s}_2, \ldots, \mathbf{s}_n$ will yield a constant surface, $y(\mathbf{s}_1) = y(\mathbf{s}_2) = \ldots = y(\mathbf{s}_n)$. Rather, mean stationarity of a process $Y(\mathbf{s})$ means that $\mu(\mathbf{s}_1) = \mu(\mathbf{s}_2) = \cdots = \mu(\mathbf{s}_n)$ at an arbitrary set of n locations, $\mathbf{s}_1, \mathbf{s}_2, \ldots, \mathbf{s}_n$, where n itself is an arbitrary positive integer. Similarly, we say that a time series, Y_t is mean stationary if $E(Y_t)$, $(= \mu_t$, say), does not depend on the value of t. A mean stationary process is rarely of interest since, often, the main interest of the study is to investigate spatial and/or temporal variation. However, we often assume a zero-mean stationary process for the underlying error distribution or a prior process in modeling.

In spatial and temporal investigations often it is of interest to study the relationships, described by covariance or correlation, between the random variables at different locations. For example, one may ask, "will the covariance between two random variables at two different locations depend on the two locations as well as the distance between the two?" A lot of simplification in analysis is afforded when it is assumed that the covariance only depends on the simple difference (given by the separation vector $\mathbf{s} - \mathbf{s}'$, $= \mathbf{h}$, say) between two locations $\mathbf{s}$ and $\mathbf{s}'$ and not on the actual locations $\mathbf{s}$ and $\mathbf{s}'$. A stochastic process $Y(\mathbf{s})$ is said to be covariance stationary if $\mathrm{Cov}(Y(\mathbf{s}), Y(\mathbf{s}')) = C(\mathbf{h})$ where C is a suitable function of the difference $\mathbf{h}$. The function $C(\mathbf{h})$ is called the covariance function of the stochastic process and plays a crucial role in many aspects of spatial analysis. The global nature of the covariance function $C(\mathbf{h})$, as it is free of any particular location in the domain $\mathbb{D}$, helps tremendously to simplify modeling and analysis and to specify joint distributions for the underlying random variables.

A stochastic process, $Y(\mathbf{s})$, is said to be *variance stationary* if its variance, $V(\mathbf{s})$, is a constant, say σ^2, over the whole domain $\mathbb{D}$. For a variance stationary process, no heterogeneity arises due to variation either in space or time. This is again a very strong assumption that may not hold in practice. However, while modeling we often assume that the underlying error distribution has a constant spatial variance. Other methods and tricks, such as data transformation and amalgamation of several processes are employed to model non-constant spatial variance.

A stochastic process, which is both mean stationary and variance stationary is called to be a *weakly stationary* (or second-order stationary) process. Note that weak stationarity does not say anything about the joint distribution of any collection of random variables, $Y(\mathbf{s}_1), \ldots, Y(\mathbf{s}_n)$ from the underlying stochastic process $Y(\mathbf{s})$. A stronger version of stationarity, called strict stationarity, goes someway to characterize the underlying joint distribution. A stochastic process $Y(\mathbf{s})$ is said to be *strictly (or strong) stationary*, if for any given $n \geq 1$, any set of n sites $\{\mathbf{s}_1, \mathbf{s}_2, \ldots, \mathbf{s}_n\}$, and any increment vector $\mathbf{h}$, the joint distribution of $Y(\mathbf{s}_1), \ldots, Y(\mathbf{s}_n)$ is the same as that of $Y(\mathbf{s}_1 + \mathbf{h}), \ldots, Y(\mathbf{s}_n + \mathbf{h})$. Thus, for a strictly stationary process, a shift in location will not result in any change in the joint distribution of the random variables. It can be shown that a strictly stationary stochastic process is also a covariance stationary stochastic process. However, the converse is not true in general since we cannot claim that two random variables will have the same distribution if respectively their means and variances happen to be equal. However, the converse is true when the underlying distribution is Gaussian and, in this case, the stochastic process is called a Gaussian Process (GP). The GPs are defined below.

The concept of *intrinsic stationarity* concerns the stationarity of the variance of the difference $Y(\mathbf{s}+\mathbf{h}) - Y(\mathbf{s})$. This property is related to the question, "is $\text{Var}(Y(\mathbf{s} + \mathbf{h}) - Y(\mathbf{s}))$ free of the location $\mathbf{s}$ and does it only depend on the separation vector $\mathbf{h}$?" If the answer is yes to both of these questions, then the process $Y(\mathbf{s})$ is said to be *intrinsically stationary*. Intrinsic stationarity and covariance stationarity are very strongly related as we discuss in the next section.

2.4   Variogram and covariogram

The quantity $\text{Var}(Y(\mathbf{s} + \mathbf{h}) - Y(\mathbf{s}))$ is called the *variogram* of the stochastic process, $Y(\mathbf{s})$ as it measures the variance of the first difference in the process at two different locations $\mathbf{s} + \mathbf{h}$ and $\mathbf{s}$. Our desire for a simplified analysis, using intrinsic stationarity, would dictate us to suppose that the variogram depends only on the separation vector $\mathbf{h}$ and not on the actual location $\mathbf{s}$.

There is a one-to-one relationship between the variogram under the assumption of mean and variance stationarity for a process $Y(\mathbf{s})$. Assuming mean and variance stationary we have $E(Y(\mathbf{s} + \mathbf{h})) = E(Y(\mathbf{s}))$ and $\mathrm{Var}(Y(\mathbf{s} + \mathbf{h})) = \mathrm{Var}(Y(\mathbf{s})) = C(\mathbf{0})$, the spatial variance. For an intrinsically stationary process, we have:

$$
\begin{aligned}
\mathrm{Var}(Y(\mathbf{s} + \mathbf{h}) - Y(\mathbf{s})) &= E\left\{Y(\mathbf{s} + \mathbf{h}) - Y(\mathbf{s}) - E(Y(\mathbf{s} + \mathbf{h}) + Y(\mathbf{s}))\right\}^2 \\
&= E\left\{(Y(\mathbf{s} + \mathbf{h}) - E(Y(\mathbf{s} + \mathbf{h}))) - (Y(\mathbf{s}) - E(Y(\mathbf{s})))\right\}^2 \\
&= E\left\{(Y(\mathbf{s} + \mathbf{h}) - E(Y(\mathbf{s} + \mathbf{h})))^2\right\} + E\left\{(Y(\mathbf{s}) - E(Y(\mathbf{s})))^2\right\} \\
&\quad - 2E\left\{(Y(\mathbf{s} + \mathbf{h}) - E(Y(\mathbf{s} + \mathbf{h})))(Y(\mathbf{s}) - E(Y(\mathbf{s})))\right\} \\
&= \mathrm{Var}(Y(\mathbf{s} + \mathbf{h})) + \mathrm{Var}(Y(\mathbf{s})) - 2\mathrm{Cov}(Y(\mathbf{s} + \mathbf{h}), Y(\mathbf{s})) \\
&= C(\mathbf{0}) + C(\mathbf{0}) - 2C(\mathbf{h}) \\
&= 2(C(\mathbf{0}) - C(\mathbf{h})).
\end{aligned}
$$

This result states that:

Variogram at separation $\mathbf{h} = 2\times$ { Spatial variance $-$ Spatial covariance function at separation $\mathbf{h}$ }.

Clearly, we can easily find the variogram if we already have a specification for the spatial covariance function for all values of its argument. However, it is not easy to retrieve the covariance function from a specification of the variogram function, $\mathrm{Var}(Y(\mathbf{s} + \mathbf{h}) - Y(\mathbf{s}))$. This needs further assumptions and limiting arguments, see e.g. Chapter 2 of Banerjee et al. (2015).

In order to study the behavior of a variogram, $2(C(\mathbf{0}) - C(\mathbf{h}))$, as a function of the covariance function $C(\mathbf{h})$, we see that the multiplicative factor 2 is only a distraction. This is why the semi-variogram, which is half of the variogram is conventionally studied in the literature. We use the notation $\gamma(\mathbf{h})$ to denote the semi-variogram, and thus $\gamma(\mathbf{h}) = C(\mathbf{0}) - C(\mathbf{h})$.

In practical modeling work we assume a specific valid covariance function $C(\mathbf{h})$ for the stochastic process and then the semi-variogram, $\gamma(\mathbf{h})$ is automatically determined. The word "valid" has been included in the previous sentence since a positive definiteness condition is required to ensure non-negativeness of variances of all possible linear combinations of the random variables $Y(\mathbf{s}_1), \ldots, Y(\mathbf{s}_n)$. The simplification provided by the assumption of intrinsic stationarity is still not enough for practical modeling work since it is still very hard to specify a valid multi-dimensional function $C(\mathbf{h})$ as a function of the separation vector $\mathbf{h}$. The crucial concept of isotropy, defined and discussed below, accomplishes this task of specifying the covariance function as a one-dimensional function.

2.5   Isotropy

So far the semi-variogram $\gamma(\mathbf{h})$, or the covariance function $C(\mathbf{h})$, has been assumed to depend on the multidimensional $\mathbf{h}$. This is very general and too broad, giving the modeler a tremendous amount of flexibility regarding the stochastic process as it varies over the spatial domain, $\mathbb{D}$. However, this flexibility throws upon a lot of burdens arising from the requirement of precise specification of the dependence structure as the process travels from one location to the next. That is, the function $C(\mathbf{h})$ needs to be specified for every possible value of multidimensional $\mathbf{h}$. Not only is this problematic from the purposes of model specification, but also it is hard to estimate all such precise features from data. Hence, the concept of *isotropy* is introduced to simplify the specification.

A covariance function $C(\mathbf{h})$ is said to be *isotropic* if it depends only on the length $\|\mathbf{h}\|$ of the separation vector $\mathbf{h}$. Isotropic covariance functions only depend on the distance but not on the angle or direction of travel. Assuming space to be in two dimensions, an isotropic covariance function guarantees that the covariance between two random variables, one at the center of a circle and the other at any point on the circumference is the same as the covariance between the two random variables one at the center and another at other point on the circumference of the same circle. Thus, the covariance does not depend on where and which direction the random variables are recorded on the circumference of the circle. Hence, such covariance functions are called omni-directional.

Abusing notations an isotropic covariance function, $C(\cdot)$ is denoted by $C(\|\mathbf{h}\|)$ or simply by $C(d)$ where $d \geq 0$ is a scalar distance between two locations. The notation $C(\cdot)$ has been abused here since earlier we talked about $C(\mathbf{h})$ where $\mathbf{h}$ is a multi-dimensional separation vector, but now the same C is used to denote the one-dimensional covariance function $C(d)$. A covariance function is called *anisotropic* if it is not isotropic.

In practice it may seem that the isotropic covariance functions are too restrictive as they are rigid in not allowing flexible covariance structure for the underlying stochastic process. For example, a pollution plume can only spread through using the prevailing wind direction, e.g. east–west. Indeed, this is true, and often, the assumption of isotropy is seen as a limitation of the modeling work. However, the overwhelming simplicity still trumps all the disadvantages, and isotropic covariance functions are used for the underlying true error process. Many mathematical constructs and practical tricks are used to build anisotropic covariance functions, see e.g. Chapter 3 of Banerjee et al. (2015).

2.6   Matèrn covariance function

In practical modeling work we need to explicitly specify a particular covariance function so that the likelihood function can be written for the purposes of parameter estimation. In this section we discuss the most commonly used covariance function, namely the Matèrn family (Matèrn, 1986) of covariance functions as an example of isotropic covariance functions. We discuss its special cases, such as the exponential and Gaussian. To proceed further recall from elementary definitions that covariance is simply variance times correlation if the two random variables (for which covariance is calculated) have the same variance. In spatial and spatio-temporal modeling, we assume equal spatial variance, which we denote by σ^2. Isotropic covariance functions depend on the distance between two locations, which we denote by d. Thus, the covariance function we are about to introduce will have the form

$$C(d) = \sigma^2 \rho(d), \ d > 0$$

where $\rho(d)$ is the correlation function. Note also that when $d = 0$, the covariance is the same as the variance and should be equal to σ^2. Indeed, we shall assume that $\rho(d) \to 1$ as $d \to 0$. Henceforth we will only discuss covariance functions in the domain when $d > 0$.

How should the correlation functions behave as the distance d increases? For most natural and environmental processes, the correlation should decay with increasing d. Indeed, the Tobler's first law of Geography (Tobler, 1970) states that, "everything is related to everything else, but near things are more related than distant things." Indeed, there are stochastic processes where we may want to assume no correlation at all for any distance d above a threshold value. Although this sounds very attractive and intuitively simple there are technical difficulties in modeling associated with this approach since an arbitrary covariance function may violate the requirement of non-negative definiteness of the variances. More about this requirement is discussed below in Section 2.7. There are mathematically valid ways to specify negligible amounts of correlations for large distances. See for example, the method of tapering discussed by Kaufman et al. (2008).

Usually, the correlation function $\rho(d)$ should monotonically decrease with increasing value of d due to the Tobler's law stated above. The particular value of d, say d_0, which is a solution of the equation $\rho(d) = 0$ is called the *range*. This implies that the correlation is exactly zero between any two random observations observed at least the range d_0 distance apart. Note that due to the monotonicity of the correlation function, it cannot climb up once it reaches the value zero for some value of the distance d. With the added assumption of Gaussianity for the data, the range d_0 is the minimum distance beyond which any two random observations are deemed to be independent. With such assumptions we claim that the underlying process does not get affected by the same process, which is taking place at least d_0 distance away.

An analytical specification for the correlation function $\rho(d)$, such as the Matèrn family defined below may not allow a finite value for the range. That is, it may not be practically possible to solve $\rho(d) = 0$ for a finite value of d. In such situations we define the *effective range* to be the minimum distance after which there is only very negligible amount of correlation, e.g. 0.05, although other values lower than 0.05 but greater than 0 can be used. Often we define the effective range as the solution of the equation $\rho(d) = 0.05$.

The Matèrn family of covariance functions provides a very general choice (satisfying all the requirements) and is given by:

$$C(d|\boldsymbol{\psi}) = \sigma^2 \rho(d|\boldsymbol{v}) \tag{2.1}$$

where

$$\rho(d|\boldsymbol{v}) = \frac{1}{2^{\nu-1}\Gamma(\nu)}(\sqrt{2\nu}\phi d)^\nu K_\nu(\sqrt{2\nu}\phi d), \quad \phi > 0, \nu > 0, d > 0 \tag{2.2}$$

where $\boldsymbol{\psi} = (\sigma^2, \boldsymbol{v})$, $\boldsymbol{v} = (\phi, \nu)$, $\Gamma(\nu)$ is the mathematical Gamma function and $K_\nu(\cdot)$ is the modified Bessel function of second kind of order ν, see e.g. Abramowitz and Stegun (1965, Chapter 9). Note that the covariance function $C(d|\boldsymbol{\psi})$ has been made to depend on two parameters ν and ϕ besides σ^2 and the new notation $C(d|\boldsymbol{\psi})$ explicitly recognizes this fact. The parameter ν in $C(d|\boldsymbol{\psi})$ determines the smoothness of the covariance function and the parameter ϕ dictates the rate of decay as d increases. Incidentally, there appears to be a typo in the formula (2.8) of Banerjee et al. (2015) where the factor 2 in the numerator should be actually $\sqrt{2}$. Otherwise, the special case of exponential covariance function discussed below does not seem to be obtainable when $\nu = 0.5$.

Higher values of the smoothness parameter ν imply more smoothness in the underlying stochastic process as can be seen from Figure 2.1. (R Code Notes 2.1 explains how this figure is drawn.) Intuitively, smoother processes are easier to predict. Hence a highly smooth stochastic process can be perfectly predicted based on only a few realizations. As an example consider predicting the smooth process, $y(\mathbf{s}) = a + bx(\mathbf{s})$, which is a straight line. Only two realizations from the straight line allows us to predict any point in the entire straight line. Note that a process having an infinite amount of smoothness is essentially a smooth deterministic process, which may not be very appropriate in practice. Thus, we often assume that the underlying process contains only a finite degree of smoothness.

Different values of ν give rise to different popular special cases of the Matèrn family of covariance function; see Figure 2.2. In case $\nu = 1/2$ the covariance function $C(d|\boldsymbol{\psi})$ simplifies to $\sigma^2 \exp(-\phi d)$ and is appropriately called the exponential covariance function. This is one of the most popular choice of the covariance function in practice because of several reasons besides its simplicity. For example, the correlation function $\rho(d|\phi) = \exp(-\phi d)$ decays exponentially as the distance d between two locations increases. The equation $\rho(d|\phi) = \exp(-\phi d) = 0$ does not yield a finite solution for any fixed given positive value of ϕ, hence there is no finite range associated with the exponential covariance function. The effective range, solution to the equation $\exp(-\phi d) = 0.05$, is $-\log(0.05)/\phi$ or $3/\phi$. This easy interpretation helps us to choose the decay parameter ϕ from our practical knowledge. We may also use this relationship to specify the domain of the prior distribution. For example, if it is only plausible for the underlying spatial process to have an effective range between 10 and 100 kilometers, then the decay parameter ϕ must lie between 0.03 and 0.3.

Other examples of closed form covariance function from the Matèrn family include the case $C(d|\psi) = \sigma^2(1 + \phi d)\exp(-\phi d)$ corresponding to $\nu = 3/2$. Assumption of this covariance function will yield a smoother process realization than that from the exponential correlation function discussed above. Another interesting example of the Matèrn family is the Gaussian covariance function when $\nu \to \infty$. The Gaussian covariance function is given by $C(d|\psi) = \sigma^2\exp(-\phi^2 d^2), d > 0$. Note that the correlation function is an exponentially decaying function of d^2. On the other hand, the exponential correlation function is a decaying function of d. Hence the correlation under the Gaussian covariance will die down at a much faster rate than the same under the exponential covariance function. As a result, the Gaussian correlation function does not provide a good fit in practical studies where spatial correlation persists over large distances.

♣ **R Code Notes 2.1. Figure 2.1** Drawing each panel in this figure is an elaborate process sketched below. In the first step, we generate $n = 500$ locations in the unit square and obtain the $n \times n$ distance matrix d between these locations. We also generate n iid standard normal random variables.

The bmstdr package contains a function called materncov which calculates the Matèrn covariance function for a given distance d and the two required parameters ϕ and ν. The parameterization of this function is as given in (2.2).

In the second step we calculate the Matèrn covariance matrix, Σ say, and use Cholesky decomposition of Σ to transform the standard normal random variables to the n-dimensional multivariate normal random variable with covariance matrix Σ.

In the third step we use the interp function from the akima library to have an image surface that can be drawn. The interpolated surface is organized as a data frame using the gather command in the tidyr library.

Finally, the functions geom_raster and stat_contour in ggplot draw the plot. We use the same set of standard normal random variates to draw all four panels in this figure.

Matern field with nu=0.5

Matern field with nu=2.5

Matern field with nu=1.5

Matern field with nu=3.5

FIGURE 2.1: Random fields generated using the Matèrn correlation function for different values of the smoothness parameter ν. The parameters ϕ and σ^2 are both kept at 1. The plot on the top left corner is the one for exponential covariance function which has been used throughout in this book for point referenced data modeling.

2.7 Gaussian processes (GP) $GP(0, C(\cdot|\boldsymbol{\psi}))$

Often Gaussian processes are assumed as components in spatial and spatio-temporal modeling. These stochastic processes are defined over a continuum, e.g. a spatial study region and specifying the resulting infinite dimensional random variable is often a challenge in practice. Gaussian processes are very convenient to work in these settings since they are fully defined by a mean function, say $\mu(\mathbf{s})$ and a valid covariance function, say $C(||\mathbf{s} - \mathbf{s}^*||) = \text{Cov}(Y(\mathbf{s}), Y(\mathbf{s}^*))$, which is required to be positive definite. A covariance function is said to be positive definite if the covariance matrix, implied by that covariance function, for a finite number of random variables belonging to that process is positive definite.

Suppose that the stochastic process $Y(\mathbf{s})$, defined over a continuous spatial region $\mathbb{D}$, is assumed to be a GP with mean function $\mu(\mathbf{s})$ and covariance function $C(\mathbf{s}, \mathbf{s}^*)$. Note that since $\mathbf{s}$ is any point in $\mathbb{D}$, the process $Y(\mathbf{s})$ defines a non-countably infinite number of random variables. However, in practice

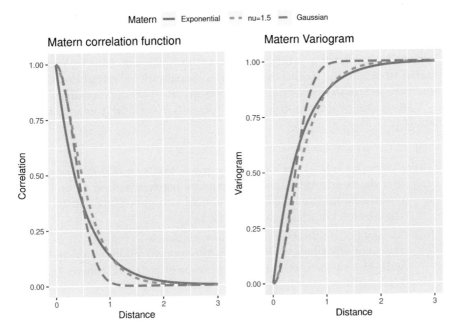

FIGURE 2.2: The Matèrn correlation function (left panel) and variogram (right panel) for different values of the smoothness parameter ν when $\phi = 1$.

the GP assumption guarantees that for any finite n and any set of n locations $\mathbf{s}_1, \ldots, \mathbf{s}_n$ within $\mathbb{D}$ the n-variate random variable $\mathbf{Y} = (Y(\mathbf{s}_1), \ldots, Y(\mathbf{s}_n))$ is normally distributed with mean $\boldsymbol{\mu}$ and covariance matrix Σ given by:

$$
\boldsymbol{\mu} = \begin{pmatrix} \mu(\mathbf{s}_1) \\ \mu(\mathbf{s}_2) \\ \vdots \\ \mu(\mathbf{s}_n) \end{pmatrix}, \quad \Sigma = \begin{pmatrix} C(0) & C(d_{12}) & \cdots & C(d_{1n}) \\ C(d_{21}) & C(0) & \cdots & C(d_{2n}) \\ \vdots & \vdots & \ddots & \vdots \\ C(d_{n1}) & C(d_{n2}) & \cdots & C(0) \end{pmatrix},
$$

where $d_{ij} = \|\mathbf{s}_i - \mathbf{s}_j\|$ is the distance between the two locations $\mathbf{s}_i$ and $\mathbf{s}_j$. From the multivariate normal distribution in Section A.1 in Appendix A, we can immediately write down the joint density of $\mathbf{Y}$ for any finite value of n. However, the unresolved matter is how do we specify the two functions $\mu(\mathbf{s}_i)$ and $C(d_{ij})$ for any i and j. The GP assumption is often made for the error process just as in usual regression modeling the error distribution is assumed to be Gaussian. Hence often a GP assumption comes with $\mu(\mathbf{s}) = 0$ for all $\mathbf{s}$. The next most common assumption is to assume the Matèrn covariance function $C(d_{ij}|\boldsymbol{\psi})$ written down in (2.1) for $C(d_{ij})$. The Matèrn family provides a valid family of positive definite covariance functions, and it is the only family used in this book.

We now introduce the GP notation $GP(\mathbf{0}, C(\cdot|\boldsymbol{\psi}))$ which will be used throughout the book. In this definition we use the notation $w(\mathbf{s})$ to denote the GP as a zero-mean spatial random variable and reserve $Y(\mathbf{s})$ for the data which will often have non-zero mean structures. We also use the lower case letter w to denote the stochastic process in keeping with the tradition of using lower case letters to denote random effects in mixed effects modeling.

A stochastic process $w(\mathbf{s})$ is said to follow $GP(\mathbf{0}, C(\cdot|\boldsymbol{\psi}))$ in a spatial domain $\mathbb{D}$ if:

(i) $E(w(\mathbf{s})) = 0$ for any $\mathbf{s} \in \mathbb{D}$;

(ii) $\text{Cov}(w(\mathbf{s}_i), w(\mathbf{s}_j)) = C(d_{ij}|\boldsymbol{\psi})$ for any $\mathbf{s}_i$ and $\mathbf{s}_j$ inside the region $\mathbb{D}$ and d_{ij} is any valid distance measure between $\mathbf{s}_i$ and $\mathbf{s}_j$ and the covariance function $C(d_{ij}|\boldsymbol{\psi})$ is a valid covariance function such as the Matèrn covariance function (2.1);

(iii) for any finite collection of random variables $\mathbf{w} = (w(\mathbf{s}_1), \ldots, w(\mathbf{s}_n))$ the joint distribution of $\mathbf{w}$ is n-dimensional multivariate normal with the mean vector $\mathbf{0}$ and covariance matrix Σ where $\Sigma_{ij} = C(d_{ij}|\boldsymbol{\psi})$ for $i, j = 1, \ldots, n$.

The joint density of $\mathbf{w}$ is given by:

$$f(\mathbf{w}|\boldsymbol{\psi}) = \left(\frac{1}{2\pi}\right)^{\frac{n}{2}} |\Sigma|^{-\frac{1}{2}} e^{-\frac{1}{2}\mathbf{w}'\Sigma^{-1}\mathbf{w}},$$

from (A.24) in Appendix A. This density will provide the likelihood function for estimating $\boldsymbol{\psi} = (\sigma^2, \phi, \nu)$.

Gaussian processes are often preferred in spatial modeling because of the above attractive distribution theory. Moreover, Kriging or spatial prediction, defined below, at yet unobserved locations conditionally on the observed data is facilitated by means of a conditional distribution, which is also normal. This convenient distribution theory is very attractive for spatial prediction in the context of modern, fully model-based spatial analysis within a Bayesian framework. The spatial predictive distributions are easy to compute and simulate from an iterative MCMC framework.

2.8   Space-time covariance functions

A particular covariance structure must be assumed for the $Y(\mathbf{s}_i, t)$ process. The pivotal space-time covariance function is defined as

$$C(\mathbf{s}_1, \mathbf{s}_2; t_1, t_2) = \text{Cov}[Y(\mathbf{s}_1, t_1), Y(\mathbf{s}_2, t_2)].$$

The zero mean spatio-temporal process $Y(\mathbf{s}, t)$ is said to be *covariance stationary* if

$$C(\mathbf{s}_1, \mathbf{s}_2; t_1, t_2) = C(\mathbf{s}_1 - \mathbf{s}_2; t_1 - t_2) = C(d; \tau),$$

where $d = \mathbf{s}_1 - \mathbf{s}_2$ and $\tau = t_1 - t_2$. The process is said to be *isotropic* if

$$C(d; \tau) = C(||d||; |\tau|),$$

that is, the covariance function depends upon the separation vectors only through their lengths $||d||$ and $|\tau|$. Processes which are not isotropic are called *anisotropic*. In the literature isotropic processes are popular because of their simplicity and interpretability. Moreover, there is a number of simple parametric forms available to model those.

A further simplifying assumption to make is the assumption of separability; see for example, Mardia and Goodall (1993). Separability is a concept used in modeling multivariate spatial data including spatio-temporal data. A *separable covariance function* in space and time is simply the product of two covariance functions one for space and the other for time.

The process $Y(\mathbf{s}, t)$ is said to be *separable* if

$$C(||d||; |\tau|) = C_s(||d||) \, C_t(|\tau|).$$

Now suitable forms for the functions $C_s(\cdot)$ and $C_t(\cdot)$ are to be assumed. A very general choice is to adopt the Matèrn covariance function introduced before.

There is a growing literature on methods for constructing non-separable and non-stationary spatio-temporal covariance functions that are useful for modeling. See for example, Gneiting (2002) who develops a class of non-separable covariance functions. A simple example is:

$$C(||d||; |\tau|) = (1 + |\tau|)^{-1} \exp\left\{-||d||/(1 + |\tau|)^{\beta/2}\right\}, \qquad (2.3)$$

where $\beta \in [0, 1]$ is a space-time interaction parameter. For $\beta = 0$, (2.3) provides a separable covariance function. The other extreme case at $\beta = 1$ corresponds to a totally non-separable covariance function. Figure 2.3 plots this function for four different values: 0, 0.25, 0.5 and 1 of β. There are some discernible differences between the functions can be seen for higher distances at the top right corner of each plot. However, it is true that it is not easy to describe the differences, and it gets even harder to see differences in model fits. The paper by Gneiting (2002) provides further descriptions of non-separable covariance functions.

There are other ways to construct non-separable covariance functions, for example, by mixing more than one spatio-temporal processes, see e.g. Sahu et al. (2006) or by including a further level of hierarchy where the covariance matrix obtained using $C(||d||; |\tau|)$ follows a inverse-Wishart distribution centred around a separable covariance matrix. Section 8.3 of the book by Banerjee et al. (2015) also lists many more strategies. For example, Schmidt and O'Hagan (2003) construct non-stationary spatio-temporal covariance structure via deformations.

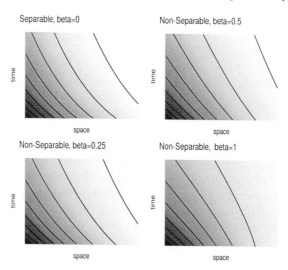

FIGURE 2.3: Space-time covariance function (2.3) at different distances in space and time for different values of β. The top left hand panel is an example of a separable covariance function. All other panels show non-separability reaching the extreme level at $\beta = 1$ in the bottom right panel.

♣ **R Code Notes 2.2. Figure 2.3**
To draw this figure we basically follow the R Code Notes 2.1. First we write a simple function

```
nonsep <- function(d, tau, beta=1) {1/(1+tau) * exp(-d/(1+
   tau)^(beta/2))}
```

We then use the `outer` function to find the values of the covariance function (2.3) over a grid of values of d and τ, the arguments of the function `nonsep`. These evaluations are repeated for four different values of β labeled in the plot.

In the next step we use the `interp` function from the `akima` library to have an image surface that can be drawn. The interpolated surface is organized as a data frame using the `gather` command in the `tidyr` library.

Finally, the functions `geom_raster` and `stat_contour` in `ggplot` draw the plot. We use the color palette generated by the command `brewer.pal` (9,"YlOrRd") using the `RColorBrewer` package.

2.9 Kriging or optimal spatial prediction

The jargon "Kriging" refers to a form of spatial prediction at an unobserved location based on the observed data. That it is a popular method is borne by the fact that "Kriging" is verbification of a method of spatial prediction named after its inventor D.G. Krige, a South African mining engineer. Kriging solves the problem of predicting $Y(\mathbf{s})$ at a new location $\mathbf{s}_0$ having observed data $y(\mathbf{s}_1), \ldots, y(\mathbf{s}_n)$.

Classical statistical theory based on squared error loss function in prediction will yield the sample mean $\bar{y}$ to be the optimal predictor for $Y(\mathbf{s}_0)$ if spatial dependency is ignored between the random variables $Y(\mathbf{s}_0), Y(\mathbf{s}_1), \ldots, Y(\mathbf{s}_n)$. Surely, because of the Tobler's law, the prediction for $Y(\mathbf{s}_0)$ will be improved if instead spatial dependency is taken into account. The observations nearer to the prediction location, $\mathbf{s}_0$, will receive higher weights in the prediction formula than the observations further apart. So, now the question is how do we determine these weights? Kriging provides the answer.

In order to proceed further we assume that $Y(\mathbf{s})$ is a GP, although some of the results we discuss below also hold in general without this assumption. In order to perform Kriging it is assumed that the best linear unbiased predictor with weights l_i, $\hat{Y}(\mathbf{s}_0)$, is of the form $\sum_{i=1}^n \ell_i Y(\mathbf{s}_i)$ and dependence between the $Y(\mathbf{s}_0), Y(\mathbf{s}_1), \ldots, Y(\mathbf{s}_n)$ is described by a covariance function, $C(d|\psi)$, of the distance d between any two locations as defined above in this chapter. The Kriging weights are easily determined by evaluating the conditional mean of $Y(\mathbf{s}_0)$ given the observed values $y(\mathbf{s}_1), \ldots, y(\mathbf{s}_n)$. These weights are "optimal" in the same statistical sense that the mean $E(X)$ minimizes the expected value of the squared error loss function, i.e. $E(X - a)^2$ is minimized at $a = E(X)$. Here we take X to be the conditional random variable $Y(\mathbf{s}_0)$ given $y(\mathbf{s}_1), \ldots, y(\mathbf{s}_n)$.

The actual values of the optimal weights are derived by partitioning the mean vector, $\boldsymbol{\mu}_{n+1}$ and the covariance matrix, Σ, of $Y(\mathbf{s}_0), Y(\mathbf{s}_1), \ldots, Y(\mathbf{s}_n)$ as follows. Let

$$\boldsymbol{\mu}_{n+1} = \begin{pmatrix} \mu_0 \\ \boldsymbol{\mu} \end{pmatrix}, \quad \Sigma = \begin{pmatrix} \sigma_{00} & \Sigma_{01} \\ \Sigma_{10} & \Sigma_{11} \end{pmatrix}$$

where $\boldsymbol{\mu}$ is the vector of the means of $\mathbf{Y} = (Y(\mathbf{s}_1), \ldots, Y(\mathbf{s}_n))'$; $\sigma_{00} = \mathrm{Var}(Y(\mathbf{s}_0))$; $\Sigma_{01} = \Sigma'_{1,0} = \mathrm{Cov} \begin{pmatrix} Y(\mathbf{s}_0) \\ \mathbf{Y}_1 \end{pmatrix}$; $\Sigma_{11} = \mathrm{Var}(\mathbf{Y}_1)$. Now standard multivariate normal distribution theory tells us that

$$Y(\mathbf{s}_0)|\mathbf{y} \sim N\left(\mu_0 + \Sigma_{01}\Sigma_{11}^{-1}(\mathbf{y} - \boldsymbol{\mu}), \sigma_{00} - \Sigma_{01}\Sigma_{11}^{-1}\Sigma_{10}\right).$$

In order to facilitate a clear understanding of the underlying spatial dependence on Kriging we assume a zero-mean GP, i.e. $\boldsymbol{\mu}_{n+1} = \mathbf{0}$. Now we have $E(Y(\mathbf{s}_0)|\mathbf{y}) = \Sigma_{01}\Sigma_{11}^{-1}\mathbf{y}$ and thus we see that the optimal Kriging weights are particular functions of the assumed covariance function. Note that the weights do not depend on the underlying common spatial variance as that is canceled in the product $\Sigma_{01}\Sigma_{11}^{-1}$. However, the spatial variance will affect the accuracy of the predictor since $\text{Var}(Y(\mathbf{s}_0)|\mathbf{y}) = \sigma_{00} - \Sigma_{01}\Sigma_{11}^{-1}\Sigma_{10}$.

It is interesting to note that Kriging is an exact predictor in the sense that $E(Y(\mathbf{s}_i)|\mathbf{y}) = y(\mathbf{s}_i)$ for any $i = 1, \ldots, n$. It is intuitively clear why this result will hold. This is because a random variable is an exact predictor of itself. Mathematically, this can be easily proved using the definition of inverse of an matrix. To elaborate further, suppose that

$$\Sigma = \begin{pmatrix} \Sigma_1' \\ \Sigma_2' \\ \vdots \\ \Sigma_n' \end{pmatrix}$$

where Σ_i' is a row vector of dimension n. Then the result $\Sigma\Sigma^{-1} = I_n$ where I_n is the identity matrix of order n, implies that $\Sigma_i\Sigma^{-1} = \mathbf{a}_i$ where the ith element of $\mathbf{a}_i$ is 1 and all others are zero.

The above discussion, with the simplified assumption of a zero mean GP, is justified since often in practical applications we only assume a zero-mean GP as a prior distribution. The mean surface of the data (or their transformations) is often explicitly modeled by a regression model and hence such models will contribute to determine the mean values of the predictions. In this context we note that in a non-Bayesian geostatistical modeling setup there are various flavors of Kriging such as simple Kriging, ordinary Kriging, universal Kriging, co-Kriging, intrinsic Kriging, depending on the particular assumption of the mean function. In our Bayesian inference set up such flavors of Kriging will automatically ensue since Bayesian inference methods are automatically conditioned on observed data and the explicit model assumptions.

2.10   Autocorrelation and partial autocorrelation

A study of time series for temporally correlated data will not be complete without the knowledge of autocorrelation. Simply put, autocorrelation means correlation with itself at different time intervals. The time interval is technically called the *lag* in the time series literature. For example, suppose Y_t is a time series random variable where $t \geq 1$ is an integer. The *autocorrelation* at lag $k(\geq 1)$ is defined as $\rho_k = \text{Cor}(Y_{t+k}, Y_t)$. It is obvious that the autocorrelation at lag $k = 0$, ρ_0, is one. Ordinarily, ρ_k decreases as k increases just as the spatial correlation decreases when the distance between two locations increases. Viewed as a function of the lag k, ρ_k is called the autocorrelation function, often abbreviated as ACF.

Sometimes high autocorrelation at any lag $k > 1$ persists because of high correlation between Y_{t+k} and the intermediate time series, $Y_{t+k-1}, \ldots, Y_{t+1}$. The *partial autocorrelation* at lag k measures the correlation between Y_{t+k} and Y_t after removing the autocorrelation at shorter lags. Formally, partial autocorrelation is defined as the conditional autocorrelation between Y_{t+k} and Y_t given the values of $Y_{t+k-1}, \ldots, Y_{t+1}$. The partial correlation can also be easily explained with the help of multiple regression. To remove the effects of intermediate time series $Y_{t+k-1}, \ldots, Y_{t+1}$ one considers two regression models: one Y_{t+k} on $Y_{t+k-1}, \ldots, Y_{t+1}$ and the other Y_t on $Y_{t+k-1}, \ldots, Y_{t+1}$. The simple correlation coefficient between two sets of residuals after fitting the two regression models is the partial auto-correlation at a given lag k. To learn more the interested reader is referred to many excellent introductory text books on time series such as the one by Chatfield (2003).

2.11   Measures of spatial association for areal data

Exploration of areal spatial data requires definition of a sense of spatial distance between all the constituting areal units within the data set. This measure of distance is parallel to the distance d between any two point referenced spatial locations discussed previously in this chapter. A blank choropleth map, e.g. Figure 1.12 without the color gradients, provides a quick visual measure of spatial distance, e.g. California, Nevada and Oregon in the west coast are spatial neighbors but they are quite a long distance away from Pennsylvania, New York and Connecticut in the east coast. More formally, the concept of spatial distance for areal data is captured by what is called a neighborhood, or a proximity, or an adjacency, matrix. This is essentially a matrix where each of its entry is used to provide information on the spatial relationship between each possible pair of the areal units in the data set.

The *proximity matrix*, denoted by W, consists of weights which are used to represent the strength of spatial association between the different areal units. Assuming that there are n areal units, the matrix W is of the order $n \times n$ where each of its entry w_{ij} contains the strength of spatial association between the units i and j, for $i, j = 1, \ldots, n$. Customarily, w_{ii} is set to 0 for each $i = 1, \ldots, n$. Commonly, the weights w_{ij} for $i \neq j$ are chosen to be binary where it is assigned the value 1 if units i and j share a common boundary and 0 otherwise. This proximity matrix can readily be formed just by inspecting a choropleth map, such as the one in Figure 1.12. However, the weighting function can instead be designed so as to incorporate other spatial information, such as the distances between the areal units. If required, additional proximity matrices can be defined for different orders, whereby the order dictates the proximity of the areal units. For instance we may have a first order proximity matrix representing the direct neighbors for an areal unit, a second order proximity matrix representing the neighbors of the first order areal units and so on. These considerations will render a proximity matrix, which is symmetric, i.e. $w_{ij} = w_{ji}$ for all i and j.

The weighting function w_{ij} can be standardized by calculating a new proximity matrix given by $\tilde{w}_{ij} = w_{ij}/w_{i+}$ where $w_{i+} = \sum_{j=1}^{n} w_{ij}$, so that each areal unit is given a sense of "equality" in any statistical analysis. However, in this case the new proximity matrix may not remain symmetric, i.e. $\tilde{w}_{ij}$ may or may not equal $\tilde{w}_{ji}$ for all i and j.

When working with grid based areal data, where the proximity matrix is defined based on touching areal units, it is useful to specify whether "queen" or "rook", in a game of chess, based neighbors are being used. In the R package spdep, "queen" based neighbors refer to any touching areal units, whereas "rook" based neighbors use the stricter criteria that both areal units must share an edge (Bivand, 2020).

There are two popular measures of spatial association for areal data which together serve as parallel to the concept of the covariance function, and equivalently variogram, defined earlier in this chapter. The first of these two measures is the Moran's I (Moran, 1950) which acts as an adaptation of Pearson's correlation coefficient and summarizes the level of spatial autocorrelation present in the data. The measure I is calculated by comparing each observed area i to its neighboring areas using the weights, w_{ij}, from the proximity matrix for all $j = 1, \ldots, n$. The formula for Moran's I is written as:

$$I = \frac{n}{\sum_{i \neq j} w_{ij}} \frac{\sum_{i=1}^{n} \sum_{j=1}^{n} w_{ij}(Y_i - \bar{Y})(Y_j - \bar{Y})}{\sum_{i=1}^{n} (Y_i - \bar{Y})^2}, \tag{2.4}$$

where $Y_i, i = 1, \ldots, n$ is the random sample from the n areal units and $\bar{Y}$ is the sample mean. It can be shown that I lies in the interval $[-1, 1]$, and its sampling variance can be found, see e.g. Section 4.1 in Banerjee et al. (2015) so that an asymptotic test can be performed by appealing to the central limit theorem. For small values of n there are permutation tests which compares the observed value of I to a null distribution of the test statistic I obtained by simulation. We shall illustrate these with a real data example in Section 3.4.

An alternative to the Moran's I is the Geary's C (Geary, 1954) which also measures spatial autocorrelation present in the data. The Geary's C is given by

$$C = \frac{(n-1)}{2 \sum_{i \neq j} w_{ij}} \frac{\sum_{i=1}^{n} \sum_{j=1}^{n} w_{ij}(Y_i - Y_j)^2}{\sum_{i=1}^{n} (Y_i - \bar{Y})^2}. \tag{2.5}$$

The measure C being the ratio of two weighted sum of squares is never negative. It can be shown that $E(C) = 1$ under the assumption of no spatial association. Small values of C away from the mean 1 indicate positive spatial association. An asymptotic test can be performed but the speed of convergence to the limiting null distribution is expected to be very slow since it is a ratio of weighted sum of squares. Monte Carlo permutation tests can be performed and those will be illustrated in Section 3.4 with a real data example.

2.12   Internal and external standardization for areal data

Internal and external standardization are two oft-quoted keywords in areal data modeling, especially in disease mapping where rates of a disease over different geographical (areal) units are compared. These two are now defined along with other relevant key words. To facilitate the comparison often we aim to understand what would have happened if all the areal units had the same uniform rate. This uniform rate scenario serves as a kind of a null hypothesis of "no spatial clustering or association". Disease incidence rates in excess or in deficit relative to the uniform rate is called the *relative risk*. Relative risk is often expressed as a ratio where the denominator corresponds to the standard dictated by the above null hypothesis. Thus, a relative risk of 1.2 will imply 20% increased risk relative to the prevailing standard rate. The relative risk can be associated with a particular geographical areal unit or even for the whole study domain when the standard may refer to an absence of the disease. Statistical models are often postulated for the relative risk for the ease of interpretation.

Return to the issue of comparison of disease rates relative to the uniform rate. Often in practical data modeling situation, the counts of number of individuals over different geographies and other categories, e.g. sex and ethnicity, are available. Standardization, internal and external, is a process by which we obtain the corresponding counts of diseased individuals under the assumption of the null hypothesis of uniform disease rates being true. We now introduce the notation n_i, for $i = 1, \ldots, k$ being the total number of individuals in region i and y_i being the observed number of individuals with the disease, often called cases, in region i. Under the null hypothesis

$$\bar{r} = \frac{\sum_{i=1}^{k} y_i}{\sum_{i=1}^{k} n_i}$$

will be an estimate of the uniform disease rate. As a result,

$$E_i = n_i \bar{r}$$

will be the expected number of individuals with the disease in region i if the null hypothesis of uniform disease rate is true. Note that $\sum_{i=1}^{k} E_i = \sum_{i=1}^{k} y_i$ so that the total number of observed and expected cases are same. Note that to find E_i we used the observations y_i, $i = 1, \ldots, k$. This process of finding the E_i's is called internal standardization. The word *internal* highlights the use of the data itself to perform the standardization.

The technique of internal standardization is appealing to the analysts since no new external data are needed for the purposes of modeling and analysis. However, this technique is often criticized since in the modeling process E_i's are treated as fixed values when in reality these are functions of the random observations y_i's of the associated random variables Y_i's. Modeling of the Y_i's while treating the E_is as fixed is the unsatisfactory aspect of this strategy. To overcome this drawback the concept of external standardization is often used and this is what is discussed next.

External standardization is an elementary and basic method in epidemiology to achieve comparability of mortality or morbidity across different subpopulations, see e.g. Lilienfeld and Lilienfeld (1980) or Bland (2000). External standardization estimates the E_i's by estimating the $\bar{r}$ using external data instead of the y_i's. An example of external data is a weighted average of a national age and sex specific death rates. The weights in the weighted average are the relative size of the age and sex specific cohorts in the population. This is also sometimes termed as *direct standardization*. Applying the rates of the reference population to the number at risk in the study population gives the *expected* number of events E which then is compared with the observed number of events E, leading to the standardized event ratio, more popularly known as the standardized mortality (or morbidity) ratio SMR $= Y/E$. Often the SMR is presented as multiplied by 100 which suggests the interpretation as percentage, which is misleading. The SMR is a rate ratio comparing the event rate in study population to the event rate in the reference population.

The previously discussed internal standardization is also known as *indirect standardization*. Indirect standardization is often considered the preferred method when the study populations have only few events so that stratification, as required by the direct method, would contribute to instability of the stratified estimates (Miettinen, 1985). Indirect standardization uses the concept of what number of events would be expected if the study population had the stratum distribution of a reference population. If this reference population is available the indirect standardization turns out to be *external*.

Statistical inference for the SMR is widely based on the assumption that Y follows a Poisson distribution with the number of expected events E as the offset, see e.g. Keiding (1987) and Clayton and Hills (1993). One difference between internal and external indirect standardization is that for the internal method the sum of the number of observed event over all study populations equals the sum of expected number of events over all study populations. This is not necessarily the case for the external indirect method.

It should be pointed out that these concepts of standardization are also widely used in demography, see e.g. Hinde (1998). A practical example of performing standardization is provided in Section 3.4.1.

2.13   Spatial smoothers

Observed spatially referenced data will not be smooth in general due to the presence of noise and many other factors, such as data being observed at a coarse irregular spatial resolution where observation locations are not on a regular grid. Such irregular variations hinder making inference regarding any dominant spatial pattern that may be present in the data. Hence researchers often feel the need to smooth the data to discover important discernible spatial trend from the data. Statistical modeling, as proposed in this book, based on formal coherent methods for fitting and prediction, is perhaps the best formal method for such smoothing needs. However, researchers often use many non-rigorous off-the shelf methods for spatial smoothing either as exploratory tools demonstrating some key features of the data or more dangerously for making inference just by "eye estimation" methods. Our view in this book is that we welcome those techniques primarily as exploratory data analysis tools but not as inference making tools. Model based approaches are to be used for smoothing and inference so that the associated uncertainties of any final inferential product may be quantified fully.

For spatially point referenced data we briefly discuss the inverse distance weighting (IDW) method as an example method for spatial smoothing. There are many other methods based on Thiessen polygons and crude application of Kriging (using ad-hoc estimation methods for the unknown parameters). These, however, will not be discussed here due to their limitations in facilitating rigorous model based inference.

To perform spatial smoothing, the IDW method first prepares a fine grid of locations covering the study region. The IDW method then performs interpolation at each of those grid locations separately. The formula for interpolation is a weighted linear function of the observed data points where the weight for each observation is inversely proportional to the distance between the observation and interpolation locations. Thus to predict $Y(\mathbf{s}_0)$ at location $\mathbf{s}_0$ the IDW method first calculates the distance $d_{i0} = ||\mathbf{s}_i - \mathbf{s}_0||$ for $i = 1, \ldots, n$. The prediction is now given by:

$$\hat{Y}(\mathbf{s}_0) = \frac{1}{\sum_{i=1}^{n} \frac{1}{d_{i0}}} \sum_{i=1}^{n} \frac{y(\mathbf{s}_i)}{d_{i0}}.$$

Variations in the basic IDW methods are introduced by replacing d_{i0} by the pth power, d_{i0}^p for some values of $p > 0$. The higher the value of p, the quicker the rate of decay of influence of the distant observations in the interpolation. Note that it is not possible to attach any uncertainty measure to the individual predictions $\hat{Y}(\mathbf{s}_0)$ since a joint model has not been specified for the random vector $Y(\mathbf{s}_0), Y(\mathbf{s}_1), \ldots, Y(\mathbf{s}_n)$. However, in practice, an overall error rate such as the root mean square prediction error can be calculated for set aside validation data sets. Such an overall error rate will fail to ascertain uncertainty for prediction at an individual location.

There are many methods for smoothing areal data as well. One such method is inspired by what are known as conditionally auto-regressive (CAR) models which will be discussed more formally later in Section 2.14. In implementing this method we first need to define a neighborhood structure. Such a structure prepares a list of areal units which are declared as neighbors of each areal unit i for $i = 1, \ldots, n$. Obviously, there is subjectivity in defining the neighborhood structure. The main consideration here is to decide what all areal units can influence the response at a given areal unit i. Often, a Markovian structure, where areal units which share a common boundary with the given areal unit i are assumed to be neighbors of the unit i. Let k_i denote the number of neighbors of areal unit i and assume that $k_i > 0$ for each $i = 1, \ldots, n$ – so that there are no islands! Then a smoothing value, $\hat{Y}(\mathbf{s}_i)$, for any areal unit i is taken as the average of the observed values at the neighboring areal units, i.e.

$$\hat{Y}(\mathbf{s}_i) = \frac{1}{k_i} \sum_{j=1}^{k_i} y(\mathbf{s}_j).$$

Again, like the IDW method there is no associated uncertainty measure for $\hat{Y}(\mathbf{s}_i)$ unless one assumes a valid joint distribution for $Y(\mathbf{s}_1), \ldots, Y(\mathbf{s}_n)$. Of course, an overall error rate can be reported. Another important point to note here is that for areal unit data the prediction problem does not often exist, i.e., it is not required to predict at a new areal unit outside of the study region. However, the above smoother can still be applied to estimate the value at areal unit i even when the $y(\mathbf{s}_i)$ value is unobserved. But this poses a problem in estimating for any areal unit j for which the areal unit i is a neighbor of. Several strategies such as ignoring the missing value or replacing all the missing values by the grand mean of the data can be adopted. Again, this method is only proposed as an exploratory tool and such ad-hoc methods of estimation may be allowed in the analysis.

2.14   CAR models

The keyword CAR in CAR models stands for **C**onditional **A**uto**R**egression. This concept is often used in the context of modeling areal data which can be either discrete counts or continuous measurements. However, the CAR models are best described using the assumption of the normal distribution although CAR models for discrete data are also available. In our Bayesian modeling for areal data CAR models are used as prior distributions for spatial effects defined on the areal units. This justifies our treatment of CAR models using the normal distribution assumption.

Assume that we have areal data Y_i for the n areal units. The conditional in CAR stands for conditioning based on all the others. For example, we like to think of Y_1 given $Y_2, \ldots, Y_n$. The **A**uto**R**egression terms stand for regression on itself (auto). Putting these concepts together the CAR models are based on regression of each Y_i conditional on the others Y_j for $j = 1, \ldots n$ but with

$j \neq i$. The constraint $j \neq i$ makes sure that we do not use Y_i to define the distribution of Y_i. Thus, a typical CAR model will be written as

$$Y_i | y_j, j \neq i ~ \sim ~ N \left(\sum_j b_{ij} y_j, \sigma_i^2 \right) \tag{2.6}$$

where the b_{ij}'s are presumed to be the regression coefficients for predicting Y_i based on all the other Y_j's. The full distributional specification for $\mathbf{Y} = (Y_1, \ldots, Y_n)$ comes from the independent product specification of the distributions (2.6) for each $i = 1, \ldots, n$. There are several key points and concepts that we now discuss to understand and we present those below as a bulleted list.

• The models (2.6) can be equivalently rewritten as

$$\mathbf{Y} = B\mathbf{Y} + \boldsymbol{\epsilon}$$

where $\boldsymbol{\epsilon} = (\epsilon_1, \ldots, \epsilon_n)$ is a multivariate normal error distribution with zero means. The appearance of $\mathbf{Y}$ on the right hand side of the above emphasizes the keywords **A**uto**R**egression in CAR.

• The CAR specification defines a valid multivariate normal probability distribution for $\mathbf{Y}$ under the additional conditions

$$\frac{b_{ij}}{\sigma_i^2} = \frac{b_{ji}}{\sigma_j^2}, ~ i, j = 1, \ldots, n,$$

which are required to ensure that the inverse covariance matrix Σ^{-1} in (A.24), is symmetric.

• To reduce influence of the spatially far away Y_j's on Y_i, the b_{ij}'s for those far away areal units are set at 0. In that case, the CAR models (2.6) provide an attractive interpretation that each Y_i has its mean determined by the values of its neighboring areal units. This naturally brings in the proximity matrix W discussed in Section 2.11. We exploit the neighborhood structure, W, in the CAR model by assuming

$$b_{ij} = \frac{w_{ij}}{w_{i+}}, ~ \sigma_i^2 = \frac{\sigma^2}{w_{i+}}, i, j = 1, \ldots, n,$$

where $w_{i+} = \sum_{j=1}^{n} w_{ij}$ as previously defined in Section 2.11. With these assumptions, the joint density of $\mathbf{Y}$ can be written as

$$f(\mathbf{y}) \propto e^{-\frac{1}{2\sigma^2} \mathbf{y}'(D_w - W)\mathbf{y}}, \tag{2.7}$$

where $D_w = \text{diag}(W\mathbf{1})$ is a diagonal matrix with w_{i+} as the ith diagonal entry.

- Let $\Sigma^{-1} = D_w - W$. Note that the row sum of Σ^{-1}, $w_{i+} - \sum_{j=1}^{n} w_{ij}$ is zero for any i. Hence Σ^{-1} is singular matrix and as a result (2.7) only defines a singular normal distribution, see (A.25).

- The singularity of the implied joint distribution of the CAR models makes it problematic to be used as a joint model for observed data. However, in Bayesian modeling of this book the CAR models are used only as prior distributions in hierarchical models, which avoids the problematic issue.

- The singularity of the CAR models can be avoided by slightly adjusting the inverse covariance matrix $\Sigma^{-1} = D_w - W$ in (2.7). The adjustment takes the form of

$$\Sigma^{-1} = D_w - \rho W$$

for a suitable value of ρ. There is a large literature on the topic of adjustment methods, see e.g. Section 4.3 of the book by Banerjee et al. (2015).

2.15   Point processes

Spatial point pattern data arise when an event of interest, e.g. outbreak of a disease, e.g. Covid-19, occurs at random locations inside a study region of interest, $\mathbb{D}$. Often the main interest in such a case lies in discovering any explainable or non-random pattern in a scatter plot of the data locations. Absence of any regular pattern in the data locations is said to correspond to the model of *complete spatial randomness*, CSR, which is also called a Poisson process. Under CSR, the number of points in any given sub-region will follow the Poisson distribution with a parameter value proportional to the area of the sub-region. Often, researchers are interested in rejecting the model of CSR in favor of their own theories of the evolution or clustering of the points. In this context the researchers have to decide what all type of clustering may possibly explain the clustering pattern of points and which one of those provides the "best" fit to the observed data. There are other obvious investigations to make, for example, are there any suitable covariates which may explain the pattern? To illustrate, a lack of trees in many areas in a city may be explained by a layer of built environment.

Spatio-temporal point process data are naturally found in a number of disciplines, including (human or veterinary) epidemiology where extensive datasets are also becoming more common. One important distinction in practice is between processes defined as a discrete-time sequence of spatial point processes, or as a spatially and temporally continuous point process. See the books by Diggle (2014) and Møller and Waagepetersen (2003) for many examples and theoretical developments.

2.16 Conclusion

The main purpose of this chapter has been to introduce the key concepts we need to pursue spatio-temporal modeling in the later chapters. Spatio-temporal modeling, as any other substantial scientific area of research, has its own unique set of keywords and concept dictionary. Not knowing some of these is a barrier to fully understanding, or more appropriately appreciating, what is going on under the hood of modeling equations. Thus, this chapter plugs the knowledge gap a reader may have regarding the typical terminology used while modeling.

It has not been possible to keep the chapter completely notation free. Notations have been introduced to keep the rigor in presentation and also as early and unique reference points for many key concepts assumed in the later chapters. For example, the concepts of Gaussian Process (GP), Kriging, internal and external standardization are defined without the data application overload. Of course, it is possible to skip reading of this chapter until a time when the reader is confronted with an un-familiar jargon.

2.17 Exercises

1. Draw a graph of the Matèrn correlation function for different values of the decay parameter ϕ and smoothness parameter ν.

2. Draw theoretical variograms based on different versions of the Matèrn correlation function. Reproduce Figure 2.2.

3. Obtain Moran's I and Geary's C statistics for different data sets introduced in Section 1.4. For spatio-temporal data obtain the two statistics at each time point and then provide a time series plot for each.

3

Exploratory data analysis methods

3.1 Introduction

Any statistical modeling must precede by a fair amount of exploratory data analysis (EDA) to understand the nature and the peculiarities in the data. This understanding directly aids in selecting the most plausible modeling methods for the data set hand. Of course, in practical research work the investigator needs to perform a literature search regarding the past and current knowledge about the practical problem they are aiming to solve. But a thorough exploratory analysis of the data enables the researcher to gain a deeper understanding regarding the possible sources which may explain the variability in the data.

The EDA is also important for other reasons. By undertaking an EDA before modeling, the researcher scrutinizes the data set for any anomalies that may be present. Now is the time to discover any occurrence of missing values, illegal values, internal representation of any categorical variable and so on. Most statistical modeling methods cannot handle any missing covariate values. Hence, for simplicity any such missing value should be dealt with at this stage. If those cannot be imputed then more advanced modeling methodologies which allow modeling of the covariates should be adopted. This is also the time to check if the types of the data columns are indeed the ones the researcher is hoping for. In other words it is important to check after the data reading step. The simple R summary command may be used for data scrutiny. Removal of all problems regarding data legality and appropriateness will help us save a lot of time in the future modeling step.

In EDA itself one of the main objectives is to discover the appropriate modeling scale and important relationships between the modeling variables. For example, for continuous response data we may have to decide on a transformation, e.g. square root or log that may encourage normality. Also, most normal distribution based modeling efforts assume homoscadasticity (constant variance). Hence the EDA may be used to choose the modeling scale that encourages homoscadasticity. Especially in spatio-temporal modeling, the EDA should be used to investigate any visible spatial and temporal patterns which, in turn, will help us in modeling.

In the remainder of this chapter we explore two spatio-temporal data sets, introduced in Sections 1.3.1 and 1.4.1, that we use as running examples in our

DOI: 10.1201/9780429318443-3

modeling chapters 6, 7 and 10. By collapsing over time we also derive two further data sets which we explore here.

The four data sets: `nyspatial`, `nysptime`, `engtotals`, and `engdeaths` are available in the accompanying R package `bmstdr` so that the reader can easily access them to reproduce all the results. Familiarity of these data sets will be advantageous in the modeling performed later on. EDA for the other data sets introduced in Chapter 1 will be performed in the respective sections where those are modeled in Chapters 8, 9 and 11.

3.2   Exploring point reference data

3.2.1   Non-spatial graphical exploration

Recall the air pollution data set `nysptime` introduced in Section 1.3.1 which contains the daily maximum ozone concentration values at the 28 sites shown in Figure 1.1 in the state of New York for the 62 days in July and August 2006. From this data set we have created a spatial data set, named `nyspatial`, which contains the average air pollution and the average values of the three covariates at the 28 sites. Figure 3.1 provides a histogram for the response, average daily ozone concentration levels, at the 28 monitoring sites. The plot does not show a symmetric bell-shaped histogram but it does admit the possibility of a unimodal distribution for the response. The R command used to draw the plot is given below:

```
p <- ggplot(nyspatial, aes(x=yo3)) +
  geom_histogram(binwidth=4.5, color="black", fill="white") +
  labs(x="Average daily ozone concentration", y = "count")
p
```

The `geom_histogram` command has been invoked with a bin width argument of 4.5. The shape of the histogram will change if a different bin width is supplied. As is well known, a lower value will provide a lesser degree of smoothing while a higher value will increase more smoothing by collapsing the number of classes. It is also possible to adopt a different scale, e.g. square root or logarithm, but we have not done so here to illustrate modeling on the original scale of the data. We shall explore different scales for the spatio-temporal version of this data set.

Figure 3.2 provides a pair-wise scatter plot of the response against the three explanatory covariates: maximum temperature, wind speed and relative humidity. The diagonal panels in this plot provides kernel density estimates of the variables. This plot reveals that wind speed correlates the most with ozone levels at this aggregated average level. As is well known, see e.g. Sahu and Bakar (2012a), the maximum temperature also positively correlates with the

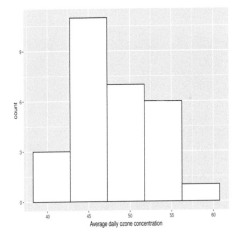

FIGURE 3.1: A simple histogram of 28 air pollution values in the data set nyspatial.

ozone levels. Relative humidity is seen to have the least amount of correlation with ozone levels. This plot has been obtained using the commands:

```
library(GGally)
p <- ggpairs(nyspatial, columns=6:9)
```

3.2.2 Exploring spatial variation

Traditionally, spatial variation is explored by drawing an empirical variogram. To obtain an empirical variogram we first obtain a variogram cloud . A variogram cloud is a simple scatter plot of the half of the squared difference between the data values at any two locations against the distance between the two locations. With the spatial coordinates supplied in Universal Transverse Mercator (UTM) co-ordinate system, the distance between two locations is simply the Euclidean distance. Usually, the UTM-X and Y coordinates are given in the unit of meter. Throughout this book we report distances in units of kilometers. The left panel of Figure 3.3 provides a variogram cloud for the nyspatial data.

To draw an empirical variogram, we divide the distance in X axis into an assumed number of bins and calculate the averages of the X and Y coordinates of the cloud of points in each of the bins. These averaged points are then plotted as the empirical variogram, see the right panel of Figure 3.3. A smooth loess fitted curve is over-laid to help make ideas regarding the true underlying variogram. This empirical plot does not look like any of the theoretical Matèrn variograms plotted in the right panel of Figure 2.2.

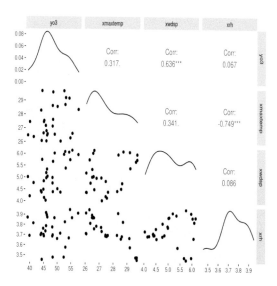

FIGURE 3.2: A pairwise scatter plot of the response and covariates in the data set `nyspatial`.

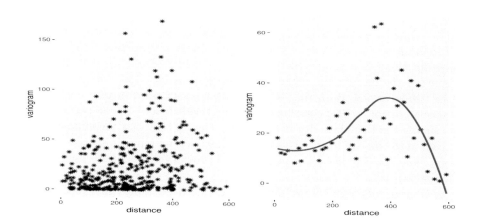

FIGURE 3.3: Variogram cloud (left panel) and an empirical variogram (right panel) for the average daily maximum ozone levels at the 28 sites in New York from the `nyspatial` data set.

♣ **R Code Notes 3.1. Figure 3.3** The `bmstdr` command

```
a <- bmstdr_variogram(data=nyspatial, formula = yo3 ~utmx
    + utmy, coordtype="utm", nb=50)
```

has been used to obtain this figure. The supplied data frame must contain the columns in the formula argument. The `coordtype` argument can take values among `"utm"`, `"lonlat"` or `"plain"`. The `nb` argument specifies the number of desired bins for the variogram. See the help file `?bmstdr_variogram`. The output a contains the two plots. The plots have been put together in a single graphsheet using the `ggarrange` command in the `ggpubr` library.

Kriging, see Section 2.9, can be performed using off-the-shelf methods by using a suitable package without going through explicit Bayesian modeling. For example, the `fields` package performs Kriging by using the function `Krig` which requires specification of at least three arguments: (i) x providing the coordinates of the locations in the data set, (ii) Y providing the response, and (iii) Z providing a matrix of covariates. The fitted object can be used to predict at new locations which can be supplied as arguments to the `predict` method. The predictions so obtained can be used to draw a heat map plot of the response surface.

In order to draw such a heat map in R we need predictions of the surface over a regular grid of point locations covering the study region. However, the `predict` method for the `Krig` function cannot predict over a regular grid unless the covariate values are also supplied for each location in the grid. In such a case, simple linear interpolation is usually performed using a package such as `akima` or `MBA`. The output of the Kriging predictions are passed on to a function such as `interp` in `akima` to obtain the interpolated predictions for each point in the grid. A raster plot can then be obtained of the output of the latest interpolation routine. See the detailed notes for coding in R Code Notes 3.2.

The kriged map in Figure 3.4 shows higher values of ozone levels in the Long Island Sound and also in the south west corner of the state. The contours of equal (interpolated) ozone concentration values are also superimposed in this plot.

However, the map is quite flat and does not show a lot of local spatial variation. One of the aims of this book is to use explicit Bayesian modeling to improve the off-the-shelf kriging methods. We will return to this in the modeling Chapter 6.

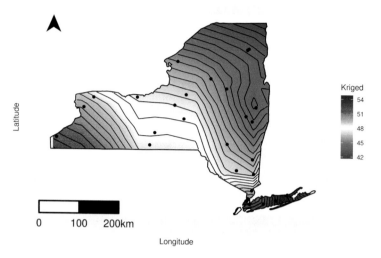

FIGURE 3.4: Kriged map of average ozone concentration values during July-August 2006 in the state of New York.

♣ **R Code Notes 3.2. Figure 3.4** As noted in the text, we fist use the fields to first perform Kriging and then predict at the 100 locations inside the state of New York contained in the data object `gridnyspatial` which is available from the `bmstdr` package. We then use the `akima` library to interpolate and then use the `gather` from the `tidyr` library to organize the interpolated data for plotting. Finally the plot is obtained using the `ggplot` function.

To draw a pretty map of the study region only we often ignore the values of interpolations outside the study region simply by setting those interpolations as 'NA' in R. We do this by using the `fnc.delete.map.XYZ` function provided in the `bmstdr` package. Contours are added using the `stat_contour` function. A distance scale and a north arrow have been placed using the `ggsn` package. The full set of code lines used to draw the plot are long and hence those are made available online from github[a].

---

[a]https://github.com/sujit-sahu/bookbmstdr.git

3.3 Exploring spatio-temporal point reference data

This section illustrates EDA methods with the nysptime data set in bmstdr. To decide the modeling scale Figure 3.5 plots the mean against variance for each site on the original scale and also on the square root scale for the response. A stronger linear mean-variance relationship with a larger value of the slope for the superimposed regression line is observed on the original scale making this less suitable for modeling purposes. This is because in linear statistical modeling we often model the mean as a function of the available covariates and assume equal variance (homoscedasticity) for the residual differences between the observed and modeled values. A word of caution here is that the right panel does not show a complete lack of mean-variance relationship. However, we still prefer to model on the square root scale to stabilize the variance and in this case the predictions we make in Chapter 7 for ozone concentration values do not become negative.

Temporal variations are illustrated in Figures 3.6 for all 28 sites and in Figure 3.8 for the 8 sites which have been used for model validation purposes in Chapters 6 and 7. Figure 3.7 shows variations of ozone concentration values for the 28 monitoring sites. Suspected outliers, data values which are at a distance beyond 1.5 times the inter quartile range from the whiskers, are plotted as red stars. Such high values of ground level ozone pollution are especially harmful to humans.

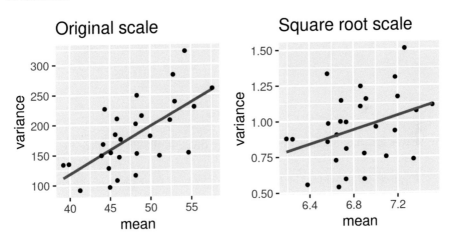

FIGURE 3.5: Mean-variance plots at the original (left panel) and square root (right panel) scales for the daily maximum ozone levels at the 28 sites in New York from the nysptime data set. A least squares fitted straight line has been superimposed in each of the plots.

♣ **R Code Notes 3.3. Figure 3.5** The site-wise summaries, mean and variance, are obtained using the `summaryBy` command in the `library(doBy)`. The function

```
fun_mean_var <- function(x) { c(mean = mean(x, na.rm=T), s2
    = var(x, na.rm=T))}
```

has been sent as the `FUN` argument in the `summaryBy` command. The `geom_smooth` command has been used with the arguments `method="lm"` and `se=F` to draw the straight line in each plot. Two plots are drawn separately and then the `ggarrange` command in the `ggpubr` library is used to put those in a single graphsheet.

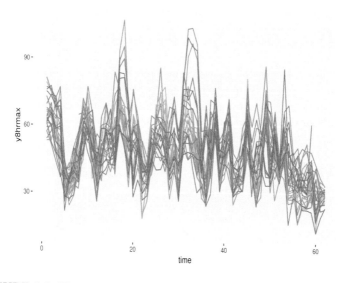

FIGURE 3.6: Time series plot of the data at the 28 monitoring sites.

Figures 3.9 provides the histogram and the scatter plots of the response on the square root (modeling) scale against the three available covariates: maximum temperature, wind speed and relative humidity. At this dis-aggregated daily temporal level a much stronger positive relationship is observed between ozone concentration values and maximum temperature. A negative relationship is observed with relative humidity. A not so strong relationship is seen between ozone values and wind speed. These exploratory plots will help us interpret the modeling results we discuss in Chapter 7. The two months identified in the plot do not show any overwhelming patterns.

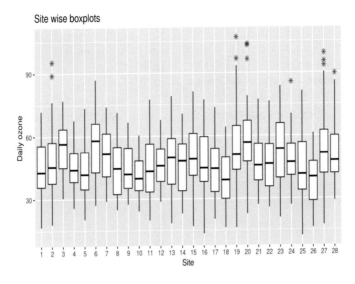

FIGURE 3.7: Box plots of daily ozone levels at the 28 monitoring sites.

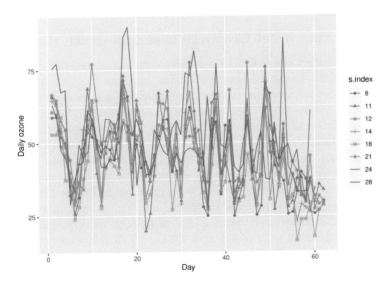

FIGURE 3.8: Time series plots of daily ozone levels for the 8 validation sites.

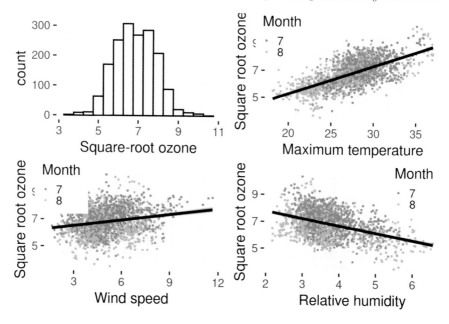

FIGURE 3.9: Four EDA plots for the `nysptime` data on the square root scale for ozone levels.

3.4 Exploring areal Covid-19 case and death data

This section explores the Covid-19 mortality data introduced in Section 1.4.1. The `bmstdr` data frame `engtotals` contains aggregated number of deaths along with other relevant information for analyzing and modeling this data set. The data frame object `engdeaths` contains the death numbers by the 20 weeks from March 13 to July 31, 2020. These two data sets will be used to illustrate spatial and spatio-temporal modeling for areal data in Chapter 10.

Typical such areal data are represented by a *choropleth map* which uses shades of color or grey scale to classify values into a few broad classes, like a histogram. Two choropleth maps have been provided in Figure 1.9.

For the `engtotals` data set the minimum and maximum number of deaths were 4 and 1223 respectively for the City of London (a very small borough within greater London with population 9721) and Birmingham with population 1,141,816 in 2019. However, the minimum and maximum death rates per 100,000 were 10.79 and 172.51 respectively for Hastings (in the South East) and Hertsmere (near Watford in greater London) respectively.

Calculation of the Moran's I for number and rate of deaths is performed by using the `moran.mc` function in the library `spdep`. This function requires the spatial adjacency matrix in a list format, which is obtained by the `poly2nb` and `nb2listw` functions in the `spdep` library. The Moran's I statistics for the raw observed death numbers and the rate are found to be 0.34 and 0.45 respectively both with a p-value smaller than 0.001 for the null hypothesis of no spatial autocorrelation. The permutation tests in statistics randomly permute the observed data and then calculates the relevant statistics for a number of replications. These replicate values of the statistics are used to approximate the null distribution of the statistics against which the observed value of the statistics for the observed data is compared and an approximate p-value is found. The tests with Geary's C statistics gave a p-value of less than 0.001 for the death rate per 100,000 but the p-value was higher, 0.025, for the un-adjusted observed Covid death numbers. Thus, the higher degree of spatial variation in the death rates has been successfully detected by the Geary's statistics. The code lines to obtain these results are given below.

```
library(bmstdr)
library(spdep)
covidrate <- engtotals$covid/engtotals$popn*100000
nbhood <- poly2nb(englad)
Wlist <- nb2listw(nbhood, style = "B", zero.policy = T)
moran.mc(engtotals$covid, Wlist, nsim=1000, zero.policy=T)
geary.mc(engtotals$covid, Wlist, nsim=1000, zero.policy=T)
## For the Covid rate per 100k
moran.mc(covidrate, Wlist, nsim=1000, zero.policy=T)
geary.mc(covidrate, Wlist, nsim=1000, zero.policy=T)
```

A glimpse of the temporal variation present in the data is observed in Figure 1.10 which plots the boxplots of the weekly death rates. Other exploratory time series plots, as in Figure 3.6 can also be considered.

3.4.1 Calculating the expected numbers of cases and deaths

To detect spatial patterns in the Covid-19 number of cases and deaths we need to perform standardization, as discussed in Section 2.12, which essentially finds the corresponding expected numbers under an 'equitable' spatial distribution. There are competing methodologies for obtaining these expected numbers, see Sahu and Böhning (2021) for a general discussion. Below we follow these authors in calculating the expected numbers.

The spatio-temporal data set `engdeaths` contains 6260 rows of data from the $K = 313$ Local Authority Districts, Counties, and Unitary Authorities (LADCUA) in England and $T = 20$ weeks. The main response variables for our spatio-temporal modeling are the number of Covid deaths, denoted by Y_{it}, and cases denoted by Z_{it}, $i = 1, \ldots, n$ and $t = 1, \ldots, T$. Also let D_{it} denote the number of deaths *from all causes*, including Covid, in LADCUA i in week t. The magnitude of Y_{it} depends on the size and demographics of the population at risk, which we adjust for by computing the expected number of Covid deaths E_{it}. There are many alternative ways to obtain the expected number of deaths, e.g. by considering the national age and sex specific death rates and then by performing internal or external indirect standardization as discussed in Section 2.12.

Estimation of mortality and morbidity requires a denominator (the number of persons at risk), which is not easily accessible in our application. However, the number of deaths from all causes is available which leads to the concept of *proportional mortality ratio* PMR, defined as the number of deaths from a specific cause (Covid) divided by the number of total deaths, see e.g. Rimm et al. (1980) or Last (2000). A direct argument shows that the PMR is identical to the ratio of the specific mortality rate to the mortality rate from all causes. The PMR is frequently used if population denominators are not readily accessible.

We now combine the philosophies of internal indirect standardization and proportional mortality ratio to an appropriate application in this example. The study populations are given by the various regions and the various weeks. We begin by calculating the number of expected deaths per week t and region i.

The availability of the number of deaths from all causes, D_{it}, in week t and region i allows us to estimate E_{it}s by indirect standardization as follows. Let

$$\bar{r} = \frac{\sum_{i=1}^{n} \sum_{t=1}^{T} Y_{it}}{\sum_{i=1}^{n} \sum_{t=1}^{T} D_{it}}$$

denote the overall proportion of Covid deaths among deaths from all causes. We consider this as an estimate of the ratio of the Covid-19 mortality to the mortality due to all causes for the entire country England, i.e. collapsed over all weeks and regions. Then, the expected number of Covid deaths is defined by:

$$E_{it} = \bar{r} \times D_{it}, \quad i = 1, \ldots, n, t = 1, \ldots, T.$$

Obviously, $\sum_{i=1}^{n} \sum_{t=1}^{T} Y_{it} = \sum_{i=1}^{n} \sum_{t=1}^{T} E_{it}$, so this can be viewed as a form of internal indirect standardization. Having obtained E_{it} we obtain the standardized mortality ratio, $\text{SMR}_{it} = Y_{it}/E_{it}$ for all values of i and t. The SMR_{it} has a simple interpretation, e.g. a value of 1.2 corresponds to a 20% increased risk compared to E_{it}. A zero value of D_{it} for some particular combinations of i and t leads to $E_{it} = 0$, which, in turn, causes problem in defining the SMR. To overcome this problem problem we replace the zero D_{it} values by 0.5 and adjust R and E_{it} accordingly.

A similar expected number of positive Covid cases is required to analyze the observed number of cases, previously denoted by Z_{it}, at the tth week in the ith LADCUA. Corresponding to R, the overall proportion of Covid deaths, we define the overall proportion of infections in the population as:

$$\bar{r}_C = \frac{\sum_{i=1}^{n} \sum_{t=1}^{T} Z_{it}}{\sum_{i=1}^{n} \sum_{t=1}^{T} P_{it}}$$

where P_{it} denotes the population at risk at time t in the ith LADCUA. Hence, the expected number of cases, corresponding to Z_{it}, denoted by C_{it} and is defined as:

$$C_{it} = \bar{r}_C \times P_{it}, \quad i = 1, \ldots, n, t = 1, \ldots, T. \tag{3.1}$$

Clearly, $\sum_{i=1}^{n} \sum_{t=1}^{T} Z_{it} = \sum_{i=1}^{n} \sum_{t=1}^{T} C_{it}$. In our implementation we assume constant population over the weeks, i.e. $P_{it} = P_i$ where the P_i is the ONS estimated population in 2019, *popn*, obtained previously at the data collection stage. Note that C_{it} is always positive since $P_{it} > 0$ for all values of i and t. Again, this can be viewed as a form of internal indirect standardization. Our focus is on the standardized Covid-19 morbidity ratio, $\text{SMCR}_{it} = \frac{Z_{it}}{C_{it}}$, which reduces to $\frac{Z_{it}}{P_i}$.

3.4.2 Graphical displays and covariate information

The number of Covid deaths, Y_{it}, varies over i and t with a mean value of 5.2 and a maximum value of 246. These values are not comparable across the different spatial units due to the differences in the at risk population size. Hence, for the purposes of exploratory analysis only, we transform Y_{it} into the rate per 100,000 people to have comparable weekly rates for the n spatial units. The death rate, so obtained, ranges from 0 to 40 with a mean value of 4.24. This rate of death is plotted in side-by-side boxplots in Figure 1.10 in Chapter 1 for the 20 weeks in our study.

It is generally believed that socio-economic deprivation leads to worse health on average and this, in turn may, affect Covid case and death rates. To investigate such issues, for each LADCUA we collected the following data from the source ONS website[1]):

(i) population data for 2019 (denoted *popn*),

(ii) population density (denoted *popden* on the log scale), defined as size of population divided by the area of the LADCUA,

(iii) percentage of the working age population receiving job-seekers allowance (denoted *jsa*) during January 2020,

(iv) median house price in March 2020 (denoted *price* on the log to the base 10 scale).

The first of these four variables, *popn*, will be used to calculate rates of cases and deaths in the subsequent sections. None of these variables vary over time (the weeks) but are spatially varying providing information regarding the long-term socio-economic nature of the spatial units. The raw values of population density have a large range from 25 to 16,427 which will cause problems in modeling. We apply the log transformation to stabilize the values of this co-variate. Similarly we apply the log transformation, but using the base 10 for ease of interpretation, for the house price data.

---

[1]https://ons.gov.uk/

To associate the Covid death rates with the temporally static but spatially varying covariates, *popden*, *price* and *jsa* we obtain an aggregated over time value of the SMR of Covid deaths for each of the *n* LADCUAs. This aggregated SMR is obtained simply as the ratio of the sum of the observed and expected number of Covid deaths over the 36 weeks. Note that this is the ratio of the sum over the weeks – not the sum of the ratios. Figure 3.10 plots a map of the SMRs and the three important covariates in different panels. The map of the log-population density, *popden*, shows the highest level of correlation among the three covariates plotted in this figure. The covariate *jsa* shows up areas in the north of England having high levels of both SMRs and *jsa*. The median house price in the log-10 scale, *price*, also correlates (but negatively) well with the SMR values – showing on average higher house price values in the south where the SMR values are lower on average.

> ♣ **R Code Notes 3.4. Figure 3.10** This figure follows on from the maps drawn in Figure 1.9. The exact same procedures have been followed for each of the plotted quantities.

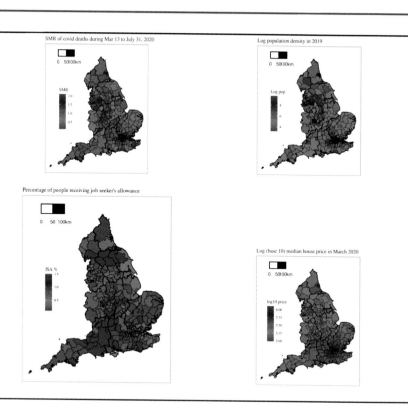

FIGURE 3.10: Average SMR of Covid deaths (top-left) and three relevant variables capturing socio-economic information.

We now investigate the rate of reported number of cases, previously denoted as Z_{it}. The number of cases, ranging from 0 to 4447, has a very high level of variability and this is not comparable across the different spatial units. Hence we apply the same transformation to convert the raw number of cases into the number of cases per 100,000 people. Figures 3.11 and 3.12 provide three important plots showing various aspects of the case numbers. The left panel in Figure 3.11 shows a clear north-south divide in the observed average weekly case numbers. The panel in the right of this figure, similar to Figure 1.10, shows the rise and fall of the case dynamics. Figure 3.12 plots the average weekly number of deaths and cases both per 100,000 people. This clearly shows a slow decline of both case and death rates.

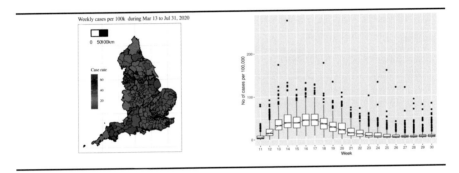

FIGURE 3.11: Graphical summaries of case rates.

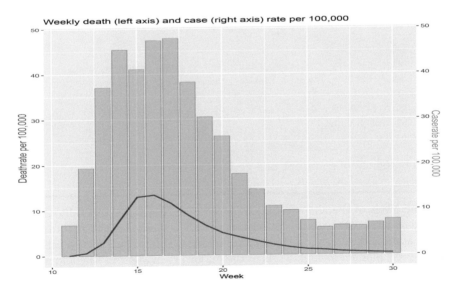

FIGURE 3.12: A plot of case rates and death rates.

♣ **R Code Notes 3.5. Figure 3.12** To obtain this plot we first
obtain the weekly mean death and case rates by aggregating over the
313 local authorities for the 20 weeks.

```
engdeaths$deathrate <- engdeaths$covid * 100000/engdeaths$popn
engdeaths$caserate <- engdeaths$noofcases * 100000/engdeaths$
    popn
u <- engdeaths[, c("Weeknumber", "deathrate", "caserate", "no2"
    )]
wkmeans <- aggregate.data.frame(u[, -1], by=list(Week=u[,1]),
    FUN=mean)
head(wkmeans)
```

In the next step we choose the colors and then issue the `ggplot` command provided below.

```
deathcolor <- "#FF0000"
casecolor <- rgb(0.2, 0.6, 0.9, 1)

twoplots <- ggplot(data=wkmeans, aes(x=Week)) +
  geom_bar( aes(y=caserate), stat="identity", size=.1, fill=
      casecolor, color="blue", alpha=.4) +
  geom_line(aes(y=deathrate), size=1, color=deathcolor) +
  scale_y_continuous(
    # Features of the first axis
    name = "Deathrate per 100,000",
    # Add a second axis and specify its features
    sec.axis = sec_axis(~.*1, name="Caserate per 100,000")
  ) +
  theme(
    axis.title.y = element_text(color = deathcolor, size=13),
    axis.title.y.right = element_text(color = casecolor, size=13)
  ) +
  ggtitle("Weekly death (left axis) and case (right axis) rate
      per 100,000")
twoplots
```

Note the use of `geom_bar` and `geom_line` commands.

The levels of the air pollutant NO_2 is the only spatio-temporally varying
environmental covariate in the `engdeaths` data set. Figure 3.13 plots the average standardized case morbidity rate and also the average NO_2 levels. The
NO_2 levels plot shows high levels in the major urban areas in England including London. The two maps in this plot also show some correlation between
the case morbidity rate and the NO_2 levels. Temporal variation in the NO_2
levels, and its association with the case rate are explored in Figure 3.14. The
boxplots in the left panel show higher levels of NO_2 after the easing of the

lockdowns after week 28 in July. The panel on the right shows that higher level of case rates are associated with higher levels of NO_2.

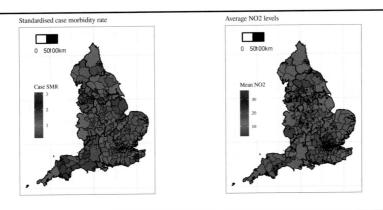

FIGURE 3.13: Average SMR for cases (left) and NO_2 levels (right) during the 20 weeks.

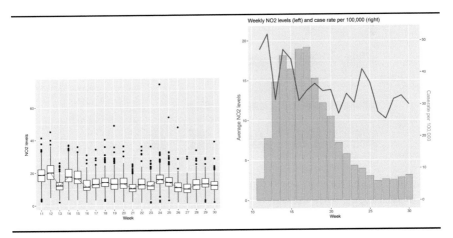

FIGURE 3.14: Boxplot of NO_2 during the 20 weeks (left panel) and the Covid-19 case and death rates (right panel).

To explore the relationships between the SMR values for death on the top-panel of Figure 3.10 and the socio-economic covariates, we provide a pairwise scatter plot in Figure 3.15. The scales of the plotted variables have been described in the caption of this figure. Various relationships between the variables present in the data set can be read from Figure 3.15. The numerical values of the correlations shown in the plot reveal significantly moderate level of positive association between the log of the SMR for cases and square-root

NO_2, *jsa, popden* and the log of the SMR of Covid deaths. The SMR of the cases are higher in the lower house price areas – mostly in the north and the SMR of the cases are lower in the areas where *jsa* is lower. Population density (on the log scale) correlates positively with the SMRs of cases and Covid deaths. Obviously, higher levels of case rates leads to higher numbers of death.

Looking at the last column of the plot labeled 'deathsmr' we see that the SMR of Covid deaths is positively associated with *jsa* and *popden* but it shows a negative association with *price*. This is because of the north-south divide in house prices in England and also a similar divide in opposite direction for the death rates as seen in Figure 3.10. However, the higher house prices in London in the south and also the higher levels of deaths in London reduces the strength of this north-south divide.

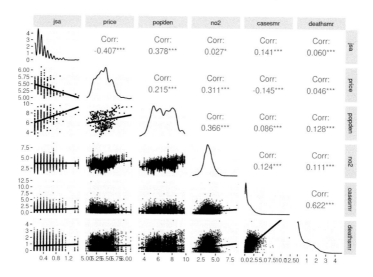

FIGURE 3.15: Pairwise scatter plots of the log SMR values of Covid cases and deaths along with various covariates. The variable *price* is on log to the base 10 scale, *popden* is on the log scale, NO_2 is on the square-root scale.

3.5　Conclusion

This chapter has illustrated various EDA plots that can be obtained to understand spatial and spatio-temporal data sets before embarking on modeling. Note that the list of plots is not exhaustive but illustrative. The most important message is that it is essential to understand the nature of data that we are planning to model. Such understanding will enable us not only to get interpretable modeling results but also to explain why a particular model may not be suitable for the data set at hand. This step is also crucial to find out if there is any potential problem larking in the data set. For example, we may see numbers in the data frame but R may have read those as a character or factor vector type. Modeling is unlikely to yield expected results if the data set is not properly read at the first place.

3.6 Exercises

1. Perform exploratory data analysis of the air pollution data set `piemonte` included in the `bmstdr` package.

2. Obtain an aggregated version of the `piemonte` data set by calculating average monthly pollution values at each monitoring site. For each month's data obtain an empirical variogram.

3. Using the fields package perform kriging for a grid of locations within the Piemonte region.

4. The `bmstdr` data set `gridnysptime` has been ontained by kriging the values of maximum temperature, wind speed and relative humidity separately at each time point. Using the `Krig` function in the library fields write code to perform Kriging to spatially interpolate each of the three covariates at the 100 locations in the grid. Obtain root mean square error by calculating the differences between your estimates and the ones already present in the `gridnysptime` data set.

5. Verify the expected number of Covid-19 death counts present in the running example data set discussed in Section 1.4.1.

6. Perform exploratory data analysis of the child poverty data set introduced in Section 1.4.5.

4

Bayesian inference methods

4.1 Introduction

The basic philosophy underlying Bayesian inference is that the only sensible measure of uncertainty is probability which can be objective as well as subjective or a combination of both. For example, we may be interested in statements such as what is the probability that a new, as yet untested, individual in a population will possess certain characteristics given the combined knowledge learned from a certain number of randomly chosen individuals and any existing historical knowledge. In a statistical modeling situation the researcher may want to evaluate the probability of a hypothesis (or a model) given the observations. Thus, in a typical Bayesian formulation of practical data analysis and modeling problem, the most important task is to formulate and then estimate probabilities of events of interest using the observed data and any prior information regarding any aspect of the scientific experiment from which data have been gathered. The actual task of integration of prior information and information from the data is facilitated through the application of the Bayes theorem due to Reverend Thomas Bayes, an amateur 18th century English mathematician.

Simply put, the Bayes theorem lets us calculate the probability of the event of interest given the occurrence of some other events in the same sample space. One classic example, where the Bayes theorem is used, is the calculation of the probability of disease given symptom. In this example, information about prevalence of a rare disease in a population is assumed and so is the effectiveness of a diagnostic test to detect the disease. Based on such information the task is to find the probability that a randomly selected individual has got the disease given that they have tested positive in the diagnostic test. This probability will not be 100% since the diagnostic test is not assumed to be 100% accurate. The Bayes theorem is a mathematical result that quantifies this probability, often called the posterior probability, based on all the available information.

In the Bayesian paradigm the observed data are assumed to come from a family of parameterized (or non-parametric) probability distributions to facilitate Bayesian inference for the parameters and the predictions. However, whereas classical statistics considers the parameters to be *fixed but unknown*, the Bayesian approach treats them as random variables in their own right.

DOI: 10.1201/9780429318443-4

Prior knowledge about any aspects of the experiment is captured through what is known as the prior distribution for the unknown parameters. Such a prior distribution, if exists, naturally enhances the strength and quality of inference. Inferences made without the appropriate use of the prior information are likely to be inferior as those are not based on the maximal available information set to solve the problem of interest.

Bayesian inference methods start by calculating the posterior probability of interest by appropriately using the Bayes theorem. In the context of making inference using parametric family for observed data the posterior probabilities are calculated using posterior distributions of the unknown parameters given the values of the observed data. Estimates of parameters and their uncertainties are calculated simply by summarizing the posterior distribution. For example, one may find the mean, median or the mode of the posterior distribution as the parameter estimate depending on the shape of the posterior distribution and also the specific learning objectives of the researcher. For example, the mean of the posterior distribution will be the correct estimate if the researcher assumes a squared error loss function, which penalizes equally for over and under estimation, as his/her objective function. On the other hand, the posterior mode will be the appropriate estimate to choose when the researcher simply wants to estimate the most likely value of the parameter given the data. The underlying loss function here is known as the 0-1 loss function as we elaborate below in this chapter. Uncertainties in the posterior estimates are simply calculated as the standard deviation of the posterior distribution or a 95% interval, called a credible interval in the Bayesian speak, capturing 95% of the posterior probability.

Bayesian methods are most attractive when it comes to performing prediction for unknown observations in either space or time in a model based setting. A simple logic works here as follows. In a model based setting, the observations are assumed to follow a parametric family of distributions. The prediction problem is to estimate a future observation from the same family conditional on the observed data values. If we assume that the parameters are known then inference for the unknown observation should simply be made by using the conditional distribution of that observation given the known parameter values and the actual observed data values. However, in practice the parameter values will be unknown. To get rid of these unknown parameters the best information we may make use of is captured by the posterior distribution of the parameters given the data values. Hence the best distribution to use for Bayesian prediction is the average of the conditional distribution of the unknown given the data values and the parameters where the averaging weights are the posterior probabilities. Thus, subsets of parameter values with higher posterior probabilities will get higher weights in the averaging process. In the extreme case when parameters are assumed to be known the averaging weight will be 100% for that known parameter value. Such an averaged distribution is called a posterior predictive distribution, and its summaries can be used for prediction and their uncertainty estimation in the same way as the posterior distribution.

The last remaining inference topics concern model comparison, traditional hypothesis testing and model adequacy checks. For hypothesis testing, the pure Bayesian answer is to calculate the posterior probability of the hypothesis of interest. In the Bayesian paradigm, this posterior probability contains all the information to make an accept/reject decision. Bayesians encounter one important practical problem using this approach when a traditional point null hypothesis such as $H_0 : \theta = 0$ is to be tested and θ is a continuous parameter capable of taking any particular value. The problem comes from the statistical convention that the probability of a continuous random variable, θ here, taking any particular value, 0 here, is exactly zero. The argument goes that the particular value, 0, is an approximation and a finer measurement scale could have recorded a different value eroding the positive probability of the zero value. There are several remedies such as to modify the point null hypothesis to an interval valued one, essentially refuting the practicality of a point null hypothesis.

The Bayesian concept of marginal likelihood is also adopted to facilitate model comparison and hypothesis testing. The premise here is to avoid comparison of parameter values in different models since those are not comparable across different models anyway. For example, in a linear regression model the constant term is the overall mean when there are no regressors but it becomes the intercept when at least one regressor is included in the model. The models are compared by the marginal likelihood of the data when the parameters have been averaged (integrated) out from the joint distribution of the data and the parameters. The marginal likelihood is evaluated using the data values for each of the competing models and intuitively, the best model is the one for which the marginal likelihood is the maximum. This is advantageous for several reasons. The marginal likelihood refers to the observed data which do not change with the model a particular researcher may want to fit. Thus, the marginal likelihoods are parameter free, although they may depend heavily on the assumption of the prior distribution. Indeed, a proper prior distribution, that guarantees total probability 1, needs to be assumed in the averaging used to obtain the marginal likelihood. Otherwise, the marginal likelihood is not well-defined since the averaging weights as induced by an improper prior distribution will not sum to a finite value leading to the arbitrariness of the average. To elaborate this, note that a weighted average $\sum_{i=1}^{n} a_i w_i$ is only well defined if the sum of the weights $\sum_{i=1}^{n} w_i$ is finite. Model comparison using the marginal likelihoods is often facilitated by the calculation of the Bayes factor, which is calculated as the ratio of two marginal likelihoods. This chapter provides further explanation regarding the interpretation of the Bayes factor and difficulties associated in calculating the marginal likelihood.

Practical difficulties in estimating the marginal likelihood have led the way to developing model comparison methods based on the so called "information criteria" such as the Akaike Information Criteria (AIC). The Bayesian literature there are many different ICs such as the Bayesian Information Criteria (BIC), Deviance Information Criteria (DIC), Watanabe Information Criteria (WAIC) and the posterior predictive model choice criteria proposed by Gelfand and Ghosh (1998). A common theme that runs across in all these model choice criteria is that each criterion is composed of two parts: (i) a measure of goodness of fit and (ii) a penalty for model complexity. The model providing the minimum value of the sum of the above two components is chosen as the best model for the data. Choosing among the different criteria is a difficult task since each criteria, based on different considerations and expectations for a model, have their own merits and demerits. The researcher, here, is asked to set their modeling objectives in the first place and then they may select the best model according to the most appropriate model selection criteria. For example, a model may only be developed for the sole purpose of out of sample predictions. In such a case the directly interpretable root mean square prediction error, or the mean absolute prediction error for hold-out data should be used instead to select the models.

Checking model adequacy is another important task that must be performed before making model based inference. What use a "best" model may have if it cannot adequately explain the full extent of variability in the data? Such a non-adequate model will fail to explain sources of variability in the data and, as a result will also fail to predict future outcomes. Bayesian posterior predictive distributions are used in many different ways to check for model adequacy. For example, comparisons can be made between the nominal and the actual achieved coverage of prediction intervals for a large number of out of sample validation predictions. A good degree of agreement between the two coverages will signal good model adequacy.

The use of prior information in the Bayesian inference paradigm has brought in controversy in statistics. Many non-Bayesian scientists often claim that they do not have any prior information and hence are not able to formulate their investigations using the Bayesian philosophy mentioned in the above paragraphs. Indeed, this may be true in some situations for which the Bayesian view would be to assume either default or non-informative prior distributions that would encourage the investigator to take decisions based on the information from data alone but still using the Bayesian paradigm which uses the above mantra that, "the only sensible measure of uncertainty is probability." However, when some prior information is available it would be philosophically sub-optimal not to use prior information which will lead to a better decision.

Another important philosophical criticism of the Bayesian view is that the Bayesian philosophy embraces the concept of "subjective" probability which has the potential to lead to "biased" inference of any sort favored by the experimenter. It is difficult to counter such hard hitting criticisms in a practical setting and human bias can never be ruled out. However, model selection, validation and posterior predictive checks mentioned above are designed to guard against human bias and un-intended errors committed while modeling. Moreover, there is a large literature on objective Bayesian inference using reference and non-informative priors. These issues are not discussed any further in this book. Instead, the reader is referred to Berger (2006) who makes a case for objective Bayesian analysis.

4.2   Prior and posterior distributions

4.3   The Bayes theorem for probability

The Bayes theorem allows us to calculate probabilities of events when additional information for some other events is available. For example, a person may have a certain disease whether or not they show any symptoms of it. Suppose a randomly selected person is found to have the symptom. Given this additional information, what is the probability that they have the disease? Note that having the symptom does not fully guarantee that the person has the disease.

To formally state the Bayes theorem, let $B_1, B_2, \ldots, B_k$ be a set of mutually exclusive and exhaustive events and let A be another event with positive probability (see illustration in Figure 4.1). The Bayes theorem states that for any i, $i = 1, \ldots, k$,

$$P(B_i|A) = \frac{P(B_i \cap A)}{P(A)} = \frac{P(A|B_i)P(B_i)}{\sum_{j=1}^{k} P(A|B_j)P(B_j)}. \tag{4.1}$$

♡ **Example 4.1.** We can understand the theorem using a simple example. Consider a rare disease that is thought to occur in 0.1% of the population. Using a particular blood test a physician observes that out of the patients with disease 99% possess a particular symptom. Also assume that 1% of the population without the disease have the same symptom. A randomly chosen person from the population is blood tested and is shown to have the symptom. What is the conditional probability that the person has the disease?

Here $k = 2$ and let B_1 be the event that a randomly chosen person has the disease and B_2 is the complement of B_1. Let A be the event that a randomly chosen person has the symptom. The problem is to determine $P(B_1|A)$.

We have $P(B_1) = 0.001$ since 0.1% of the population has the disease, and $P(B_2) = 0.999$. Also, $P(A|B_1) = 0.99$ and $P(A|B_2) = 0.01$. Now

$$
\begin{aligned}
P(\text{disease} \mid \text{symptom}) = P(B_1|A) &= \frac{P(A|B_1)\,P(B_1)}{P(A|B_1)\,P(B_1) + P(A|B_2)\,P(B_2)} \\
&= \frac{0.99 \times 0.001}{0.99 \times 0.001 + 0.999 \times 0.01} \\
&= \frac{99}{99 + 999} = 0.09.
\end{aligned}
$$

The probability of disease given symptom here is very low, only 9%, since the disease is a very rare disease and there will be a large percentage of individuals in the population who have the symptom but not the disease, highlighted by the figure 999 as the second last term in the denominator above.

It is interesting to see what happens if the same person is found to have the same symptom in another independent blood test. In this case, the prior probability of 0.1% would get revised to 0.09 and the revised posterior probability is given by:

$$P(\text{disease} \mid \text{twice positive}) = \frac{0.99 \times 0.09}{0.99 \times 0.09 + 0.91 \times 0.01} = 0.908.$$

As expected, this probability is much higher since it combines the evidence from two independent tests. This illustrates an aspect of the Bayesian world view: the prior probability gets continually updated in the light of new evidence.

□

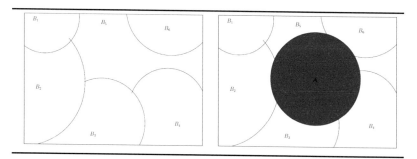

FIGURE 4.1: The left panel shows the mutually exclusive and exhaustive events $B_1, \ldots, B_6$ (they form a partition of the sample space); the right figure shows a possible event A.

4.4   Bayes theorem for random variables

The Bayes theorem stated above is generalized for two random variables instead of two events A and B_i's as noted above. In the generalization, B_i's will be replaced by the generic parameter θ which we want to estimate and A will be replaced by the observation random variable denoted by Y. Also, the probabilities of events will be replaced by the probability (mass or density) function of the argument random variable. Thus, $P(A|B_i)$ will be substituted by $f(y|\theta)$ where $f(\cdot)$ denotes the probability (mass or density) function of the random variable X given a particular value of θ. The replacement for $P(B_i)$ is $\pi(\theta)$, which is the prior distribution of the unknown parameter θ. If θ is a discrete parameter taking only finite many, k say, values, then the summation in the denominator of the above Bayes theorem will stay as it is since $\sum_{j=1}^{k} \pi(\theta_j)$ must be equal to 1 as the total probability. If, however, θ is a continuous parameter then the summation in the denominator of the Bayes theorem must be replaced by an integral over the range of θ, which is generally taken as the whole of the real line.

The Bayes theorem for random variables is now stated as follows. Suppose that two random variables Y and θ are given with probability density functions (pdfs) $f(y|\theta)$ and $\pi(\theta)$, then

$$\pi(\theta|y) = \frac{f(y|\theta)\pi(\theta)}{\int_{-\infty}^{\infty} f(y|\theta)\pi(\theta)d\theta}, \quad -\infty < \theta < \infty. \tag{4.2}$$

The probability distribution given by $\pi(\theta)$ captures the prior beliefs about the unknown parameter θ and is the prior distribution in the Bayes theorem. The posterior distribution of θ is given by $\pi(\theta|y)$ after observing the value y of the random variable Y. We illustrate the theorem with the following example.

♡ **Example 4.2. Binomial** Suppose $Y \sim$ binomial(n, θ) where n is known and we assume Beta(α, β) prior distribution for θ. Here the likelihood function is

$$f(y|\theta) = \binom{n}{y}\theta^y(1 - \theta)^{n-y}$$

for $0 < \theta < 1$. The function $f(y|\theta)$ is to be viewed as a function of θ for a given value of y, although its argument is written as $y|\theta$ instead of $\theta|y$. This is because we use the probability density function of Y, which is more widely known, and we avoid introducing further notation for the likelihood function e.g. $L(\theta; y)$.

Suppose that the prior distribution is the beta distribution (A.22) having density

$$\pi(\theta) = \frac{1}{B(\alpha, \beta)}\theta^{\alpha-1}(1 - \theta)^{\beta-1}, \quad 0 < \theta < 1. \tag{4.3}$$

Hence, the posterior distribution is given by:

$$\begin{aligned}
\pi(\theta|y) &= \frac{f(y|\theta)\pi(\theta)}{\int_0^1 f(y|\theta)\pi(\theta)d\theta} \\
&= \frac{\theta^y(1-\theta)^{n-x}\theta^{\alpha-1}(1-\theta)^{\beta-1}}{\int_0^1 \theta^y(1-\theta)^{n-y}\theta^{\alpha-1}(1-\theta)^{\beta-1}d\theta} \\
&= \frac{\theta^{y+\alpha-1}(1-\theta)^{n-y+\beta-1}}{\int_0^1 \theta^{y+\alpha-1}(1-\theta)^{n-y+\beta-1}d\theta} \\
&= \frac{1}{B(y+\alpha, n-y+\beta)}\theta^{y+\alpha-1}(1 - \theta)^{n-y+\beta-1},
\end{aligned}$$

where $0 < \theta < 1$. Note that the term $\binom{n}{y}\frac{1}{B(\alpha,\beta)}$ in the product $f(y|\theta)\pi(\theta)$ has been canceled in the ratio defining the Bayes theorem. Thus, the posterior distribution is recognized to be the beta distribution (4.3) but with revised parameters $y + \alpha$ and $n - y + \beta$. □

The Bayes theorem for random variables (4.2) also holds if the single random variable Y is replaced by a joint distribution of an n-dimensional random variable, $Y_1, \ldots, Y_n$ as in the case of a random sample from the probability distribution of Y. The above Bayes theorem now delivers the posterior distribution:

$$\pi(\theta|y_1, \ldots, y_n) = \frac{f(y_1, \ldots, y_n|\theta)\pi(\theta)}{\int_{-\infty}^{\infty} f(y_1, \ldots, y_n \mid \theta)\pi(\theta)d\theta}, \quad -\infty < \theta < \infty.$$

♡ **Example 4.3. Exponential** Suppose $Y_1, \ldots, Y_n$ is a random sample from the distribution with pdf $f(y|\theta) = \theta e^{-\theta y}$, for $\theta > 0$. Suppose the prior distribution for θ is given by $\pi(\theta) = \mu e^{-\mu\theta}$ for some known value of $\mu > 0$.

The joint distribution of $Y_1, \ldots, Y_n$, or equivalently the likelihood function, is given by:

$$f(y_1, y_2, \ldots, y_n|\theta) = \theta e^{-\theta y_1} \cdots \theta e^{-\theta y_n} = \theta^n e^{-\theta \sum_{i=1}^{n} y_i}.$$

Hence the posterior distribution is given by:

$$\pi(\theta|y) = \frac{\theta^n e^{-\theta \sum_{i=1}^{n} y_i} \mu e^{-\mu\theta}}{\int_0^{\infty} \theta^n e^{-\theta \sum_{i=1}^{n} y_i} \mu e^{-\mu\theta} d\theta}$$

$$= \frac{\theta^n e^{-\theta(\mu + \sum_{i=1}^{n} y_i)}}{\int_0^{\infty} \theta^n e^{-\theta(\mu + \sum_{i=1}^{n} y_i)} d\theta}, \quad \theta > 0.$$

This is recognized to be the pdf of the gamma distribution $G(n + 1, \mu + \sum_{i=1}^{n} y_i)$, see Section A.1. □

4.5 Posterior ∝ Likelihood × Prior

Consider the denominator in the posterior distribution. The denominator, given by, $\int_{-\infty}^{\infty} f(y|\theta)\pi(\theta)d\theta$ or $\int_{-\infty}^{\infty} f(y_1, \ldots, y_n|\theta)\pi(\theta)d\theta$ is free of the unknown parameter θ since θ is only a dummy in the integral, and it has been integrated out in the expression. The posterior distribution $\pi(\theta|y_1, \ldots, y_n)$ is to be viewed as a function of θ and the denominator is merely a constant. That is why, we often ignore the constant denominator and write the posterior distribution $\pi(\theta|y_1, \ldots, y_n)$ as

$$\pi(\theta|y_1, \ldots, y_n) \propto f(y_1, \ldots, y_n \mid \theta) \times \pi(\theta).$$

By noting that $f(y_1, \ldots, y_n \mid \theta)$ provides the likelihood function of θ and $\pi(\theta)$ is the prior distribution for θ, we write:

> Posterior ∝ Likelihood × Prior.

Hence we always know the posterior distribution up-to a normalizing constant. Often we are able to identify the posterior distribution of θ just by looking at the numerator as in the two preceding examples.

4.6 Sequential updating of the posterior distribution

The structure of the Bayes theorem allows sequential updating of the posterior distribution. By Bayes theorem we "update" the prior belief $\pi(\theta)$ to $\pi(\theta|\mathbf{y})$. Note that $\pi(\theta|y_1) \propto f(y_1|\theta)\pi(\theta)$ and if Y_2 is independent of Y_2 given the parameter θ, then:

$$
\begin{aligned}
\pi(\theta|y_1, y_2) &\propto f(y_2|\theta)f(y_1|\theta)\pi(\theta) \\
&\propto f(y_2|\theta)\pi(\theta|y_1).
\end{aligned}
$$

Thus, at the second stage of data collection, the first stage posterior distribution, $\pi(\theta|y_1)$ acts as the prior distribution to update our belief about θ after. Thus, the Bayes theorem shows how the knowledge about the state of nature represented by θ is continually modified as new data becomes available. There is another strong point that jumps out of this sequential updating. It is possible to start with a very weak prior distribution $\pi(\theta)$ and upon observing data sequentially the prior distribution gets revised to a stronger one, e.g. $\pi(\theta|y_1)$ when just one observation has been recorded – of course, assuming that data are informative about the unknown parameter θ.

4.7 Normal-Normal example

The theory learned so far can be used to derive the posterior distribution of the mean parameter θ of a normal distribution when the variance parameter, σ^2 is assumed to be known. In the derivation we also assume that a normal prior distribution with known mean μ and variance τ^2 is assumed for θ. This example is to be used in many discussions to illustrate many appealing properties of the Bayesian methods. However, the following derivation of the posterior distribution is non-trivial and can be skipped in the first reading. In this case the reader can skip to the text discussion at the end of this section.

Suppose $Y_1, \ldots, Y_n \sim N(0, \sigma^2)$ independently, where σ^2 is known. Let us assume the prior distribution $\theta \sim N(\mu, \tau^2)$ for known values of μ and τ^2. The likelihood function is:

$$
\begin{aligned}
f(y_1, y_2, \ldots, y_n|\theta) &= \prod_{i=1}^{n} \frac{1}{\sqrt{2\pi\sigma^2}} \exp\left\{-\frac{1}{2}\frac{(y_i-\theta)^2}{\sigma^2}\right\} \\
&= \left(\frac{1}{2\pi\sigma^2}\right)^{\frac{n}{2}} \exp\left\{-\frac{1}{2}\sum_{i=1}^{n}\frac{(y_i-\theta)^2}{\sigma^2}\right\}.
\end{aligned}
$$

The prior distribution is:

$$
\pi(\theta) = \frac{1}{\sqrt{2\pi\tau^2}} \exp\left\{-\frac{1}{2}\frac{(\theta-\mu)^2}{\tau^2}\right\}.
$$

The posterior distribution is proportional to the Likelihood × Prior. Hence in the product for the posterior distribution we keep the terms involving θ only.

$$\pi(\theta|y_1,\ldots,y_n) \propto \exp\left\{-\frac{1}{2}\left[\sum_{i=1}^{n}\frac{(y_i-\theta)^2}{\sigma^2} + \frac{(\theta-\mu)^2}{\tau^2}\right]\right\}.$$

The above quantity inside the box brackets, $\mathbb{M}$ say, is a quadratic in θ and hence we use the method of completion of squares below to complete the square in θ.

$$
\begin{aligned}
\mathbb{M} &= \sum_{i=1}^{n}\frac{(y_i-\theta)^2}{\sigma^2} + \frac{(\theta-\mu)^2}{\tau^2}\\
&= \frac{\sum_{i=1}^{n}y_i^2 - 2\theta\sum_{i=1}^{n}y_i + n\theta^2}{\sigma^2} + \frac{\theta^2 - 2\theta\mu + \mu^2}{\tau^2}\\
&= \theta^2\left(\frac{n}{\sigma^2} + \frac{1}{\tau^2}\right) - 2\theta\left(\frac{\sum_{i=1}^{n}y_i}{\sigma^2} + \frac{\mu}{\tau^2}\right) + \frac{\sum_{i=1}^{n}y_i^2}{\sigma^2} + \frac{\mu^2}{\tau^2}\\
&= \theta^2 a - 2\theta b + c
\end{aligned}
$$

where

$$a = \frac{n}{\sigma^2} + \frac{1}{\tau^2}, \quad b = \frac{n\bar{y}}{\sigma^2} + \frac{\mu}{\tau^2}, \quad c = \frac{\sum_{i=1}^{n}y_i^2}{\sigma^2} + \frac{\mu^2}{\tau^2},$$

and $\bar{y} = \frac{1}{n}\sum_{i=1}^{n}y_i$. Note that none of a, b and c involves θ. These are introduced just for writing convenience. Now

$$
\begin{aligned}
\mathbb{M} &= a\left(\theta^2 - 2\theta\frac{b}{a}\right) + c\\
&= a\left(\theta^2 - 2\theta\frac{b}{a} + \frac{b^2}{a^2} - \frac{b^2}{a^2}\right) + c\\
&= a\left(\theta - \frac{b}{a}\right)^2 + \frac{b^2}{a} + c.
\end{aligned}
$$

In the above the first term only involves θ since none of a, b and c involves θ. Hence the last two terms when exponentiated can be absorbed in the proportionality constant. Hence after discarding the last two terms we write:

$$\pi(\theta|y_1,\ldots,y_n) \propto \exp\left\{-\frac{1}{2}a\left(\theta - \frac{b}{a}\right)^2\right\}$$

which is recognized to be the pdf of a normal distribution with mean $\frac{b}{a}$ and variance $\frac{1}{a}$. More explicitly,

$$\pi(\theta|\mathbf{y}) = N\left(\mu_p \equiv \sigma_p^2\left(\frac{n\bar{y}}{\sigma^2} + \frac{\mu}{\tau^2}\right), \quad \sigma_p^2 \equiv \left(\frac{n}{\sigma^2} + \frac{1}{\tau^2}\right)^{-1}\right). \quad (4.4)$$

For notational convenience, we have defined μ_p and σ_p^2 to denote the mean and variance of the posterior distribution of θ given $\mathbf{y}$. The above posterior distribution reveals several points worth noting.

1. The posterior distribution only involves the data through the sample mean $\bar{y}$. A more general result is true for Bayesian posterior distributions. The posterior distribution for a parameter, θ here, will only depend on the sufficient statistics, (Bernardo and Smith, 1994) $\bar{y}$ here.

2. The posterior precision, defined as the inverse of the posterior variance σ_p^2, is sum of the data precision $\frac{n}{\sigma^2}$ and prior precision $\frac{1}{\tau^2}$. Thus, the Bayes theorem allows us naturally to add precisions from the data and prior information.

3. The posterior mean, μ_p is a convex combination of the data mean $\bar{y}$ and the prior mean μ. The weights of the data and prior means are respectively proportional to their precisions.

4. Assume that $\tau^2 = \frac{\sigma^2}{m}$, so that the prior distribution is $N\left(\mu, \frac{\sigma^2}{m}\right)$. Note that if $X_1, \ldots, X_m \sim N(\mu, \sigma^2)$ independently then $\bar{X} \sim \left(\mu, \frac{\sigma^2}{m}\right)$. Thus, the $N\left(\mu, \frac{\sigma^2}{m}\right)$ prior distribution for θ can be thought to have come from m fictitious (or imaginary) past observations before collecting the n new observations $y_1, \ldots, y_n$.

   (a) The posterior mean, μ_p, simplifies to $\frac{n\bar{y}+m\mu}{n+m}$, which is the aggregated mean from $n+m$ combined observations

   (b) The posterior variance, σ_p^2 simplifies to $\frac{\sigma^2}{n+m}$ which can be interpreted as the variance of the sample mean from $n+m$ combined observations.

   (c) When the data and prior distributions are of equal weight, i.e. $n = m$, then μ_p is the simple average of the data and prior mean. The posterior precision is then twice that of the data precision as can be intuitively expected.

5. Suppose $\tau^2 \to 0$ corresponding to a very precise prior, then the posterior mean μ_p is equal to the prior mean μ and the posterior variance $\sigma_p^2 = 0$. In this case, there is no need to collect any data for making inference about θ. The prior belief will never change.

6. Suppose $\tau^2 \to \infty$ corresponding to a very imprecise limiting prior distribution. In this case the limiting posterior distribution is $N\left(\bar{y}, \frac{\sigma^2}{n}\right)$, which is based completely from the information obtained from data alone.

♡ **Example 4.4.** We consider the New York air pollution example to illustrate the prior and posterior distributions. In this illustration we take the mean ozone concentration levels from the $n = 28$ sites as independent observations. For the sake of illustration we assume $\sigma^2 = 22$ which has been guided by the sample variance $s_y^2 = 22.01$. Also, we take the prior mean to be data mean $\bar{y} = 47.8$ plus $\sigma/\sqrt{n}$ so that it is one standard deviation (of the data mean) above the data mean. Also, to have an informative prior distribution the prior variance τ^2 is thought to come from $m = 10$ hypothetical past observations. Thus, we take:

$$\mu = \bar{y} + \frac{\sigma}{\sqrt{n}}, \quad \tau^2 = \frac{\sigma^2}{10}.$$

The resulting prior and posterior distributions in the left panel of Figure 4.2 show disagreement between the data and prior as expected. The prior distribution becomes non-informative for smaller values of m. For example, the prior distribution is effectively "flat" relative to the likelihood when $m = 1$, see the right panel of Figure 4.2. More discussion regarding the choice of the prior distribution is provided below in Section 4.10.

Warning: The prior mean μ has been parameterized here with the data mean $\bar{y}$ for illustration purposes only. In the Bayesian philosophy prior distributions are formed before observing the data and our strategy for adopting the prior distribution is not legal according to the Bayesian rule book. Hence, we do not recommend this method for practical modeling work that we will present in later chapters. We however still use these choice of the hyper-parameters to illustrate the theory with informative prior distributions and to study sensitivity.   □

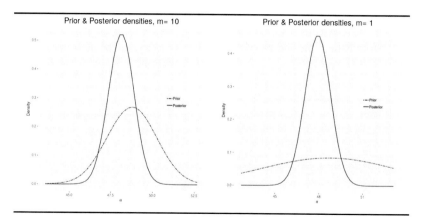

FIGURE 4.2: Prior and posterior distribution corresponding to two prior distribution s with $m = 10$ and $m = 1$ for the New York air pollution example.

4.8   Bayes estimators

Given the posterior distribution $\pi(\theta|y_1, \ldots, y_n)$, we require a mechanism to choose a reasonable estimator $\hat{\theta}$. Suppose the true parameter is θ_0, which is unknown. Let a be our estimate for it. In real life we may not have $a = \theta_0$. Then it is sensible to measure the penalty we have to pay for estimating incorrectly. The penalty may be measured by $(a - \theta_0)^2$ or $|a - \theta_0|$ or some other function. Intuitively, we should choose that value of a which minimizes the expected loss $E[L(a, \theta)]$, sometimes called the **risk**, where the expectation is taken with respect to the posterior distribution $\pi(\theta|y_1, \ldots, y_n)$ of θ. Note that a should not be a function of θ, rather it should be a function of $y_1, \ldots, y_n$, the random sample. The minimiser, $\hat\theta$ say, is called the *Bayes estimator* of θ.

4.8.1 Posterior mean

Suppose we consider the squared error loss function:

$$L(a, \theta) = (a - \theta)^2.$$

Let $b = E_{\pi(\theta|y_1, y_2, \ldots, y_n)}(\theta) = \int_{-\infty}^{\infty} \theta\, \pi(\theta|y_1, y_2, \ldots, y_n)d\theta$. Now, we have,

$$
\begin{aligned}
E[L(a, \theta)] &= \int_{-\infty}^{\infty} L(a, \theta)\, \pi(\theta|y_1, \ldots, y_n)d\theta \\
&= \int_{-\infty}^{\infty} (a - b + b - \theta)^2 \pi(\theta|y_1, \ldots, y_n)d\theta \\
&= (a - b)^2 + \int (b - \theta)^2 \pi(\theta|y_1, \ldots, y_n)d\theta \\
&\geq \int_{-\infty}^{\infty} (b - \theta)^2 \pi(\theta|y_1, \ldots, y_n)d\theta,
\end{aligned}
$$

for any value of a. In the above derivation, the cross-product term is

$$\int_{-\infty}^{\infty} 2(a - b)(b - \theta)\, \pi(\theta|y_1, y_2, \ldots, y_n)d\theta = 0$$

since

$$b = \int_{-\infty}^{\infty} \theta\, \pi(\theta|y_1, y_2, \ldots, y_n)d\theta.$$

When will the above inequality be an equality? The answer is when $a = b = E_{\pi(\theta|y_1, y_2, \ldots, y_n)}$. Hence we say that:

the Bayes estimator under squared error loss is the posterior mean.

♡ **Example 4.5. Binomial** The Bayes estimator under a squared error loss function is

$$\hat\theta = \frac{y + \alpha}{y + \alpha + n - y + \beta} = \frac{y + \alpha}{n + \alpha + \beta}.$$

□

♡ **Example 4.6. Exponential** The Bayes estimator under a squared error loss function is

$$\hat\theta = \frac{n + 1}{\mu + \sum_{i=1}^{n} y_i}.$$

Note that μ is a known constant, and hence $\hat\theta$ can be evaluated numerically.

□

♡ **Example** 4.7. Let $y_1, \ldots, y_n \sim Poisson(\theta)$ and suppose that the prior distribution is $\pi(\theta) = e^{-\theta}, \theta > 0$. The likelihood function is:

$$f(y_1, \ldots, y_n | \theta) = \prod_{i=1}^{n} \frac{1}{y_i!} e^{-\theta} \theta^{y_i} = \frac{1}{y_1! y_2! \cdots y_n!} \exp\{-n\theta\} \, \theta^{\sum_{i=1}^{n} y_i}$$

In order to derive the posterior distribution of θ we only collect the terms involving θ only from the likelihood times the prior.

$$\pi(\theta | y_1, \ldots, y_n) \quad \propto \exp\{-n\theta\} \, \theta^{\sum_{i=1}^{n} y_i} \exp\{-\theta\}$$
$$\propto \exp\{-(n+1)\theta\} \, \theta^{\sum_{i=1}^{n} y_i}$$

which is the pdf of $G\left(1 + \sum_{i=1}^{n} y_i, \frac{1}{n+1}\right)$. Hence the Bayes estimator of θ under squared error loss is:

$$\hat{\theta} = \text{Posterior mean} = \frac{1 + \sum_{i=1}^{n} y_i}{1 + n}.$$

□

4.8.2 Posterior median

Now suppose that we assume the absolute error loss function, $L(a, \theta) = |a - \theta|$. Then it can be proved that $E(|a - \theta|)$ is minimized by taking a to be the median of the posterior distribution of θ.

The median of a random variable X with pdf $g(x)$ is defined as the value m which solves:

$$\int_{-\infty}^{m} g(y) dy = \frac{1}{2}.$$

For symmetric posterior distributions the median is the point of symmetry. In general, it is hard to solve the above equation for finding m, the median. However, numerical solutions may be obtained.

4.8.3 Posterior mode

Suppose, we consider the loss function:

$$L(a, \theta) = \begin{cases} 0 & \text{if } |a - \theta| \le \delta \\ 1 & \text{if } |a - \theta| > \delta \end{cases}$$

where δ is a given small positive number. Thus, $L(a, \theta) = I(|a - \theta| > \delta)$ where $I(\cdot)$ is the indicator function of its argument, i.e. $I(A) = 1$ if A is true and $I(A)$ takes the value 0 otherwise. Now we find the expected value of the loss function where the expectation is to be taken with respect to the posterior distribution.

$$
\begin{aligned}
E[L(a,\theta)] &= \int_{-\infty}^{\infty} I(|a-\theta| > \delta)\pi(\theta|\mathbf{x})d\theta \\
&= \int_{-\infty}^{\infty} \left(1 - I(|a-\theta| \le \delta)\right)\pi(\theta|\mathbf{y})d\theta \\
&= 1 - \int_{a-\delta}^{a+\delta} \pi(\theta|\mathbf{y})d\theta \\
&\approx 1 - 2\delta\pi(a|\mathbf{x}).
\end{aligned}
$$

In order to minimize the risk we need to maximize $\pi(a|\mathbf{y})$ with respect to a and the Bayes estimator is the maximiser. Therefore, the Bayes estimator is that value of θ which maximize the posterior density function, i.e. the modal value. This estimator is called the maximum a-posteriori (MAP) estimator.

♡ **Example 4.8. New York air pollution data** The Bayes estimator under all three loss functions is

$$
\hat\theta = \left(\frac{n}{\sigma^2} + \frac{1}{\tau^2}\right)^{-1} \left(\frac{n\bar y}{\sigma^2} + \frac{\mu}{\tau^2}\right).
$$

Here $\bar y = 47.88$ and we continue to assume $\sigma^2 = 22$. Now we parameterize,

$$
\mu = \bar y + k\frac{\sigma}{\sqrt{n}} \quad \text{and} \quad \tau^2 = \frac{\sigma^2}{m}. \tag{4.5}
$$

Again, the reader is reminded of the warning noted in Example 4.7 regarding the choice of the hyper-parameters for the prior distribution. We intentionally use the parameterization (4.5) to study prior sensitivity and data-prior conflict as is done now.

Table 4.1 provides values of $\hat\theta$ for different values of k and m. Notice that $m = 0.1$ corresponds to a large prior variance relative to data variance and the first column of the table shows that the posterior mean being very close to the data mean does not get affected by the prior distribution even if it has a very "wrong" prior mean corresponding to $k = 16$. However, a strong prior distribution with $m = 100$ overwhelms the information from the data and results in a very different posterior mean.

	m			
k	0.1	1	10	100
1	47.88	47.91	48.11	48.57
2	47.88	47.94	48.34	49.26
4	47.89	48.00	48.81	50.65
16	47.93	48.36	51.61	58.96

TABLE 4.1: Table showing effect of assuming different prior distributions.

□

4.9   Credible interval

A point estimate, such as the posterior summaries above, does not convey the uncertainty associated with it. A quick way to assess uncertainty of the posterior estimate is to calculate the posterior standard deviation. However, more precise information can be obtained since the entire posterior distribution is available to us. For example, by continuing to assume that θ is a scalar parameter we can obtain an interval, say (a, b) where a and b may depend on the actual data values $y_1, \ldots, y_n$, such that

$$P(a \leq \theta \leq b | y_1, \ldots, y_n) = 1 - \alpha \tag{4.6}$$

for any given value $0 < \alpha < 1$. Such an interval (a, b) is called a credible or probability interval for θ. The associated interpretation of a credible interval is really appealing and sound since $1 - \alpha$ is the probability that the true θ lies inside the interval (a, b). Recall the basic Bayesian philosophy that probability is the only measure of uncertainty. Contrast this with the long-run frequency interpretation attached to a traditional confidence interval obtained in classical statistics. A calculated 95% confidence interval, say $(29.5, 51.3)$ does not allow us to say $P(29.5 \leq \theta \leq 51.3) = 0.95$ since θ is non-random. However, if we obtain $a = 29.5$ and $b = 51.3$ satisfying (4.6) then we can clearly write $P(29.5 \leq \theta \leq 51.3 | y_1, \ldots, y_n) = 0.95$.

The credible interval defined in (4.6) does not guarantee that it includes all the highest probability regions of the posterior distribution. Such an interval may leave out high density areas. To guard against such problems we define the *highest posterior density*, or HPD, intervals. A credible interval satisfying (4.6) is an HPD credible interval if $\pi(\theta | \mathbf{y}) \geq \pi(\psi | \mathbf{y})$ for all $\theta \in (a, b)$ and $\psi \notin (a, b)$. That is, for all ψ values outside the HPD the posterior density will be lower than that for all θ values inside the credible region. Intuitively, the HPD credible intervals are the preferred ones as those do not leave out high probability areas of the posterior distribution. However, finding an HPD interval requires more knowledge of the posterior distribution.

The concepts of credible intervals and HPD credible intervals can easily be generalized to credible sets and HPD regions when there is more than one parameter. For example, a set A is called a $100(1 - \alpha)\%$ credible set for θ if $P(\theta \in A | \mathbf{y}) = 1 - \alpha$. A set A is called a *Highest Posterior Density* (HPD) credible region if $\pi(\theta | \mathbf{x}) \geq \pi(\psi | \mathbf{x})$ for all $\theta \in A$ and $\psi \notin A$.

♡ **Example 4.9. Normal–Normal**   The 95% HPD credible region for θ is given by:

$$\left(\frac{n}{\sigma^2} + \frac{1}{\tau^2} \right)^{-1} \left(\frac{n\bar{y}}{\sigma^2} + \frac{\mu}{\tau^2} \right) \pm 1.96 \left(\frac{n}{\sigma^2} + \frac{1}{\tau^2} \right)^{-2}.$$

For the New York air pollution data example suppose that μ and τ^2 are as in (4.5). For $k = 1$ and $m = 10$, the 95% HPD is found to be (46.20 49.61) while for $k = 1$ and $m = 1$ the 95% HPD is (46.62, 49.60).

□

4.10 Prior Distributions

4.10.1 Conjugate prior distribution

Suppose that we have a hierarchical model where $f(\mathbf{y}|\theta)$ provides the likelihood function and $\pi(\theta|\eta)$ is the prior distribution where η are hyperparameters. If the posterior distribution $\pi(\theta|\mathbf{y}, \eta)$ belongs to the same parametric family as $\pi(\theta|\eta)$, then we say that $\pi(\theta|\eta)$ is a conjugate prior distribution for θ. With conjugate prior distributions the posterior analysis becomes much easier if in addition we assume the hyper parameters η to be known. Here is a table showing Natural conjugacies:

Likelihood	Prior
Binomial	Beta
Poisson	Gamma
Normal	Normal
Exponential	Gamma

If the hyper-parameters η are unknown then a suitable prior distribution, $\pi(\eta)$ must be provided as part of the hierarchical model specification so that we can write down the joint posterior distribution as:

$$\pi(\theta, \eta|\mathbf{y}) \propto f(\mathbf{y}|\theta)\,\pi(\theta|\eta)\,\pi(\eta).$$

Bayesian inference can proceed by evaluating the marginal posterior distribution: $\pi(\theta|\mathbf{y} = \int_{-\infty}^{\infty} \pi(\theta, \eta|\mathbf{y})d\eta$. Section 4.13 provides an example of this.

There are other suggested alternatives in the literature. For example, it may be possible to evaluate the marginal likelihood of η given by:

$$\pi(\eta; \mathbf{y}) = \int_{-\infty}^{\infty} f(\mathbf{y}|\theta)\,\pi(\theta|\eta)d\theta.$$

It is then intuitively suggested that one finds the maximum likelihood estimate of η from this marginal likelihood and use that estimate in the posterior distribution of $\pi(\theta|\mathbf{y})$. Such a prior distribution is called an ML-II prior, see page 232 of Berger (1985).

4.10.2   Locally uniform prior distribution

Often, a locally uniform prior distribution is suggested. Such a prior distribution puts equal prior mass for all values of θ, i.e.

$$\pi(\theta) = k, \; k > 0, \; -\infty < \theta < \infty.$$

That is, the prior distribution is uniform over the real line. Such a prior is often synonymously called a non-informative, vague, diffuse, or flat prior distribution. These default prior distributions are assumed to conduct Bayesian analysis in case there is no prior information with which to choose $\pi(\theta)$.

There is a potential problem in using a flat prior distribution that arises from the fact that $\int_{-\infty}^{\infty} \pi(\theta)d\theta$ is not finite for such a prior distribution. Hence, $\pi(\theta) = k$ does not provide a well-defined prior distribution. This is called an **improper** prior distribution since it is not a proper probability density function which can integrate to one. However, recall that all statistical inferences, which have been discussed so far, are made using the posterior distribution $\pi(\theta|\mathbf{y})$. Hence it is "legal" to assume an improper prior distribution if only it is guaranteed that the posterior distribution is proper so that the inferences made are all valid.

Improper prior distributions are also justified on another theoretical ground. Suppose that we assume that $\pi(\theta) = k$ only for values of θ where the likelihood function has appreciable value, and $\pi(\theta) = 0$ otherwise. This $\pi(\theta)$ will then define a proper probability density function and no theoretical problem arises.

4.10.3   Non-informative prior distribution

Often in practical settings researchers are faced with having no prior distribution to go by. That is, they do not have access to any prior information such as which values of θ are more likely than others; which values may not occur at all and so on. In some of these situations non-informative prior distributions are used. By definition, a *non-informative prior* should not contain any "information" for θ.

The locally uniform prior distribution defined previously may still contain information regarding θ since information here is interpreted relative to the sampling experiment. More precisely information is measured as the following property of the likelihood function $f(\mathbf{y}|\theta)$:

$$I(\theta) = -E\left[\frac{\partial^2}{\partial\theta^2} \log f(\mathbf{Y}|\theta)\right],$$

where the expectation is taken with respect to the joint distribution of $Y_1, \ldots, Y_n$. This $I(\theta)$ is known as the Fisher information and is used heavily in developing likelihood based statistical theory for inference.

The most widely used non-informative prior distribution is provided by Jeffreys (1961) and is given by:

$$\pi(\theta) = \sqrt{I(\theta)}. \tag{4.7}$$

As an example, for the binomial distribution it can be shown that the non-informative prior distribution is given by:

$$\pi(\theta) \propto \{\theta(1-\theta)\}^{-\frac{1}{2}}.$$

This prior distribution is a beta distribution with parameters $\frac{1}{2}$ and $\frac{1}{2}$ and thus is a proper distribution. However, in most situations the non-informative prior distribution 4.7) yields an improper prior distribution. In such cases we have to guarantee that the resulting posterior distribution is a proper probability distribution.

Why does the above prior distribution (4.7) give a non-informative prior? The answer in brief is the following. The above prior distribution induces a one-to-one function $g(\theta)$, $= \phi$ say, for which the pdf of ϕ, $\pi(\phi) \propto 1$. That is, for the transformed parameter ϕ the induced prior distribution is locally uniform or non-informative. Hence the prior distribution (4.7) for θ (which is a one-to-one transformation function of ϕ) is also non-informative.

There is a huge literature on prior selection and prior elicitation, see e.g. Berger (1985). A systematic study of that literature is beyond the scope of the current book. In our modeling development we will mostly use either conjugate or flat prior distributions to demonstrate methodology. Where possible, a limited amount of sensitivity study with respect to the choice of the prior distributions will also be conducted to demonstrate robustness in the drawn inference.

4.11 Posterior predictive distribution

"What is the probability that the sun will rise tomorrow, given that it has risen without fail for the last n days?" In order to answer questions like these we need to learn what are called predictive distributions.

Let $Y_1, \ldots, Y_n$ be an i.i.d. sample from the distribution $f(y|\theta)$. Let $\pi(\theta)$ be the prior distribution and $\pi(\theta|\mathbf{y})$ be the posterior distribution. We want to evaluate the probability distribution (pdf or pmf) of $Y_0|Y_1, \ldots, Y_n$. The given notation is to denote that $Y_1, \ldots, Y_n$ have already been observed, like the sun has risen for the last n days. We define the **posterior predictive distribution** to be:

$$f(y_0|y_1, \ldots, y_n) = \int_{-\infty}^{\infty} f(y_0|\theta, y_1, \ldots, y_n) \, \pi(\theta|y_1, \ldots, y_n) \, d\theta. \tag{4.8}$$

This density simplifies to

$$f(y_0|y_1,\ldots,y_n) = \int_{-\infty}^{\infty} f(y_0|\theta)\,\pi(\theta|y_1,\ldots,y_n)\,d\theta, \tag{4.9}$$

when Y_0 is conditionally independent of the random sample $Y_1,\ldots,Y_n$ given the parameter θ. For spatially or temporally correlated data modeling the first version (4.8) will be the appropriate one to use. The second simpler version (4.9) is used for modeling random sample data.

The above posterior predictive density is the density of a future observation, Y_0 given everything else, i.e., the "model" and the observations, $y_1,\ldots,y_n$. (The model is really the function $f(y|\theta)$ in the case of random sampling.) In such a case, intuitively, if θ is known then y_0 will follow $f(y_0|\theta)$ since it is from the same population as $y_1,\ldots,y_n$ are. However, we do not know θ but the posterior distribution $\pi(\theta|y)$ contains all the information (from the likelihood and the prior) that we know about θ. Therefore, the predictive distribution is obtained as an average over $\pi(\theta|y)$. We now derive some predictive distributions.

♡ **Example** 4.10. We return to the sun example. Let

$$Y_i = \begin{cases} 1 & \text{if its sunny on the } i\text{th day,} \\ 0 & \text{otherwise} \end{cases}$$

for $i = 0, 1, \ldots, n$. Note that Y_0 will be binary as well. We would like to evaluate $P(Y_0 = 1|y = (1,1,\ldots,1))$. Assume that $f(y_i|\theta) = \theta^{y_i}(1-\theta)^{1-y_i}$, and Y_i are independent for $i = 1,\ldots,n+1$. Therefore, the likelihood function is

$$\begin{aligned} f(y|\theta) &= \theta^{\sum_{i=1}^n y_i}(1-\theta)^{n-\sum_{i=1}^n y_i}, \\ &= \theta^n \text{ since } y = (1,1,\ldots,1). \end{aligned}$$

Let us assume a uniform prior distribution for θ, i.e. $\pi(\theta) = 1$ if $0 < \theta < 1$. Now the posterior distribution is:

$$\pi(\theta|y) = \frac{\theta^n}{\int_0^1 \theta^n d\theta} = (n+1)\theta^n.$$

Note that $f(Y_0 = 1|\theta) = \theta$. Finally we can evaluate the posterior predictive distribution using (4.9).

$$\begin{aligned} P(Y_0 = 1|y) &= \int_0^1 \theta(n+1)\theta^n d\theta \\ &= (n+1)\int_0^1 \theta^{n+1} d\theta \\ &= \frac{n+1}{n+2}. \end{aligned}$$

Intuitively, this probability goes to 1 as $n \to \infty$. Also note that $P(Y_0 = 0|y) = \frac{1}{n+2}$. Thus, we have obtained the posterior predictive distribution given the observations y. □

4.11.1 Normal-Normal example

Return to the Normal-Normal example introduced in Section 4.7. We now derive the posterior predictive distribution of a future observation Y_0 given the observations $y_1, \ldots, y_n$.

Recall that the posterior distribution (4.4) is given by:

$$\pi(\theta|\mathbf{y}) = N\left(\mu_p = \sigma_p^2\left(\frac{n\bar{y}}{\sigma^2} + \frac{\mu}{\tau^2}\right), \ \sigma_p^2 = \left(\frac{n}{\sigma^2} + \frac{1}{\tau^2}\right)^{-1}\right).$$

Below we prove that the posterior predictive distribution

$$Y_0|\mathbf{y} \sim N\left(\mu_0 \equiv \mu_p, \ \sigma_0^2 \equiv \sigma^2 + \sigma_p^2\right). \tag{4.10}$$

The above posterior predictive distribution reveals several points worth noting.

1. Note that the mean of the posterior predictive distribution is same as the mean of the posterior distribution. This is intuitive and justified by the hierarchical specifications that

   $$Y_0|\theta, \mathbf{y} \sim N(\theta, \sigma^2) \text{ and } \theta|\mathbf{y} \sim N(\mu_p, \sigma_p^2).$$

   This hierarchical specification implies

   $$E(Y_0|\mathbf{y}) = E(E(Y_0|\mathbf{y}, \theta))$$

   where the outer expectation is with respect to the posterior distribution of θ. The inner expectation, $E(Y_0|\mathbf{y}, \theta)$ is simply θ and the outer expectation is the posterior mean – giving the result.

2. The variance of the posterior predictive distribution is interpreted as the sum of the data and posterior variances. This is again justified by the hierarchical specification and the formula (which can be proved in general) that

   $$\text{Var}(Y_0|\mathbf{y}) = E\left(\text{Var}(Y_0|\mathbf{y}, \theta)\right) + \text{Var}\left(E\left(Y_0|\mathbf{y}, \theta\right)\right).$$

3. The variance of the prediction is the sum of data variance, σ^2 and the posterior variance σ_p^2. This addition of the variances is justified since the posterior predictive distribution aims to predict a future observation (with its own variability) – not just the overall mean.

4. By taking $\tau^2 = \frac{\sigma^2}{m}$ we have $\sigma_p^2 = \frac{\sigma^2}{n+m}$. Now:

   $$
   \begin{aligned}
   \sigma_0^2 &= \sigma^2 + \frac{\sigma^2}{n+m} \\
   &= \frac{n+m+1}{n+m}\sigma^2 \\
   &= (n+m+1)\sigma_p^2.
   \end{aligned}
   $$

This shows that the variance of the posterior predictive distribution is $n + m + 1$ times the variance of the posterior distribution. This result also holds for the unknown variance case as we shall see in Section 4.14.

5. The posterior predictive distribution (4.10) is inference-ready! No further derivation is required to assess uncertainty of predictions unlike in the classical inference case. For example, this distribution is ready to be used for making point and interval predictions for future observations just the same way as Bayes estimators were found in Sections 4.8 and 4.9.

6. The limiting behaviors as $n \to \infty$ and/or $\tau^2 \to \infty$ discussed in Section 4.7 are also relevant here.

♡ **Example** 4.11. **New York Air pollution data** Returning to the example with the prior distribution $k = 1$ and $m = 1$, we have $\mu_p = 47.91$ and $\sigma_p^2 = 0.58$. Hence, any new observation Y_0 will have the $N(47.91, 22.58)$ distribution. Contrast this with the conditional distribution of $Y_0|\theta$, which is $N(\theta, 22)$. The added variability in the distribution of $Y_0|\mathbf{y}$ is due to the uncertainty in estimating θ. Prediction estimates and intervals are easily obtained from the summaries of the posterior predictive distribution $N(47.91, 22.58)$.   □

We provide a proof of the main result (4.10) below for the sake of completeness. As in Section 4.7 this derivation is a bit tedious and can be skipped in the first reading.

Recall that $Y_0|\theta \sim N(\theta, \sigma^2)$ and we are using the notation $\theta|\mathbf{y} \sim N(\mu_p, \sigma_p^2)$. We would like to prove that $Y_0|\mathbf{y} \sim N(\mu_p, \sigma^2 + \sigma_p^2)$. By definition (4.9),

$$
\begin{aligned}
f(y_0|\mathbf{y}) &= \int_{-\infty}^{\infty} f(y_0|\theta)\, \pi(\theta|\mathbf{y}) d\theta \\
&= \int_{-\infty}^{\infty} \frac{1}{\sqrt{2\pi\sigma^2}} e^{-\frac{1}{2\sigma^2}(y_0-\theta)^2} \frac{1}{\sqrt{2\pi\sigma_p^2}} \exp\left\{ -\frac{1}{2\sigma_p^2}(\theta - \mu_p)^2 \right\} d\theta \\
&= \int_{-\infty}^{\infty} \frac{1}{2\pi\sqrt{\sigma^2\sigma_p^2}} \exp\left\{ -\frac{1}{2}\left[\frac{(y_0-\theta)^2}{\sigma^2} + \frac{(\theta-\mu_p)^2}{\sigma_p^2} \right] \right\} d\theta \\
&= \int_{-\infty}^{\infty} \frac{1}{2\pi\sqrt{\sigma^2\sigma_p^2}} \exp\left\{ -\tfrac{1}{2}\mathbb{M} \right\} d\theta,
\end{aligned}
$$

where

$$
\begin{aligned}
\mathbb{M} &= \frac{(y_0-\theta)^2}{\sigma^2} + \frac{(\theta-\mu_p)^2}{\sigma_p^2} \\
&= \theta^2\left\{ \frac{1}{\sigma^2} + \frac{1}{\sigma_p^2} \right\} - 2\theta\left\{ \frac{y_0}{\sigma^2} + \frac{\mu_p}{\sigma_p^2} \right\} + \left\{ \frac{y_0^2}{\sigma^2} + \frac{\mu_p^2}{\sigma_p^2} \right\} \\
&= \theta^2 a - 2\theta b + c, \quad \text{say} \\
&= a\left(\theta - \frac{b}{a} \right)^2 - \frac{b^2}{a} + c,
\end{aligned}
$$

where

$$
a = \frac{1}{\sigma^2} + \frac{1}{\sigma_p^2}, \quad b = \frac{y_0}{\sigma^2} + \frac{\mu_p}{\sigma_p^2}, \quad c = \frac{y_0^2}{\sigma^2} + \frac{\mu_p^2}{\sigma_p^2}.
$$

Now

$$
\begin{aligned}
-\frac{b^2}{a} + c &= -\left\{\frac{y_0}{\sigma^2} + \frac{\mu_p}{\sigma_p^2}\right\}^2 \frac{\sigma^2 \sigma_p^2}{\sigma^2 + \sigma_p^2} + \frac{y_0^2}{\sigma^2} + \frac{\mu_p^2}{\sigma_p^2} \\
&= -\frac{\sigma^2 \sigma_p^2}{\sigma^2 + \sigma_p^2}\left(\frac{y_0^2}{\sigma^4} + 2\frac{y_0 \mu_p}{\sigma^2 \sigma_p^2} + \frac{\mu_p^2}{\sigma_0^4}\right) + \frac{y_0^2}{\sigma^2} + \frac{\mu_p^2}{\sigma_p^2} \\
&= -\frac{y_0^2 \sigma_p^2}{\sigma^2 (\sigma^2 + \sigma_p^2)} - 2\frac{y_0 \mu_p}{\sigma^2 + \sigma_p^2} - \frac{\mu_p^2 \sigma^2}{\sigma_p^2 (\sigma^2 + \sigma_p^2)} + \frac{y_0^2}{\sigma^2} + \frac{\mu_p^2}{\sigma_p^2} \\
&= \frac{y_0^2}{\sigma^2}\left(1 - \frac{\sigma_p^2}{\sigma^2 + \sigma_p^2}\right) - 2\frac{y_0 \mu_p}{\sigma^2 + \sigma_p^2} + \frac{\mu_p^2}{\sigma_p^2}\left(1 - \frac{\sigma^2}{\sigma^2 + \sigma_p^2}\right) \\
&= \frac{y_0^2}{\sigma^2}\frac{\sigma^2}{\sigma^2 + \sigma_p^2} - 2\frac{y_0 \mu_p}{\sigma^2 + \sigma_p^2} + \frac{\mu_p^2}{\sigma_p^2}\frac{\sigma_p^2}{\sigma^2 + \sigma_p^2} \\
&= \frac{1}{\sigma^2 + \sigma_p^2}(y_0 - \mu_p)^2
\end{aligned}
$$

Therefore,

$$
\begin{aligned}
\pi(y_0|\mathbf{y}) &= \int_{-\infty}^{\infty} \frac{1}{2\pi\sqrt{\sigma^2 \sigma_p^2}} \exp\left\{-\frac{1}{2}\mathbb{M}\right\} d\theta \\
&= \int_{-\infty}^{\infty} \frac{1}{2\pi\sqrt{\sigma^2 \sigma_p^2}} \exp\left\{-\frac{1}{2}\left\{a\left(\theta - \frac{b}{a}\right)^2 + \frac{1}{\sigma^2 + \sigma_p^2}(y_0 - \mu_p)^2\right\}\right\} d\theta \\
&= \frac{1}{\sqrt{2\pi\sigma^2\sigma_p^2}} \exp\left\{-\frac{1}{2}\frac{1}{\sigma^2 + \sigma_p^2}(y_0 - \mu_p)^2\right\} \int_{-\infty}^{\infty} \frac{1}{\sqrt{2\pi}} \exp\left\{-\frac{a}{2}\left(\theta - \frac{b}{a}\right)^2\right\} d\theta \\
&= \frac{1}{\sqrt{2\pi\sigma^2\sigma_p^2}} \exp\left\{-\frac{1}{2}\frac{1}{\sigma^2 + \sigma_p^2}(y_0 - \mu_p)^2\right\} \sqrt{\frac{1}{a}} \\
&= \frac{1}{\sqrt{2\pi(\sigma^2 + \sigma_p^2)}} \exp\left\{-\frac{1}{2}\frac{1}{\sigma^2 + \sigma_p^2}(y_0 - \mu_p)^2\right\},
\end{aligned}
$$

since

$$
\int_{-\infty}^{\infty} \frac{1}{\sqrt{2\pi}} \exp\left\{-\frac{a}{2}\left(\theta - \frac{b}{a}\right)^2\right\} d\theta = \sqrt{\frac{1}{a}}
$$

and $a\sigma^2\sigma_p^2 = \sigma^2 + \sigma_p^2$.

4.12 Prior predictive distribution

The posterior predictive distributions discussed in Section 4.11 derived the distribution of a future observation conditional on n independent observations. In addition, and somewhat in contrast, the Bayesian methods also allow prediction of observations conditional on no observation at all based on what are called prior predictive distributions, provided that a proper prior distribution has been assumed for the unknown parameter. The prior predictive distribution is defined as

$$
f(y) = \int_{-\infty}^{\infty} f(y|\theta)\pi(\theta)\,d\theta. \tag{4.11}
$$

The assumption of the proper prior distribution ensures that the averaging of $f(y|\theta)$ with respect to $\pi(\theta)$ is well defined. Note that $f(y)$, as defined in (4.11) is simply the normalizing constant in $\pi(\theta|y)$ in (4.2). The distribution $f(y)$ is also interpreted as the marginal distribution of the random variable Y. Lastly, note that the prior predictive distribution (4.11) is of the same form as the posterior predictive distribution (4.9) where the averaging density $\pi(\theta|y_1,\ldots,y_n)$ is replaced by the prior distribution $\pi(\theta)$.

With n independent samples, we define the (**joint**) prior predictive distribution of $\mathbf{y} = y_1,\ldots,y_n$ as

$$f(\mathbf{y}) = \int_{-\infty}^{\infty} f(\mathbf{y}|\theta)\,\pi(\theta)\,d\theta. \tag{4.12}$$

♡ **Example** 4.12. For the Normal-Normal example, the prior predictive distribution is,

$$f(\mathbf{y}) = \int_{-\infty}^{\infty} \prod_{i=1}^{n} N(y_i|\theta,\sigma^2) N(\theta|\mu,\tau^2) d\theta.$$

This integral can be evaluated using the same technique as in Section 4.11.1 to conclude that the above $f(\mathbf{y})$ defines a multivariate normal probability density function. However, the means, variances and the covariances can be derived using the hierarchical specifications that $Y_i|\theta,\sigma^2 \sim N(\theta,\sigma^2)$ and $\theta \sim N(\mu,\tau^2)$, and using the result that, for any two random variable with finite variances:

$$E(X) = E(E(X|Y)), \quad \text{Var}(X) = E(\text{Var}(X|Y)) + \text{Var}(E(X|Y)).$$

For this example, it can be shown that $E(Y_i) = \mu$, $\text{Var}(Y_i) = \tau^2 + \sigma^2$ and $\text{Cov}(Y_i, Y_j) = \tau^2$ when $i \neq j$. Thus, marginally Y_i and Y_j will not be independent once θ has been integrated out. □

The prior predictive distribution $f(\mathbf{y})$ provides the marginal likelihood of the data and this evaluates to be a number when evaluated at the observations $\mathbf{y} = (y_1,\ldots,y_n)$. This number comes in handy when it is desired to compare two competing models for the same data. Intuitively, out of the two, the one with a higher marginal likelihood should be preferred. Hence Bayesian model choice measures based on the ratio of marginal likelihoods are defined in the next section. These marginal likelihoods save us from having to compare parameter estimates from parametric models when parameters have different interpretations under different models as have been noted previously.

4.13   Inference for more than one parameter

The discussion so far has intentionally treated the unknown parameter θ to be a scalar to gain a simpler understanding of the basic methods for Bayesian inference. However, in most realistic applications of statistical models, there are more than one unknown parameters. In principle, everything proceeds as before, but by treating θ as a vector of parameters instead. Thus, instead of writing θ we write the vectorised parameter $\boldsymbol{\theta}$ which contain a number, p say, of parameters. In this case $\boldsymbol{\theta} = (\theta_1, \ldots, \theta_p)$ and the posterior distribution is a multivariate distribution written as:

$$\pi(\boldsymbol{\theta}|\mathbf{y}) = \frac{f(\mathbf{y}|\boldsymbol{\theta})\,\pi(\boldsymbol{\theta})}{\int_{-\infty}^{\infty} \cdots \int_{-\infty}^{\infty} f(\mathbf{y}|\boldsymbol{\theta})\,\pi(\boldsymbol{\theta})d\theta_1 \cdots d\theta_p}.$$

Notice that the integral for $\boldsymbol{\theta}$ in the denominator is a multivariate integral, but the posterior distribution is still seen to be of the general likelihood $\times$ prior form

$$\pi(\boldsymbol{\theta}|\mathbf{y}) \propto f(\mathbf{y}|\boldsymbol{\theta}) \times \pi(\boldsymbol{\theta}).$$

Posterior inference can proceed by calculating the marginal posterior distribution for the parameter of interest. For example, if we are interested in θ_1, we can obtain the marginal posterior distribution of θ_1 as

$$\pi(\theta_1|\mathbf{y}) = \int_{-\infty}^{\infty} \cdots \int_{-\infty}^{\infty} \pi(\boldsymbol{\theta}|\mathbf{y})\,d\theta_2 d\theta_3 \ldots d\theta_p. \tag{4.13}$$

Note that θ_1 is not integrated out in the above integration. Now we can calculate features of the above distribution, for example $E(\theta_1|\mathbf{y})$ and $\text{Var}(\theta_1|\mathbf{y})$, for making inference.

The multivariate posterior distribution allows us to study properties of the joint distribution such as $P(\theta_1 < a, \theta_2 > b|\mathbf{y})$ for specified values of a and b. In a similar vein we can study correlations between any two parameters conditional on the observed data, $\mathbf{y}$. Thus, simultaneous inference, such as simultaneous credible sets, can be obtained using the multivariate posterior distribution without much extra effort. The integrals involved in these definitions will be estimated using Bayesian computation. Hence it is relatively simple to think of the inferential extensions to the multi-parameter case.

4.14 Normal example with both parameters unknown

We re-formulate the previous Normal-Normal example in Section 4.7 but now assume that σ^2 is unknown. Continue to assume that $Y_1, Y_2, \ldots, Y_n$ are i.i.d. $N(\theta, \sigma^2)$. We parameterize this by the mean θ and the precision $\lambda^2 = \frac{1}{\sigma^2}$, instead of σ^2, for ease of both analytical and numerical investigations. However, no generality will be lost by doing this and inference regarding σ^2 will be made easily by calculating the appropriate integral as demonstrated below.

Following Berger (1985), page 288 we assume a hierarchical specification for the joint prior distribution of θ and σ^2 as $\pi(\theta|\lambda^2) \times \pi(\lambda^2)$. This comes from natural hierarchical Bayesian modeling machinery and also helps in obtaining analytical results that we aim to verify by Bayesian computation methods. We assume that $\theta|\lambda^2 \sim N\left(\mu, \frac{1}{m\lambda^2}\right)$ and $\lambda^2 \sim G(a,b)$ for known values of μ, $V > 0$, $a > 0$ and $b > 0$. The gamma distribution $G(a,b)$ is parameterized so that it has mean a/b and variance a/b^2, see (A.18). Notice that m here has the same interpretation as before, viz. it is the fictitious number of past observations used to obtain the prior distribution $\pi(\theta|\lambda^2)$.

We now derive the likelihood function, joint posterior distribution and the conditional and marginal posterior distributions which will enable us to make inference. The long derivations below assume some familiarity with the gamma integrals but can be skipped at the first reading.

The joint posterior distribution for θ and λ^2 is written as likelihood $\times$ prior

$$(\lambda^2)^{\frac{n}{2}} \exp\left\{-\frac{\lambda^2}{2}\sum_{i=1}^{n}(y_i - \theta)^2\right\}(\lambda^2)^{\frac{1}{2}}\exp\left\{-\frac{m\lambda^2}{2}(\theta - \mu)^2\right\}(\lambda^2)^{a-1}\exp\left\{-b\lambda^2\right\}$$

for $-\infty < \theta < \infty$ and $\lambda^2 > 0$. After re-arranging the terms we write:

$$\pi(\theta, \lambda^2|\mathbf{y}) \propto (\lambda^2)^{\frac{n+1}{2}+a-1}\exp\left\{-\frac{\lambda^2}{2}\left[2b + \sum_{i=1}^{n}(y_i - \theta)^2 + m(\theta - \mu)^2\right]\right\}.$$
(4.14)

By comparing this with the density of the gamma distribution, it is easy to recognize that the conditional posterior distribution

$$\lambda^2|\theta, \mathbf{y} \sim G\left(\frac{n+1}{2} + a, \frac{1}{2}\left[2b + \sum_{i=1}^{n}(y_i - \theta)^2 + m(\theta - \mu)^2\right]\right).$$
(4.15)

We also complete the square in θ to write:

$$\sum_{i=1}^{n}(y_i - \theta)^2 + m(\theta - \mu)^2 = (n-1)s_y^2 + (n+m)(\theta - \mu_p)^2 + \frac{mn}{n+m}(\bar{y} - \mu)^2,$$
(4.16)

where

$$s_y^2 = \frac{1}{n-1}\sum_{i=1}^{n}(y_i - \bar{y})^2 \quad \text{and} \quad \mu_p = \frac{n\bar{y} + m\mu}{n+m}.$$

Note that s_y^2 is the sample variance of y and μ_p is the same as the posterior mean in the Normal-Normal example for known variance σ^2 in Section 4.11.1 and prior variance $\tau^2 = \frac{\sigma^2}{m}$. Now the joint posterior distribution (4.14) is re-written as:

$$\pi(\theta, \lambda^2|\mathbf{y}) \propto (\lambda^2)^{\frac{n+1}{2}+a-1}$$

$$\times \exp\left\{-\frac{\lambda^2}{2}\left[2b + (n-1)s_y^2 + (n+m)(\theta - \mu_p)^2 + \frac{mn}{n+m}(\bar{y} - \mu)^2\right]\right\}.$$
(4.17)

Hence, it is easy to recognize that the conditional posterior distribution

$$\theta|\lambda^2, \mathbf{y} \sim N\left(\mu_p = \frac{n\bar{y} + m\mu}{n + m}, \ \sigma_p^2 = \frac{1}{\lambda^2(n+m)}\right), \qquad (4.18)$$

where note that σ_p^2 here coincides with the definition before when $\tau^2 = \frac{\sigma^2}{m}$.

The marginal posterior distribution, as generally defined in (4.13), of $\lambda^2|\mathbf{y}$ is found as follows:

$$\pi(\lambda^2|\mathbf{y}) \propto \int_{-\infty}^{\infty} \pi(\theta, \lambda^2|\mathbf{y})d\theta$$

$$\propto (\lambda^2)^{\frac{n+1}{2}+a-1} \exp\left\{-\frac{\lambda^2}{2}\left[2b + (n-1)s_y^2 + \frac{mn}{n+m}(\bar{y}-\mu)^2\right]\right\}$$

$$\int_{-\infty}^{\infty} \exp\left\{-\frac{\lambda^2}{2}(n+m)(\theta - \mu_p)^2\right\}d\theta$$

$$\propto (\lambda^2)^{\frac{n+1}{2}+a-1} \exp\left\{-\frac{\lambda^2}{2}\left[2b + (n-1)s_y^2 + \frac{mn}{n+m}(\bar{y}-\mu)^2\right]\right\}$$

$$\times (\lambda^2(n+m))^{-\frac{1}{2}}$$

$$\propto (\lambda^2)^{\frac{n}{2}+a-1} \exp\left\{-\frac{\lambda^2}{2}\left[2b + (n-1)s_y^2 + \frac{mn}{n+m}(\bar{y}-\mu)^2\right]\right\}$$

by noting that the last definite integral is proportional to an integral over the whole domain the normal density function with mean μ_p and variance $\{\lambda^2(n+m)\}^{-1}$. By comparing the above with that of the density of a gamma distribution we conclude that

$$\lambda^2|\mathbf{y} \sim G\left(\frac{n}{2}+a, \ \frac{1}{2}\left[2b + (n-1)s_y^2 + \frac{mn}{n+m}(\bar{y}-\mu)^2\right]\right).$$

The result that if $X \sim G(a,b)$ then $E\left(\frac{1}{X}\right) = \frac{b}{a-1}$, gives us:

$$E(\sigma^2|\mathbf{y}) = \frac{1}{n+2a-2}\left[2b + (n-1)s_y^2 + \frac{mn}{n+m}(\bar{y}-\mu)^2\right].$$

Similarly, the result that if $X \sim G(a,b)$ then $\mathrm{Var}\left(\frac{1}{X}\right) = \frac{b^2}{(a-1)^2(a-2)}$ gives us:

$$\mathrm{Var}(\sigma^2|\mathbf{y}) = \frac{2}{(n+2a-2)^2(n+2a-4)}\left[2b + (n-1)s_y^2 + \frac{mn}{n+m}(\bar{y}-\mu)^2\right]^2.$$

A limiting joint flat prior distribution for θ and σ^2 is obtained when $m \to 0$ and both a and b approach zero. In such a case note that we have $E(\sigma^2|\mathbf{y}) = s_y^2$, as in the case of classical inference. By applying the integral (A.19), we now obtain the marginal posterior distribution of $\theta|\mathbf{y}$.

$$
\begin{aligned}
\pi(\theta|\mathbf{y}) &\propto \int_0^\infty \pi(\theta, \lambda^2|\mathbf{y})d\lambda^2 \\
&\propto \int_0^\infty (\lambda^2)^{\frac{n+1}{2}+a-1} \exp\left\{-\frac{\lambda^2}{2}[2b_0]\right\} d\lambda^2 \\
&\propto \int_0^\infty (\lambda^2)^{a_0-1} \exp\left\{-\lambda^2 b_0\right\} d\lambda^2 \\
&\propto \frac{\Gamma(a_0)}{b_0^{a_0}}
\end{aligned}
$$

where

$$
a_0 = \frac{n+2a+1}{2}, \text{ and } b_0 = b+\frac{1}{2}(n-1)s_y^2+\frac{1}{2}\frac{mn}{n+m}(\bar{y}-\mu)^2+\frac{1}{2}(n+m)(\theta-\mu_p)^2.
$$

Let $b_p = b + \frac{1}{2}(n-1)s_y^2 + \frac{1}{2}\frac{mn}{n+m}(\bar{y}-\mu)^2$. Now,

$$
\begin{aligned}
\pi(\theta|\mathbf{y}) &\propto \left[b_p + \frac{1}{2}(n+m)(\theta-\mu_p)^2\right]^{-a_0} \\
&\propto \left[1 + \frac{1}{2b_p}(n+m)(\theta-\mu_p)^2\right]^{-a_0} \\
&\propto \left[1 + \frac{(\theta-\mu_p)^2}{\alpha\gamma^2}\right]^{-\frac{\alpha+1}{2}},
\end{aligned}
$$

where $\alpha = n + 2a$ and

$$
\gamma^2 = \frac{1}{\alpha}\frac{2b_p}{n+m} = \frac{1}{(n+2a)(n+m)}\left(2b+(n-1)s_y^2+\frac{mn}{n+m}(\bar{y}-\mu)^2\right).
\tag{4.19}
$$

Now by comparing $\pi(\theta|\mathbf{y})$ with the density (A.26) of the t-distribution we recognize that $\theta|\mathbf{y}$ follows the $t(\mu_p, \gamma^2, n+2a)$. From the properties of the t-distribution, we have $E(\theta|\mathbf{y}) = \mu_p$ and

$$
\begin{aligned}
\text{Var}(\theta|\mathbf{y}) &= \frac{n+2a}{n+2a-2}\gamma^2 \\
&= \frac{n+2a}{n+2a-2}\frac{1}{n+2a}\frac{2b_p}{n+m} \\
&= \frac{1}{(n+2a-2)(n+m)}\left(2b+(n-1)s_y^2+\frac{nm}{n+m}(\bar{y}-\mu)^2\right).
\end{aligned}
$$

For a limiting flat prior (as above), $E(\theta|\mathbf{y}) = \mu_p$ approaches the sample mean $\bar{y}$ and $\text{Var}(\theta|\mathbf{y})$ approaches $\frac{(n-1)s_y^2}{n(n-2)}$.

Lastly, we find the posterior predictive distribution (4.9) of a new observation Y_0 given the data $\mathbf{y}$. Note that $Y_0|\theta, \sigma^2 \sim N(\theta, \sigma^2)$ and $\lambda^2 = \frac{1}{\sigma^2}$, we have

$$
\begin{aligned}
f(y_0|\mathbf{y}) &\propto \int_{-\infty}^\infty \int_0^\infty f(y_0|\theta, \lambda^2)\pi(\theta, \lambda^2|\mathbf{y})d\theta d\lambda^2 \\
&\propto \int_{-\infty}^\infty \int_0^\infty (\lambda^2)^{\frac{1}{2}}\exp\left\{-\frac{\lambda^2}{2}(y_0-\theta)^2\right\}(\lambda^2)^{\frac{n+1}{2}+a-1} \\
&\qquad \exp\left\{-\frac{\lambda^2}{2}\left[2b+\sum_{i=1}^n(y_i-\theta)^2+m(\theta-\mu)^2\right]\right\}d\theta d\lambda^2 \\
&\propto \int_{-\infty}^\infty \int_0^\infty (\lambda^2)^{\frac{n+2}{2}+a-1}\exp\left\{-\frac{\lambda^2}{2}[\mathbb{M}]\right\}d\theta d\lambda^2,
\end{aligned}
$$

say, where

$$
\begin{aligned}
\text{M} &= 2b + \sum_{i=1}^{n}(y_i - \theta)^2 + m(\theta - \mu)^2 + (y_0 - \theta)^2 \\
&= 2b + (n-1)s_y^2 + \tfrac{nm}{n+m}(\bar{y} - \mu)^2 + (n+m)(\theta - \mu_p)^2 + (y_0 - \theta)^2 \\
&= 2b_p + (n+m)(\theta - \mu_p)^2 + (y_0 - \theta)^2 \\
&= 2b_p + \tfrac{n+m}{n+m+1}(y_0 - \mu_p)^2 + (n+m+1)(\theta - \mu_p)^2.
\end{aligned}
$$

In the above, we have first applied the identity (4.16) and then completed the square in θ. Now we can integrate θ out in the double integral above defining $f(y_0|\mathbf{y})$ using the conditional normal distribution $\theta \sim N(\mu_p, \lambda^2/(1+n+m))$ as dictated by the expression M. This integration will absorb $(\lambda^2)^{\frac{1}{2}}$ and hence

$$
\begin{aligned}
f(y_0|\mathbf{y}) &\propto \int_0^\infty (\lambda^2)^{\frac{n+1}{2}+a-1} \exp\left\{ -\tfrac{\lambda^2}{2}\left[2b_p + \tfrac{n+m}{n+m+1}(y_0 - \mu_p)^2 \right] \right\} d\lambda^2 \\
&\propto \left[2b_p + \tfrac{n+m}{n+m+1}(y_0 - \mu_p)^2 \right]^{-\frac{n+2a+1}{2}} \\
&\propto \left[1 + \tfrac{1}{2b_p}\tfrac{n+m}{n+m+1}(y_0 - \mu_p)^2 \right]^{-\frac{n+2a+1}{2}} \\
&\propto \left[1 + \tfrac{1}{\alpha\xi^2}(y_0 - \mu_p)^2 \right]^{-\frac{\alpha+1}{2}},
\end{aligned}
$$

where

$$
\alpha = n + 2a \quad \text{and} \quad \xi^2 = \frac{2b_p}{\alpha}\frac{n+m+1}{n+m} = (n+m+1)\gamma^2.
$$

This proves that $Y_0|\mathbf{y}$ follows $t(\mu_p, (n+m+1)\gamma^2, n+2a)$ distribution with the above values of the parameters. These theoretical results will be numerically illustrated in the Bayesian computation Chapter 5. Note that the mean of the $Y_0|\mathbf{y}$ is still the posterior mean μ_p, and

$$
\begin{aligned}
\text{Var}(Y_0|\mathbf{y}) &= \tfrac{n+2a}{n+2a-2}(n+m+1)\gamma^2 \\
&= \tfrac{(n+2a)(n+m+1)}{n+2a-2}\tfrac{1}{(n+2a)(n+m)}\left(2b + (n-1)s_y^2 + \tfrac{mn}{n+m}(\bar{y} - \mu)^2 \right). \\
&= \tfrac{n+m+1}{(n+m)(n+2a-2)}\left(2b + (n-1)s_y^2 + \tfrac{mn}{n+m}(\bar{y} - \mu)^2 \right) \\
&= (n + m + 1)\text{Var}(\theta|\mathbf{y}).
\end{aligned}
$$

Thus, the variance of the posterior predictive distribution is $n + m + 1$ times the variance of the posterior distribution. This is the same result as we have seen in the known variance case earlier in Section 4.11.1.

♡ **Example 4.13. New York air pollution data** We now relax the previous assumption of known variance. First we consider the posterior distribution $\pi(\theta|\mathbf{y})$. For the known variance case, $\theta|\mathbf{y} \sim N(\mu_p, \sigma_p^2)$. Recall that the prior mean of θ, μ is parameterized as $\mu = \bar{y} + k\frac{\sigma}{\sqrt{n}}$ and the prior variance is taken as $\tau^2 = \frac{\sigma^2}{m}$. For $k = 1$ and $m = 1$ the posterior distribution is evaluated to be $\pi(\theta|\mathbf{y})$ is $N(\mu_p = 47.91, \sigma_p^2 = 0.58)$ where

$$
\mu_p = \frac{n\bar{y} + m\mu}{n + m} \quad \text{and} \quad \sigma_p^2 = \frac{\sigma^2}{n + m}.
$$

In the unknown σ^2 case we let

$$\mu = \bar{y} + k\frac{s_y}{\sqrt{n}}$$

to have a comparable choice for the prior mean. Now we consider the special case with $k = 1$ and $m = 1$ as above. The posterior distribution is $t(n + 2a, \mu_p, \gamma^2)$ where μ_p is above ($= 47.91$) and

$$\begin{aligned}
\gamma^2 &= \frac{1}{(n+2a)(n+m)}\left(2b + (n-1)s_y^2 + \frac{mn}{n+m}(\bar{y} - \mu)^2\right) \\
&= \frac{1}{(n+2a)(n+m)}(2b + 595.03)
\end{aligned}$$

resulting in

$$\text{Var}(\theta|\mathbf{y}) = \frac{n + 2a}{n + 2a - 2}\gamma^2 = \frac{2b + 595.03}{(n + 2a - 2)(n + m)}.$$

Thus, the posterior mean 47.91 remains the same as before but the posterior variance depends on the hyper-parameters a and b of the $G(a, b)$ prior distribution for $\lambda^2 = \frac{1}{\sigma^2}$. For the sake of illustration with a proper prior distribution with $a = 2$ and $b = 1$ we have $\gamma^2 = 0.64$ (note that $n = 28$ here) and

$$\text{Var}(\theta|\mathbf{y}) = \frac{n + 2a}{n + 2a - 2}\gamma^2 = 0.69.$$

This larger variance (than earlier variance of 0.58) includes the penalty we must pay for not knowing σ^2. For the non-informative prior distribution where both a and b approach 0, $\text{Var}(\theta|\mathbf{y}) = 0.79$ – showing an even greater increase in the posterior variance.

The posterior mean of $\sigma^2|\mathbf{y}$ is obtained to be 19.26 which compares well with $s_y^2 = 22.01$. Now consider the earlier posterior predictive distribution $N(47.91, 22.58)$ for the known σ^2 case. That distribution is revised to $t(\mu_p, (n + m + 1)\gamma^2, n + 2a)$, which is evaluated to be $t(47.91, 19.3, 32)$ for $a = 2$ and $b = 1$. This distribution has the variance 20.58, which is smaller than the variance 22.58 in the known σ^2 case. This is an effect of the assumed proper prior distribution. For the non-informative prior where both a and b approach 0, $\text{Var}(Y_0|\mathbf{y}) = 23.67$ which larger than the earlier variance 22.58. This last result is also intuitive since a more informative prior distribution is expected to yield a more precise posterior distribution.   □

4.15   Model choice

4.15.1   The Bayes factor

Suppose that we have to choose between two hypotheses H_0 and H_1 corresponding to assumptions of alternative models M_0 and M_1 for data $\mathbf{y}$. The likelihood functions are denoted by $f_i(\mathbf{y}|\theta_i)$ and the prior distributions are $\pi_i(\theta_i)$, for $i = 0, 1$. In many cases, the competing models have a common set of parameters, but this is not necessary in the general development below; hence the general notations f_i, π_i and θ_i are justified.

Recall that the prior predictive distribution (4.12) for model i is,

$$f(\mathbf{y}|M_i) = \int_{-\infty}^{\infty} f_i(\mathbf{y}|\theta_i)\pi_i(\theta_i)d\theta_i.$$

Simple intuition suggests that we should choose the model for which $f(\mathbf{y}|M_i)$ is the largest. The **Bayes factor**, defined by

$$B_{01}(\mathbf{y}) = \frac{f(\mathbf{y}|M_0)}{f(\mathbf{y}|M_1)}, \qquad (4.20)$$

captures the intuition and the model M_0 is preferred if $B_{01}(\mathbf{y}) > 1$. Thus, the Bayes factor is simply the ratio of the marginal likelihoods under two different models.

♡ **Example** 4.14. **Geometric versus Poisson** Suppose that:

$$M_0 : Y_1, Y_2, \ldots, Y_n|\theta_0 \sim f_0(y|\theta_0) = \theta_0(1 - \theta_0)^y, \quad y = 0, 1, \ldots.$$

$$M_1 : Y_1, Y_2, \ldots, Y_n|\theta_1 \sim f_1(y|\theta_1) = \exp\{-\theta_1\}\theta_1^y/y!, \quad y = 0, 1, \ldots.$$

Further, assume that θ_0 and θ_1 are known. How should we decide between the two models based on $y_1, y_2, \ldots, y_n$?

Since the parameters are known under the models, we do not need to assume any prior distributions for them. Consequently,

$$f(\mathbf{y}|M_0) = \theta_0^n(1 - \theta_0)^{n\bar{y}}.$$

and

$$f(\mathbf{y}|M_1) = \exp\{-n\theta_1\}\theta_1^{n\bar{y}}/\prod_{i=1}^{n} y_i!.$$

Now the Bayes factor is just the ratio of the above two. To illustrate, let $\theta_0 = 1/3$ and $\theta_1 = 2$ (then the two distributions have same mean). Now if $n = 2$ and $y_1 = y_2 = 0$ then $B_{01}(\mathbf{y}) = 6.1$, however if $n = 2$ and $y_1 = y_2 = 2$ then $B_{01}(\mathbf{y}) = 0.3$. □

This example illustrates that the Bayes factor can be used to compare arbitrary models (for the same data set) which do not necessarily need to have any similarities or built-in nestedness. Moreover, calibration tables are available for the Bayes factor. For example, Raftery (1996) provides the following calibration table.

B_{01}	$2\log(B_{01})$	Evidence for M_0
<1	<0	Negative (supports M_1)
1 to 3	0 to 2	Barely worth mentioning
3 to 12	2 to 5	Positive
12 to 150	5 to 10	Strong
>150	>0	Very strong

TABLE 4.2: Calibration of the Bayes factor.

4.15.2   Posterior probability of a model

The reader may have wondered why the Bayes factor is called a factor in the first place? Moreover, does not the use of the Bayes factor violate the Bayesian philosophy of assessing uncertainty by probability only? The answer to both of these two questions lies in interpreting the Bayes factor by explicitly calculating the "probability" of a model – very unlike in the case of classical inference. Here each of the competing models is given a prior probability, which will define a discrete probability distribution of models assuming that there is only a finite number of models that researcher would like to entertain for their data. In order to interpret the Bayes factor we assume that there are only two models M_0 and M_1 and their prior probabilities are $P(M_0)$ and $P(M_1)$ with $P(M_0) + P(M_1) = 1$.

The posterior probability of model M_i given the observed data $\mathbf{y}$, $P(M_i|\mathbf{y})$ is the one we seek for each $i = 0, 1$. The Bayes theorem is applied again to calculate the posterior probability as follows:

$$P(M_0|\mathbf{y}) = \frac{P(M_0)f(\mathbf{y}|M_0)}{P(M_0)f(\mathbf{y}|M_0) + P(M_1)f(\mathbf{y}|M_1)}.$$

Hence the posterior odds for model M_0 is given by:

$$\frac{P(M_0|\mathbf{y})}{P(M_1|\mathbf{y})} = \frac{P(M_0)f(\mathbf{y}|M_0)}{P(M_1)f(\mathbf{y}|M_1)} = \frac{P(M_0)}{P(M_1)} \times \frac{f(\mathbf{y}|M_0)}{f(\mathbf{y}|M_1)}.$$

Now we recognize that the posterior odds for M_0 has been factored into two parts: (i) the prior odds and (ii) the Bayes factor. Thus for M_0,

$$\boxed{posterior\ odds = prior\ odds \times the\ Bayes\ factor.}$$

In other words, the Bayes factor is the multiplicative factor that must be used to convert the prior odds to posterior odds. Intuitively, the Bayes factor provides a measure of whether the data $\mathbf{y}$ have increased or decreased the odds on M_0 relative to M_1. Thus, $B_{01}(\mathbf{y}) > 1$ signifies that M_0 is more relatively plausible in the light of $\mathbf{y}$. That is why the Bayes factor is called a factor.

We do not need to know the prior probabilities $P(M_i), i = 0, 1$ to calculate the Bayes factor. Those are needed if we wish to calculate the posterior probability $P(M_i|\mathbf{y})$. If two models are equally likely a-priori, then Bayes factor is equal to the posterior odds.

4.15.3   Hypothesis testing

The Bayes factor can be used to perform testing of hypotheses in traditional statistical inference settings. Suppose that we have a random sample $Y_1, \ldots, Y_n$ from a parametric family $f(y|\theta)$ and we wish to test

$$H_0 : \theta \in \Theta_0 \text{ against } H_1 : \theta \in \Theta_1,$$

where Θ_0 and Θ_1 are subsets of the whole parameter space Θ. The Bayesian solution to this testing problem is to treat each of the two hypotheses as a separate model and then use the Bayes factor to compare the models. In order to calculate the marginal likelihoods under the two models the Bayesian method will need to assume a suitable prior distribution for each of the two models. Let $\pi_0(\theta)$ and $\pi_1(\theta)$ be the prior distributions. Note that each of these must satisfy the constraint: $\int_{\Theta_i} \pi_i(\theta)d\theta = 1$, for $i = 0, 1$. The marginal likelihood for M_i is defined as:

$$f(\mathbf{y}|M_i) = \int_{\Theta_i} f(\mathbf{y}|\theta)\pi_i(\theta)d\theta$$

for $i = 0, 1$. Now to test H_0 against H_1 we simply calculate the Bayes factor (4.20). The Bayes factor will provide simpler versions depending on the nature of the hypotheses we would like to test. For example, consider a point null hypothesis $H_0 : \theta = \theta_0$ where θ_0 is a particular value in the parameter space. In this case, the prior distribution on θ under H_0 will put the total probability one at the value θ_0, i.e. $\pi_0(\theta) = 1$ when $\theta = \theta_0$. Hence $f(\mathbf{y}|M_0)$ is simply taken as $f(\mathbf{y}|\theta_0)$. Thus, we obtain the following special forms of the Bayes factor:

$$B_{01}(\mathbf{y}) = \frac{f(\mathbf{y}|\theta_0)}{f(\mathbf{y}|\theta_1)} \qquad \text{when } H_0 : \theta = \theta_0, \quad H_1 : \theta = \theta_1.$$

$$B_{01}(\mathbf{y}) = \frac{f(\mathbf{y}|\theta_0)}{\int_{\Theta_1} f(\mathbf{y}|\theta)\pi_1(\theta)d\theta} \qquad \text{when } H_0 : \theta = \theta_0, \quad H_1 : \theta \in \Theta_1.$$

$$B_{01}(\mathbf{y}) = \frac{\int_{\Theta_0} f(\mathbf{y}|\theta)\pi_0(\theta)d\theta}{\int_{\Theta_1} f(\mathbf{y}|\theta)\pi_1(\theta)d\theta} \qquad \text{when } H_0 : \theta = \Theta_0, \quad H_1 : \theta \in \Theta_1.$$

Thus, it is easy to note that for testing the two simple hypotheses $H_0 : \theta = \theta_0$ against $H_1 : \theta = \theta_1$ the Bayes factor provides the test, which is the likelihood ratio test that will be the optimal test according to the classical statistical theories such as the Neyman-Pearson Lemma. However, the Bayesian approach differs fundamentally from traditional hypothesis testing using P-values. The difference arises because the Bayesian approach is based on conditioning the observed data values, $\mathbf{y}$ in our notation, whereas the p-values are calculated using the full sampling distribution of the random sample, $\mathbf{Y}$. Recall the definition of the p-value for a test of hypothesis using a test statistic $T(\mathbf{y})$. The p-value is defined as the probability that $T(Y)$ is the observed value $T(\mathbf{y})$ or something more extreme in the direction of the alternative hypothesis assuming that the null hypothesis is true. This is often incorrectly interpreted as the *probability* that H_0 is true is smaller than the p-value. We illustrate the issues with the following example.

♡ **Example** 4.15. **Taste-test** In an experiment to determine whether an individual possesses discriminating powers, she has to identify correctly which of the two brands of a consumer product (such as tea or shampoo) she is provided with, over a series of trials.

Let θ denote the probability of her choosing the correct brand in any trial and Y_i be the Bernoulli random variable taking the value 1 for correct guess in the ith trial. Suppose that in first 6 trials the results are 1, 1, 1, 1, 1, 0.

We wish to test that the tester does not have any discriminatory power against the alternative that she does. So our problem is:

$$H_0 : \theta = \frac{1}{2} \text{ versus } H_1 : \theta > \frac{1}{2}.$$

It is a simple versus composite case and we have $\Theta_0 = \frac{1}{2}$ and $\Theta_1 = (\frac{1}{2}, 1)$. Let us assume uniform prior distribution on θ under the alternative. So the prior $\pi_1(\theta) = 2$ if $\frac{1}{2} < \theta < 1$. For the point null hypothesis the implied prior distribution is $\pi_0(\theta = \frac{1}{2}) = 1$.

Defining $\mathbf{y} = (1, 1, 1, 1, 1, 0)$, we calculate the Bayes factor as

$$B_{01}(\mathbf{y}) = \frac{\frac{1}{2}^6}{\int_{\frac{1}{2}}^1 \theta^5 (1 - \theta) 2 \, d\theta} = \frac{1}{2.86}.$$

This suggests that she does appear to have some discriminatory power but not a lot (refer to the calibration table for the Bayes factor 4.2).

In order to calculate the p-value for this example we need to obtain the distribution of the test statistic $Y = Y_1 + \ldots + Y_n$, where $n = 6$ denote the number of trials. Two cases arise depending on the sampling design. In the first case, suppose that n is fixed in advance so that the sampling distribution of Y is binomial with parameter $n = 6$ and $\theta = \frac{1}{2}$ under H_0. Hence,

$$
\begin{aligned}
\text{p-value} \quad &= \quad P(Y = 5 \text{ or something more extreme } |\theta = \tfrac{1}{2}) \\
&= \quad P(Y = 5 \text{ or } Y = 6 | \theta = \tfrac{1}{2}) \\
&= \quad 7 \times \left(\tfrac{1}{2}\right)^6 = 0.109.
\end{aligned}
$$

In the other case suppose that the sampling design is to continue with the trials until the first time she fails to identify the correct brand. In this case the test statistic Y will be the number of trials until the first failure. So now

Y will be geometrically distributed with parameter θ and we have:

$$
\begin{aligned}
\text{p-value} \quad &= \quad P(Y = 5 \text{ or something more extreme } |\theta = \tfrac{1}{2}) \\
&= \quad P(Y = 5, 6, 7, \ldots |\theta = \tfrac{1}{2}) \\
&= \quad \left(\tfrac{1}{2}\right)^6 + \left(\tfrac{1}{2}\right)^7 + \cdots \\
&= \quad 0.031.
\end{aligned}
$$

Note that the first p-value does not allow the experimenter to declare a significant result whereas the second p-value *does* at any level of significance between 3.1% and 10%. Thus, different inferences are made despite observing exactly the same sequence of events.   □

The above example demonstrates that the classical statistical inference methods using the p-values violate the *likelihood principle* which states that the same likelihood function for a parameter θ should provide the same inference for θ. In the above example, $\theta^5(1 - \theta)$ is the likelihood function for both binomial and geometric sampling distribution. Bayesian methods do not, however, violate the likelihood principle since the likelihood times prior representation of the posterior distribution. However, the frequentist procedure typically violates the principle, since long run behavior under hypothetical repetitions depends on the entire distribution $\{f(\mathbf{y}|\boldsymbol{\theta}), \mathbf{y} \in \mathcal{Y})\}$ where $\mathcal{Y}$ is the sample space and not only on the likelihood.

4.16   Criteria-based Bayesian model selection

The pure Bayesian ways of comparing models and hypotheses using the Bayes factor and the associated posterior probabilities are the only criteria to be used according to orthodox Bayesian views. However, in modern day practical modeling situations with large data sets and a large number of parameters it is difficult to calculate the Bayes factor. The difficulty comes from two sources: (i) the requirement that proper prior distributions are used to calculate the marginal likelihood and (ii) the large dimension of the parameter space and hence the large multi-dimensional integral that must be performed to calculate the marginal likelihood. These problems have prohibited the routine use of the Bayes factor in practical modeling work and this explains why the Bayes factor is not a standard output in Bayesian model fitting software packages. To remedy such computational problems we now discuss some popularly used model choice criteria.

Historically the Akaike Information Criterion (AIC) (Akaike, 1973) and another modification of it, known as the Bayesian information criteria (BIC) (Schwartz, 1978) have enjoyed the most popularity. These criteria are based on two components: (i) a goodness-of-fit term and (ii) a penalty term for criticizing model complexity. The resolution always is to choose the model which has the minimum value of the criteria from among competing models. These model choice criteria often do not have any natural scale and, hence their differences matter in model selection – not their absolute values.

The AIC is defined as

$$\text{AIC} = -2\log f(\mathbf{y}|\hat{\boldsymbol{\theta}}) + 2p$$

where $\hat{\boldsymbol{\theta}}$ is the maximum likelihood estimate (mle) of $\boldsymbol{\theta}$ maximizing the likelihood function of the p parameters in the model denoted by $\boldsymbol{\theta}$. The BIC, given by:

$$\text{BIC} = -2\log f(\mathbf{y}|\hat{\boldsymbol{\theta}}) + p\log(n)$$

gives a larger penalty (when $\log n > 2$) per parameter compared to AIC.

♡ **Example** 4.16. To gain insights we will illustrate all the model choice criteria with exact calculations using the Normal-Normal example introduced in Section 4.7. In this example, $Y_1, \ldots, Y_n \sim N(\theta, \sigma^2)$, where σ^2 is known and the prior distribution is $\pi(\theta) \sim N(\mu, \tau^2)$ for known μ and τ^2. Recall that the log-likelihood function is:

$$\log f(\mathbf{y}|\theta) = -\frac{n}{2}\log\left(2\pi\sigma^2\right) - \frac{1}{2}\sum_{i=1}^{n}\frac{(y_i - \theta)^2}{\sigma^2}.$$

The maximum likelihood estimate of θ is $\hat{\theta} = \bar{y}$, the sample mean and $p = 1$. Hence:

$$\begin{aligned}
\text{AIC} &= n\log\left(2\pi\sigma^2\right) + \frac{1}{\sigma^2}\sum_{i=1}^{n}(y_i - \bar{y})^2 + 2 \\
&= n\log\left(2\pi\sigma^2\right) + \frac{1}{\sigma^2}(n-1)s_y^2 + 2,
\end{aligned}$$

where $s_y^2 = \frac{1}{n-1}\sum_{i=1}^{n}(y_i - \bar{y})^2$ is the sample variance of the observations. The BIC is given by:

$$\text{BIC} = n\log\left(2\pi\sigma^2\right) + \frac{1}{\sigma^2}(n-1)s_y^2 + \log(n).$$

□

In Bayesian modeling with hierarchical models and with prior distributions providing information it is difficult to identify the number of parameters and hence it is difficult to use the AIC and BIC for model choice purposes. Also the Bayesian model choice criteria that we will describe below are motivated by the need to measure predictive accuracy for a single data point. Thus, the model specified distribution $f(y_0|\theta)$ and the posterior predictive distribution $f(y_0|\mathbf{y})$ will play crucial roles in the development below. The Bayesian criteria are also influenced by the Bayesian philosophy of averaging over the full posterior distribution $\pi(\boldsymbol{\theta}|\mathbf{y})$ for the unknown parameter $\boldsymbol{\theta}$ rather than plugging in the maximum likelihood estimate $\hat{\boldsymbol{\theta}}$ as has been used in the definition of AIC and BIC given above. In this section we discuss three different model choice criteria: (i) Deviance Information Criteria (DIC), (ii) Watanabe Akaike Information Criteria (WAIC) and (iii) Predictive model choice criteria (PMCC).

4.16.1   The DIC

The DIC (Spiegelhalter et al., 2002) is a generalization of the AIC to the case of hierarchical models. It mimics the AIC and BIC by taking the first (goodness-of-fit) term as the -2 times the log-likelihood function evaluated at the posterior estimate $\hat{\boldsymbol{\theta}}_{\text{Bayes}}$ of $\boldsymbol{\theta}$ instead of the maximum likelihood estimate. The second term is twice the number of parameters, which is defined as:

$$p_{\text{DIC}} = 2\left[\log\left\{f\left(\mathbf{y}|\hat{\boldsymbol{\theta}}_{\text{Bayes}}\right)\right\} - E_{\theta|\mathbf{y}}\left\{\log\left(f(\mathbf{y}|\boldsymbol{\theta})\right)\right\}\right]. \tag{4.21}$$

Note that the second term in p_{DIC} is the posterior average of the log of the predictive density $f(\mathbf{y}|\boldsymbol{\theta})$. The posterior mean, $E(\boldsymbol{\theta}|\mathbf{y})$, of $\boldsymbol{\theta}$ as $\hat{\boldsymbol{\theta}}_{\text{Bayes}}$ will produce the maximum log-predictive density, the first term inside the box brackets in p_{DIC}, when it happens to be same as the posterior mode. If this does not happen, i.e., the posterior mean is far away from the posterior mode then negative p_{DIC} can occur. For fixed effects linear models, i.e. linear regression models, the p_{DIC} will produce the same number of parameters as in the model. We shall see this later. Now the DIC is defined as:

$$\text{DIC} = -2\log\left\{f\left(\mathbf{y}|\hat{\boldsymbol{\theta}}_{\text{Bayes}}\right)\right\} + 2\,p_{\text{DIC}}. \tag{4.22}$$

For the Normal-Normal example, we take $\hat{\boldsymbol{\theta}}_{\text{Bayes}} = \mu_p$ the posterior mean. To calculate p_{DIC} we evaluate

$$
\begin{aligned}
E_{\theta|\mathbf{y}}(y_i - \theta)^2 &= E_{\theta|\mathbf{y}}(\theta - \mu_p + \mu_p - y_i)^2 \\
&= E_{\theta|\mathbf{y}}\left\{(\theta - \mu_p)^2 + (\mu_p - y_i)^2 + 2(\theta - \mu_p)(\mu_p - y_i)\right\} \\
&= E_{\theta|\mathbf{y}}\left\{(\theta - \mu_p)^2\right\} + (\mu_p - y_i)^2 + 2(\mu_p - y_i)E_{\theta|\mathbf{y}}(\theta - \mu_p) \\
&= \sigma_p^2 + (y_i - \mu_p)^2
\end{aligned}
$$

since $E_{\theta|\mathbf{y}}(\theta) = \mu_p$. Hence

$$
\begin{aligned}
p_{\text{DIC}} &= -n\log\left(2\pi\sigma^2\right) - \sum_{i=1}^{n}\frac{(y_i-\mu_p)^2}{\sigma^2} + n\log\left(2\pi\sigma^2\right) + \sum_{i=1}^{n}\frac{E_{\theta|\mathbf{y}}(y_i-\theta)^2}{\sigma^2} \\
&= \sum_{i=1}^{n}\left[-\frac{(y_i-\mu_p)^2}{\sigma^2} + \frac{\sigma_p^2+(y_i-\mu_p)^2}{\sigma^2}\right] \\
&= n\frac{\sigma_p^2}{\sigma^2}.
\end{aligned}
$$

Now we have

$$\text{DIC} = n\log\left(2\pi\sigma^2\right) + \sum_{i=1}^{n}\frac{(y_i - \mu_p)^2}{\sigma^2} + 2n\frac{\sigma_p^2}{\sigma^2}. \tag{4.23}$$

If we assume $\tau^2 = \frac{\sigma^2}{m}$, then $\sigma_p^2 = \frac{\sigma^2}{n+m}$ and $p_{DIC} = \frac{n}{n+m}$. For a very informative prior distribution (i.e. when m is very large relative to n) the penalty is very small as expected.

To compare this with the AIC obtained earlier, we let $\tau^2 \to \infty$ corresponding to a flat prior for θ. For this flat prior we have seen that $\mu_p = \bar{y}$ and $\sigma_p^2 = \frac{\sigma^2}{n}$. Hence

$$\text{DIC} = n \log \left(2\pi\sigma^2\right) + \sum_{i=1}^{n} \frac{(y_i - \bar{y})^2}{\sigma^2} + 2,$$

which is exactly equal to the AIC. Moreover, p_{DIC}, the effective number of parameters is 1 as before. However, for an informative prior distribution the DIC and AIC will differ. For example, when the prior distribution and the data are of same strength, i.e. $\tau^2 = \frac{\sigma^2}{n}$, then $\sigma_p^2 = \frac{\sigma^2}{2n}$ and consequently, $p_{DIC} = \frac{1}{2}$. That is, the number of parameters is halved.

Gelman et al. (2014) suggest an alternative p_{DIC} as

$$p_{DIC \; alt} = 2 \, \text{Var}_{\theta|\mathbf{y}} \left[\log\{ f(\mathbf{y}|\theta)\} \right].$$

This alternative form guarantees a positive number of effective parameters since this is twice the sum of predictive variances. For the Normal-Normal example with a flat prior, for which $\tau^2 \to \infty$, the $p_{DIC \; alt}$ is same as p_{DIC}. The general expression is given by:

$$
\begin{aligned}
p_{DIC \; alt} &= 2 \, \text{Var}_{\theta|\mathbf{y}} \left[\log\{f(\mathbf{y}|\theta)\} \right] \\
&= 2 \, \text{Var}_{\theta|\mathbf{y}} \left[-\frac{n}{2} \log\left(2\pi\sigma^2\right) - \frac{1}{2} \sum_{i=1}^{n} \frac{(y_i - \theta)^2}{\sigma^2} \right] \\
&= 2 \frac{1}{4\sigma^4} \text{Var}_{\theta|\mathbf{y}} \left[\sum_{i=1}^{n} (y_i - \theta)^2 \right] \\
&= \frac{1}{2\sigma^4} \text{Var}_{\theta|\mathbf{y}} \left[\sum_{i=1}^{n} (y_i - \bar{y})^2 + n(\theta - \bar{y})^2 \right] \\
&= \frac{n^2}{2\sigma^4} \text{Var}_{\theta|\mathbf{y}} \left[(\theta - \bar{y})^2 \right] \\
&= \frac{n^2}{2\sigma^4} \text{Var}_{\theta|\mathbf{y}} \left[\left(\frac{\sigma_p(\theta - \bar{y})}{\sigma_p} \right)^2 \right] \\
&= \frac{n^2}{2\sigma^4} \text{Var} \left[\sigma_p^2 Z^2 \right], \quad \text{where } Z = \frac{\theta - \bar{y}}{\sigma_p} \sim N\left(\frac{\mu_p - \bar{y}}{\sigma_p}, 1 \right) \\
&= \frac{n^2}{2\sigma^4} \sigma_p^4 \text{Var} \left[Z^2 \right], \quad \text{where } Z^2 \sim \text{ non-central} \\
&\quad \chi^2 \left(df = 1, \lambda = \left(\frac{\mu_p - \bar{y}}{\sigma_p} \right)^2 \right) \\
&= \frac{n^2}{2\sigma^4} \sigma_p^4 \, 2 \left(1 + 2 \left(\frac{\mu_p - \bar{y}}{\sigma_p} \right)^2 \right) \\
&= \frac{n^2}{\sigma^4} \sigma_p^2 \left(\sigma_p^2 + 2 (\mu_p - \bar{y})^2 \right).
\end{aligned}
$$

For the flat prior (as $\tau^2 \to \infty$) we have $\mu_p = \bar{y}$ and $\sigma_p^2 = \frac{\sigma^2}{n}$. Hence

$$p_{DIC \; alt} = \frac{n^2}{\sigma^4} \frac{\sigma^2}{n} \left(\frac{\sigma^2}{n} + 0 \right) = 1,$$

which is same as p_{DIC}. However, when $\tau^2 = \frac{\sigma^2}{n}$, then $\sigma_p^2 = \frac{\sigma^2}{2n}$ and $\mu_p = \frac{\bar{y}+\mu}{2}$.
Then

$$
\begin{aligned}
p_{\text{DIC alt}} &= \frac{n^2}{\sigma^4}\frac{\sigma^2}{2n}\left(\frac{\sigma^2}{2n} + \frac{(\mu-\bar{y})^2}{2}\right) \\
&= \frac{n}{4\sigma^2}\left(\frac{\sigma^2}{n} + (\mu-\bar{y})^2\right).
\end{aligned}
$$

Here the penalty also depends on the prior mean μ.

Both of the components of DIC are easy to calculate based on the posterior distribution $\pi(\boldsymbol{\theta}|\mathbf{y})$ unlike the calculation of the marginal likelihoods. This ease of computation has led to the inclusion of DIC as a standard model comparison tool in the Bayesian modeling software package, BUGS. Usually "bigger" models are favored by the DIC.

4.16.2 The WAIC

The WAIC have been developed, see e.g. Watanabe (2010) and Gelman et al. (2014), by using the log of the predictive density for a single data point $f(y|\boldsymbol{\theta})$. Thus, $\log(f(y|\boldsymbol{\theta}))$ is the contribution to the log-likelihood function from a single data point y, but in the following discussion we view it as log of the predictive density for a new data point so that we can judge the predictive capability of the fitted model to replicate the data. This predictive density is same as the $f(y_0|\boldsymbol{\theta})$ inside the integral defining the posterior predictive distribution in (4.9).

The WAIC is defined similar to the DIC in (4.22) but instead of plugging-in $\hat{\boldsymbol{\theta}}_{\text{Bayes}}$ we average over the posterior distribution $\pi(\boldsymbol{\theta}|\mathbf{y})$ as follows:

$$
\text{WAIC} = -2\sum_{i=1}^{n}\log\left\{\int_{-\infty}^{\infty}f(y_i|\boldsymbol{\theta})\pi(\boldsymbol{\theta}|\mathbf{y})d\boldsymbol{\theta}\right\} + 2\,p_{\text{waic}} \tag{4.24}
$$

where p_{waic} is the penalty defined below in two ways as for the DIC. Notice that the term inside the parenthesis in (4.24) is really the posterior predictive density $f(y_i|\mathbf{y})$ as given in (4.9). Hence we write:

$$
\text{WAIC} = -2\sum_{i=1}^{n}\log\left\{f(y_i|\mathbf{y})\right\} + 2\,p_{\text{waic}}.
$$

The first form of p_{waic} is defined similarly as in (4.21) but with the modification of replacing the point estimate $\hat{\boldsymbol{\theta}}_{\text{Bayes}}$ by posterior averaging:

$$
\begin{aligned}
p_{\text{waic 1}} &= 2\sum_{i=1}^{n}\left[\log\left\{E_{\boldsymbol{\theta}|\mathbf{y}}f(y_i|\boldsymbol{\theta})\right\} - E_{\theta|\mathbf{y}}\left\{\log(f(y_i|\boldsymbol{\theta}))\right\}\right] \\
&= 2\sum_{i=1}^{n}\left[\log\left\{f(y_i|\mathbf{y})\right\} - E_{\theta|\mathbf{y}}\left\{\log(f(y_i|\boldsymbol{\theta}))\right\}\right].
\end{aligned}
$$

The second alternative form of the penalty term is of the same form as in the DIC

$$
p_{\text{waic 2}} = \sum_{i=1}^{n}\text{Var}_{\boldsymbol{\theta}|\mathbf{y}}\left[\log\{f(y_i|\boldsymbol{\theta})\}\right]. \tag{4.25}
$$

Note that this penalty differs in two ways from the $p_{\text{DIC alt}}$:

    (i)   it does not have the factor 2 upfront and

    (ii) it computes the variance for each data point separately and then sums.

Gelman et al. (2014) recommend this to achieve stability arising due to the summing.

Like the DIC, WAIC also balances between goodness-of-fit and model complexity. Moreover, WAIC is easily generalisable for estimating model choice criteria in the predictive space, e.g. for leave one out cross-validation and $k-$ fold cross-validation (see Section 6.8). These generalizations are postponed to the Bayesian Computation Chapter 5 so that we can discuss computation of these criteria simultaneously.

We now return to the Normal-Normal example to illustrate the WAIC. For this example, from (4.10) we have

$$\log\{f(y_i|\mathbf{y})\} = -\frac{1}{2}\log\{2\pi(\sigma^2 + \sigma_p^2)\} - \frac{1}{2}\frac{(y_i - \mu_p)^2}{\sigma^2 + \sigma_p^2}, i = 1, \ldots, n.$$

Thus, the first term in (4.24) is now written as:

$$-2\sum_{i=1}^{n}\log\{f(y_i|\mathbf{y})\} = n\log\{2\pi(\sigma^2 + \sigma_p^2)\} + \frac{\sum_{i=1}^{n}(y_i - \mu_p)^2}{\sigma^2 + \sigma_p^2}. \qquad (4.26)$$

We further need to calculate $E_{\theta|\mathbf{y}}\log\{f(y_i|\theta)\}$ and $\text{Var}_{\theta|\mathbf{y}}[\log\{f(y_i|\theta)\}]$ to obtain the WAIC along with the two forms of the penalty function. We have the posterior distribution $\theta|\mathbf{y} \sim N(\mu_p, \sigma_p^2)$ according to (4.4).

$$\log\{f(y_i|\theta)\} = -\frac{1}{2}\log\left(2\pi\sigma^2\right) - \frac{1}{2}\frac{(y_i - \theta)^2}{\sigma^2}.$$

Now

$$E_{\theta|\mathbf{y}}\log\{f(y_i|\theta)\} = -\frac{1}{2}\log\left(2\pi\sigma^2\right) - \frac{1}{2\sigma^2}E_{\theta|\mathbf{y}}(y_i - \theta)^2.$$

$$\begin{aligned}
E_{\theta|\mathbf{y}}(y_i - \theta)^2 &= E_{\theta|\mathbf{y}}(\theta - \mu_p + \mu_p - y_i)^2 \\
&= E_{\theta|\mathbf{y}}(\theta - \mu_p)^2 + (\mu_p - y_i)^2, \quad \text{since } E_{\theta|\mathbf{y}}(\theta) = \mu_p, \\
&= \sigma_p^2 + (y_i - \mu_p)^2.
\end{aligned}$$

Hence,

$$E_{\theta|\mathbf{y}}\log\{f(y_i|\theta)\} = -\frac{1}{2}\log\left(2\pi\sigma^2\right) - \frac{1}{2\sigma^2}\left(\sigma_p^2 + (y_i - \mu_p)^2\right).$$

Now

$$\text{Var}_{\theta|\mathbf{y}}[\log\{f(y_i|\theta)\}] = \frac{1}{4\sigma^4}\text{Var}_{\theta|\mathbf{y}}\left[(y_i - \theta)^2\right].$$

$$\text{Var}_{\theta|\mathbf{y}}\left[(y_i - \theta)^2\right] = \text{Var}_{\theta|\mathbf{y}}\left[\left(\frac{\sigma_p(\theta - y_i)}{\sigma_p}\right)^2\right]$$

$$= \text{Var}\left[\sigma_p^2 Z^2\right], \quad \text{where } Z = \frac{\theta - y_i}{\sigma_p} \sim N\left(\frac{\mu_p - y_i}{\sigma_p}, 1\right)$$

$$= \sigma_p^4 \text{Var}\left[Z^2\right], \quad \text{where } Z^2 \sim \text{ non-central}$$

$$\chi^2\left(df = 1, \lambda = \left(\frac{\mu_p - y_i}{\sigma_p}\right)^2\right)$$

$$= \sigma_p^4 2\left(1 + 2\left(\frac{\mu_p - y_i}{\sigma_p}\right)^2\right)$$

$$= 2\sigma_p^2\left(\sigma_p^2 + 2\left(\mu_p - y_i\right)^2\right).$$

$$p_{\text{waic 1}} = 2\sum_{i=1}^n\left[\log\left\{f(y_i|\mathbf{y})\right\} - E_{\theta|\mathbf{y}}\left\{\log\left(f(y_i|\boldsymbol{\theta})\right)\right\}\right]$$

$$= \sum_{i=1}^n\left[-\log\{2\pi(\sigma^2 + \sigma_p^2)\} - \frac{(y_i - \mu_p)^2}{\sigma^2 + \sigma_p^2} + \log\left(2\pi\sigma^2\right) + \frac{\sigma_p^2 + (y_i - \mu_p)^2}{\sigma^2}\right]$$

$$= \sum_{i=1}^n\left[\log\left\{\frac{\sigma^2}{\sigma^2 + \sigma_p^2}\right\} - \frac{(y_i - \mu_p)^2}{\sigma^2 + \sigma_p^2} + \frac{\sigma_p^2 + (y_i - \mu_p)^2}{\sigma^2}\right]$$

$$= n\log\left\{\frac{\sigma^2}{\sigma^2 + \sigma_p^2}\right\} + n\frac{\sigma_p^2}{\sigma^2} + \frac{\sigma_p^2}{\sigma^2(\sigma^2 + \sigma_p^2)}\sum_{i=1}^n(y_i - \mu_p)^2.$$

Now,

$$p_{\text{waic 2}} = \sum_{i=1}^n \text{Var}_{\boldsymbol{\theta}|\mathbf{y}}\left[\log\{f(y_i|\boldsymbol{\theta})\}\right]$$

$$= \sum_{i=1}^n \frac{1}{4\sigma^4} 2\sigma_p^2\left(\sigma_p^2 + 2\left(\mu_p - y_i\right)^2\right)$$

$$= \frac{n}{2}\frac{\sigma_p^4}{\sigma^4} + \frac{\sigma_p^2}{\sigma^4}\sum_{i=1}^n\left(y_i - \mu_p\right)^2.$$

Assume that $\sigma^2 = 1$ and let $\tau^2 = \frac{1}{m}$. Now $\sigma_p^2 = \frac{1}{m+n}$ and $\mu_p = \frac{n\bar{y} + m\mu}{m+n}$. Hence the first tern in WAIC (4.26) reduces to:

$$n\log(2\pi) + \log\left(1 + \frac{1}{m+n}\right) + \frac{\sum_{i=1}^n(y_i - \mu_p)^2}{1 + \frac{1}{m+n}}$$

$$= n\log(2\pi) + \log\left(\frac{m+n+1}{m+n+1}\right) + \frac{m+n}{m+n+1}\sum_{i=1}^n\left(y_i - \frac{n\bar{y} + m\mu}{m+n}\right)^2$$

$$= n\log(2\pi) + \log\left(\frac{m+n+1}{m+n}\right) + \frac{m+n}{m+n+1}\left[\sum_{i=1}^n\left(y_i - \bar{y}\right)^2 + \frac{m^2 n}{(m+n)^2}(\bar{y} - \mu)^2\right].$$

Now

$$p_{\text{waic 1}} = n\log\left\{\frac{\sigma^2}{\sigma^2 + \sigma_p^2}\right\} + n\frac{\sigma_p^2}{\sigma^2} + \frac{\sigma_p^2}{\sigma^2(\sigma^2 + \sigma_p^2)}\sum_{i=1}^n(y_i - \mu_p)^2$$

$$= n\log\left\{\frac{m+n}{m+n+1}\right\} + \frac{n}{m+n} + \frac{1}{m+n+1}$$
$$\left[\sum_{i=1}^n\left(y_i - \bar{y}\right)^2 + \frac{m^2 n}{(m+n)^2}(\bar{y} - \mu)^2\right].$$

$$p_{\text{waic 2}} = \frac{n}{2}\frac{1}{(m+n)^2} + \frac{1}{m+n}\left[\sum_{i=1}^n\left(y_i - \bar{y}\right)^2 + \frac{m^2 n}{(m+n)^2}(\bar{y} - \mu)^2\right].$$

In the case of a flat prior, i.e. when $m = 0$, the above expressions simplifies considerably as follows: The first term in WAIC (4.26) reduces to:

$$-2\sum_{i=1}^n\log\{f(y_i|\mathbf{y})\} = n\log(2\pi) + \log\left(\frac{n+1}{n}\right) + \frac{n(n-1)}{n+1}s_y^2.$$

Also,

$$p_{\text{waic 1}} = n\log\left\{\frac{n}{n+1}\right\} + 1 + \frac{n-1}{n+1}s_y^2,$$

and
$$p_{\text{waic 2}} = \frac{1}{2n} + \frac{n-1}{n} s_y^2.$$

In case the data and prior are of equal strength, i.e. $m = n$ we obtain the following simplified forms. The first term in WAIC (4.26) reduces to:

$$n\log(2\pi) + \log\left(\frac{2n+1}{2n}\right) + \frac{2n}{2n+1}\left[(n-1)s_y^2 + \frac{n}{4}(\bar{y}-\mu)^2\right].$$

$$p_{\text{waic 1}} = n\log\left\{\frac{2n}{2n+1}\right\} + \frac{1}{2} + \frac{1}{2n+1}\left[(n-1)s_y^2 + \frac{n}{4}(\bar{y}-\mu)^2\right].$$

$$p_{\text{waic 2}} = \frac{1}{8n} + \frac{1}{2n}\left[(n-1)s_y^2 + \frac{n}{4}(\bar{y}-\mu)^2\right].$$

These expressions match with various special cases reported in Gelman et al. (2014). Also for large n, treating $S_y^2 = \frac{1}{n-1}\sum_{i=1}^{n}(Y_i - \bar{Y})$ as random and thereby taking $E(S_y^2) = \sigma^2$, we can show that $p_{\text{waic}} \to 1$. For small sample sizes, the WAIC will depend on the relative strength of the data and the prior distribution and must be computed numerically for model comparison. This concludes the discussion here. In the next sub-section we introduce another popular model choice criteria based on minimizing a posterior predictive loss function.

4.16.3 Posterior predictive model choice criteria (PMCC)

The PMCC is based on the idea of a loss function measuring discrepancies between a hypothetical replication of the data and an action of guessing the hypothetical data set. The expected value of the loss function with respect to the posterior predictive distribution of the hypothetical replicated data set is calculated and minimize over the action space to obtain the PMCC (Gelfand and Ghosh, 1998). Note that here the observed data are treated as fixed as in the Bayesian paradigm. We introduce the criteria below with a simpler version of a squared error loss function, which is often used and justified when the top-level data distribution is assumed to be Gaussian.

Let Y_{i0} denote the model based prediction for observed data y_i for $i = 1,\ldots,n$. In the Bayesian paradigm, Y_{i0} will be a random variable conditional on the observed data $y_1,\ldots,y_n$ and will have the probability distribution given by the posterior prediction distribution (4.9) discussed before. The PMCC criteria will then be sum of two components given by:

$$\text{PMCC} = \sum_{i=1}^{n} \{y_i - E(Y_{i0}|\mathbf{y})\}^2 + \sum_{i=1}^{n} \text{Var}(Y_{i0}|\mathbf{y}). \qquad (4.27)$$

The first term above is a goodness-of-fit term since for a well fitted model the posterior predictive mean $E(Y_{i0}|\mathbf{y})$ will be very close to the observed data point y_i. The $E(Y_{i0}|\mathbf{y})$ values are usually known as the fitted values in the linear model literature, see Section 6.4. The second term is a penalty term for prediction of the observed y_i expressed through the predictive variance

$\text{Var}(Y_{i0}|\mathbf{y})$ for $i = 1, \ldots, n$. According to PMCC the best model is the one which minimizes the PMCC. Thus, the best model has to strike a balance between the goodness-of-fit and the predictive penalty.

Returning to the Normal-Normal example, we know from (4.10) that $E(Y_{i0}|\mathbf{y}) = \mu_0$ and $\text{Var}(Y_{i0}|\mathbf{y}) = \sigma_0^2 = \sigma^2 + \sigma_p^2$. Hence the PMCC for this example is given by:

$$\text{PMCC} = \sum_{i=1}^{n} \{y_i - \mu_0\}^2 + n\sigma_0^2.$$

For the special case that $\tau^2 = \frac{1}{m}$ and $\sigma^2 = 1$, we have $\mu_0 = \frac{n\bar{y}+m\mu}{m+n}$, $\sigma_0^2 = 1 + \frac{1}{m+n}$, and

$$\begin{aligned} \text{PMCC} &= \sum_{i=1}^{n}\left\{y_i - \frac{n\bar{y}+m\mu}{m+n}\right\}^2 + n\left(1 + \frac{1}{m+n}\right) \\ &= (n-1)s_y^2 + \frac{m^2 n}{(m+n)^2}(\bar{y}-\mu)^2 + n\frac{m+n+1}{m+n}. \end{aligned}$$

For a flat prior we take $m = 0$, in which case PMCC $= (n-1)s_y^2 + n + 1$. For a very informative prior distribution when $m \to \infty$, the limiting value of the PMCC $= (n-1)s_y^2 + n(\bar{y}-\mu)^2 + n$ which highlights the effect of possible conflict between the data mean $\bar{y}$ and the prior mean μ.

♡ **Example 4.17. New York air pollution data**

All the above model choice criteria are now illustrated for this running example with σ^2 assumed to be known. We continue to use: $\mu = \bar{y} + k\frac{\sigma}{\sqrt{n}}$ and $\tau^2 = \frac{\sigma^2}{m}$. Recall that for $k = 1$ and $m = 1$ the posterior distribution $\pi(\theta|\mathbf{y})$ is $N(\mu_p = 47.91, \sigma_p^2 = 0.58)$.

AIC & BIC The relevant formulae, as given in Example 4.16, are given below along with their numerical values:

$$\begin{aligned} \text{AIC} &= n\log(2\pi\sigma^2) + \frac{1}{\sigma^2}(n-1)s_y^2 + 2 = 167.02, \\ \text{BIC} &= n\log(2\pi\sigma^2) + \frac{1}{\sigma^2}(n-1)s_y^2 + \log(n) = 168.35. \end{aligned}$$

DIC The formula for DIC is given in (4.23) as:

$$\text{DIC} = n\log(2\pi\sigma^2) + \sum_{i=1}^{n}\frac{(y_i-\mu_p)^2}{\sigma^2} + 2n\frac{\sigma_p^2}{\sigma^2} = 166.95$$

with

$$p_{\text{DIC}} = n\frac{\sigma_p^2}{\sigma^2} = 0.97.$$

The alternative penalty is

$$p_{\text{DIC alt}} = \frac{n^2}{\sigma^4}\sigma_p^2\left(\sigma_p^2 + 2(\mu_p - \bar{y})^2\right) = 0.93$$

and hence the corresponding value of the alternative DIC is 166.89.

WAIC The formula for WAIC is

$$\text{WAIC} = n \log\{2\pi(\sigma^2 + \sigma_p^2)\} + \frac{\sum_{i=1}^{n}(y_i - \mu_p)^2}{\sigma^2 + \sigma_p^2} + 2\,p_{\text{waic}}$$

where p_{waic} has two forms:

$$p_{\text{waic 1}} = n \log\left\{\frac{\sigma^2}{\sigma^2 + \sigma_p^2}\right\} + n\frac{\sigma_p^2}{\sigma^2} + \frac{\sigma_p^2}{\sigma^2(\sigma^2 + \sigma_p^2)} \sum_{i=1}^{n}(y_i - \mu_p)^2$$

and

$$p_{\text{waic 2}} = \frac{n}{2}\frac{\sigma_p^4}{\sigma^4} + \frac{\sigma_p^2}{\sigma^4} \sum_{i=1}^{n}(y_i - \mu_p)^2.$$

For the current example, the first term in WAIC is evaluated to be 165.07 and $p_{\text{waic 1}} = 0.92$, $p_{\text{waic 2}} = 0.96$ and the corresponding values of the WAIC are respectively 166.91 and 167.

PMCC The PMCC is given by:

$$\text{PMCC} = \sum_{i=1}^{n} \{y_i - \mu_p\}^2 + n(\sigma^2 + \sigma_p^2) = 594.30 + 637.24 = 1231.54.$$

This shows that the PMCC is on a completely different scale than all the previous information criteria. The different information criteria values are not much different for this simple example as can be expected. However, these exact numerical values will be useful when we try to estimate these criteria values for the unknown σ^2 case using the Bayesian computation methods of the next chapter.

□

We conclude this section with the observation that all the model choice criteria have been defined for conditionally independent univariate observations, denoted by $\mathbf{y} = (y_1, \ldots, y_n)$. When spatial models are assumed, e.g. in Chapter 6, these definitions need to be changed. One solution that we adopt in the later chapters, e.g. in Section 4.16, for multivariate model specification is that we replace $f(y_i|\theta)$ by the conditional density $f(y_i|\mathbf{y}_{-i}, \theta)$ where $\mathbf{y}_{-i}$ denotes the $n-1$ dimensional vector obtained from $\mathbf{y}$ by deleting the ith component y_i. This, however, is not the only possible solution for the multivariate models. There are other possibilities, e.g. one can treat the $\mathbf{y}$ as a single observation and proceed with $f(\mathbf{y}|\theta)$, effectively treating $n = 1$. We do not investigate such a scenario here.

4.17 Bayesian model checking

Checking for model adequacy is one of the major tasks for any modeling exercise. The modeler is asked to verify if the chosen models are able to capture all sources of variation present in the data. Moreover, the assumptions made in the modeling stage must also be addressed in the light of the nature and the quality of the model-fit. Traditionally for linear models, model checking and diagnostics proceed by calculating the residuals, observed − fitted, and then plotting the residuals in various displays to check particular assumptions or lack-of-fit in different ways.

The main idea behind Bayesian model checking is that simulations from the fitted Bayesian model should look similar to the observed data. The simulations from the fitted model are particular realizations from the posterior predictive distribution for the full data set. That is, for each of the data values y_i we simulate a number of future replicates $y_i^{(j)}$ for $j = 1, \ldots, N$. In other words, $Y_i^{(j)} \sim f(y|\mathbf{y})$ and $y_i^{(j)}$ is an observed sample value from this distribution. This sampling gives us N replicated data sets, $\mathbf{y}^{(j)} = \left(y_1^{(j)}, \ldots, y_n^{(j)} \right)$, for $j = 1, \ldots, N$ from the fitted model and each of these data sets corresponds to the observed data set $\mathbf{y}$. The Bayesian computation Chapter 5 will detail how to obtain these simulated replicated data.

Bayesian model checking can now proceed by obtaining informal graphical displays of various data summaries or discrepancy measures of interests and then superimposing the observed value of the same measure by using the observed data values $\mathbf{y}$. The replicated data sets are used to construct the "null" distribution under the fitted model and then the observed value is positioned within the null distribution to assess the quality of the fit of the model. The observed value is expected to be seen as an extreme value if the data $\mathbf{y}$ are in conflict with the fitted model. These considerations have led to the definition of the Bayesian p-value for model checking. The Bayesian p-value is simply estimated as the proportion of replications which are equal to or more extreme than the observed value of the summary statistics or discrepancy measures. (See the definition of the p-value above in Section 4.15.3.) Informal model checking is then performed by using the estimated p-values in the usual way.

A notable distinction here is that the discrepancy measures in the above paragraph may also involve parameters of the model $\boldsymbol{\theta}$ here unlike the p-value setup in the classical inference which are based solely on pivotal statistics whose sampling distribution must not depend on the parameters. The Bayesian method to handle the parameters is to use simulation replicates for the parameters from their posterior distributions just as the same way as the simulation replications from the posterior predictive distributions. Thus, the Bayesian model checking methods are able to perform using the any possible statistics formed using data and parameters – not only the pivotal statistics.

This is seen to be an advantage of the Bayesian methods when those are implemented in conjunction with Bayesian computation.

4.17.1   Nuisance parameters

While discussing multi-parameter situation it is worthwhile to afford some discussion on how to handle nuisance parameters. For example, for data obtained from a normal distribution with unknown mean and variance we may want to infer about the unknown mean and treat the unknown variance as a nuisance parameter. The most obvious Bayesian solution to this is to obtain the joint posterior distribution by assuming a joint prior distribution on all the parameters. Inference for the parameters of interest then proceeds by using the marginal posterior distribution as noted above. However, this requires a bit of extra work in terms of specifying a prior for the nuisance parameters and then integrating those out from the joint posterior distribution. This also increases the dimension of the posterior distribution that we need to explore. Hence other solutions have been proposed. Some of those are described below. Suppose we partition $\boldsymbol{\theta} = (\boldsymbol{\gamma}, \boldsymbol{\eta})$ and we are interested in $\boldsymbol{\gamma}$.

In the "marginal likelihood" technique we integrate the nuisance parameter $\boldsymbol{\eta}$ from the likelihood function, i.e. we obtain

$$\int_{-\infty}^{\infty} f(\mathbf{y}|\boldsymbol{\gamma}, \boldsymbol{\eta}) d\boldsymbol{\eta}$$

and then use this as the likelihood times prior calculation to derive the posterior distribution. This avoids the extra task of specifying prior distribution for the nuisance parameter $\boldsymbol{\eta}$. However, this may be hard to perform the above integration. Even when it is possible to integrate the marginal likelihood may turn out to be a difficult function to analyze and work with.

In another technique called the "profile likelihood" we are first tasked with finding the maximum likelihood estimate $\hat{\boldsymbol{\eta}}(\boldsymbol{\gamma})$ of $\boldsymbol{\eta}$ conditional on $\boldsymbol{\gamma}$ from the full likelihood function $f(\mathbf{y}|\boldsymbol{\gamma}, \boldsymbol{\eta})$. The maximum likelihood is plugged in the full likelihood and then one proceeds to calculate the posterior distribution using the familiar likelihood × prior construct. As in the last marginal likelihood case, there is no need to specify a prior distribution for the nuisance parameter $\boldsymbol{\eta}$.

As a third and final approach one finds a sufficient statistics $\mathbf{t}$ for $\boldsymbol{\theta}$ such that its distribution only depends on the parameter of interest $\boldsymbol{\gamma}$. That is, we may assume

$$\mathbf{y}|\mathbf{t}, \boldsymbol{\theta} \sim f(\mathbf{y}|\mathbf{t})$$

so that given $\mathbf{t}$, $\mathbf{y}$ does not bring any additional information about $\boldsymbol{\theta}$. Now if the distribution of $\mathbf{t}$ only depends on $\boldsymbol{\gamma}$ then we can make inference about $\boldsymbol{\gamma}$ solely based on the distribution of $\mathbf{t}$. This is called conditional inference.

4.18   The pressing need for Bayesian computation

In Bayesian hierarchical modeling for practical problems of interest the dimension of the set of parameters is often very large. In such cases it is necessary to solve a high dimensional integration to normalize the joint posterior density. High dimensional integration is also required to compute any desired expectation. In fact, when working with un-normalized posterior distributions a ratio of two integrations is required to evaluation any expectation. For example, suppose we have posterior $\propto$ likelihood $\times$ the prior, i.e., $\pi(\boldsymbol{\theta}|\mathbf{y}) \propto f(\mathbf{y}|\boldsymbol{\theta})\pi(\boldsymbol{\theta})$. Here the posterior mean is given by:

$$E(\boldsymbol{\theta}|\mathbf{y}) = \frac{\int_{-\infty}^{\infty} \boldsymbol{\theta} f(\mathbf{y}|\boldsymbol{\theta})\pi(\boldsymbol{\theta})d\boldsymbol{\theta}}{\int_{-\infty}^{\infty} f(\mathbf{y}|\boldsymbol{\theta})\pi(\boldsymbol{\theta})d\boldsymbol{\theta}}.$$

If $\boldsymbol{\theta}$ is multidimensional then the two integrations above will be multi-dimensional as well. Affording full flexibility in modeling will mean that the integrand, likelihood $\times$ the prior, will be a complex multivariate function of the multivariate parameter $\boldsymbol{\theta}$. As a result direct integration will not be possible for all but very simple models. Indeed, such computational limitations limited the use of Bayesian inference only to toy problems.

Modern Bayesian computation methods view this integration problem as yet another statistical problem. Here the posterior distribution $\pi(\boldsymbol{\theta}|\mathbf{y})$ is viewed as an population distribution whose features are unknown to us. This however is the main problem in statistics where the researcher is interested in discovering unknown features of a population. The statistical solution to this problem is to draw samples from the population and then make inference about the unknown population characteristics by forming suitable statistical averages of the sampled values. Sampling based Bayesian computation methods exactly do this by drawing samples from the posterior distribution. This sampling has been facilitated by the wide availability of inexpensive and high-speed computing machines and software development shared by research communities from across different parts of the world. Now tables are turned; through hierarchical modeling in a Bayesian framework, models that are inaccessible in a classical framework can be handled within a Bayesian framework. This is particularly the case for spatial and temporal data models due to concerns with asymptotics and with getting the correct uncertainty in inferential statements. The Bayesian computation chapter is devoted to discussing the key methods and techniques for sampling based approaches.

The sampling based approaches have also liberated statistical modeling from the clutches of only mathematically gifted and able statisticians. Applied scientists with a reasonable amount of numeracy skills and some limited exposure to high school level mathematics can appreciate and apply advanced Bayesian modeling to solve their practical problems. By avoiding hard core mathematics and calculus, this book aims to be a bridge that removes the statistical knowledge gap from among the applied scientists.

4.19 Conclusion

This chapter has introduced all the Bayesian inference keywords we require for data analysis in the later chapters. The chapter also illustrated these concepts with a normal distribution example that serves as the foundation stone to build the structured spatial and spatio-temporal models. The example includes most of the intermediate steps to enable a typical beginner reader to master and experience the richness of scientific understanding offered by the Bayesian methods. The exercise section below contains several examples which the reader may wish to attempt in order to test their understanding of Bayesian inference.

An applied reader without sufficient background in mathematical statistics is able to skip the detailed theoretical derivations and proceed straight Bayesian modeling using software packages. The mathematical derivations help keep the rigor in modeling and may be more appealing to students pursuing a degree with a large statistics content.

4.20 Exercises

1. A certain disease affects 0.1% of the population. A diagnostic test for the disease gives a positive response with probability 0.95 if the disease is present. It also gives a positive response with probability 0.02, however, if the disease is not present. If the test gives a positive response for a given person, what is the probability that the person has the disease?

2. The Japanese car company Maito make their Professor model in three countries, Japan, England and Germany, with one half of the cars being built in Japan, two tenths in England and three tenths in Germany. One percent of the cars built in Japan have to be returned to the dealer as faulty while the figures for England and Germany are four percent and two percent respectively. What proportion of Professors are faulty? If I buy a Professor and find it to be faulty, what is the chance that it was made in England?

3. Suppose that the number of defects on a roll of magnetic recording tape has a Poisson distribution for which the mean θ is unknown and that the prior distribution of θ is a gamma distribution with

parameters $\alpha = 3$ and $\beta = 1$. When five rolls of this tape are selected at random and inspected, the number of defects found on the rolls are 2, 2, 6, 0, and 3. If the squared error loss function is used what is the Bayes estimate of θ?

4. Suppose that $Y_1, \ldots, Y_n$ is a random sample from the distribution with pdf

$$f(y|\theta) = \begin{cases} \theta y^{\theta-1} & \text{if } 0 < y < 1, \\ 0 & \text{otherwise.} \end{cases}$$

Suppose also that the value of the parameter θ is unknown $(\theta > 0)$ and that the prior distribution of θ is a gamma distribution with parameters α and β $(\alpha > 0$ and $\beta > 0)$. Determine the posterior distribution of θ and hence obtain the Bayes estimator of θ under a squared error loss function.

5. Suppose that we have a random sample of normal data

$$Y_i \sim N(\mu, \sigma^2), \quad i = 1, \ldots, n$$

where σ^2 is known but μ is unknown. Thus, for μ the likelihood function comes from the distribution

$$\bar{Y} \sim N(\mu, \sigma^2/n),$$

where $\bar{Y} = \frac{1}{n} \sum_{i=1}^{n} Y_i$. Assume the prior distribution for μ is given by $N(\gamma, \sigma^2/n_0)$ where γ and n_0 are known constants.

(i) Show that the posterior distribution for μ is normal with

$$\text{mean} = E(\mu|\bar{y}) = \frac{n_0\gamma + n\bar{y}}{n_0 + n}, \quad \text{variance} = \text{var}(\mu|\bar{y}) = \frac{\sigma^2}{n_0 + n}.$$

(ii) Provide an interpretation for each of $E(\mu|\bar{y})$ and $\text{var}(\mu|\bar{y})$ in terms of the prior and data means and prior and data sample sizes.

(iii) By writing a future observation $\tilde{Y} = \mu + \epsilon$ where $\epsilon \sim N(0, \sigma^2)$ independently of the posterior distribution $\pi(\mu|\bar{y})$ explain why the posterior predictive distribution of $\tilde{Y}$ given $\bar{y}$ is normally distributed. Obtain the mean and variance of this posterior predictive distribution.

Suppose that in an experiment $n = 2$, $\bar{y} = 130$, $n_0 = 0.25$, $\gamma = 120$ and $\sigma^2 = 25$. Obtain:

a. the posterior mean, $E(\mu|\bar{y})$ and variance, $\text{var}(\mu|\bar{y})$,
b. a 95% credible interval for μ given $\bar{y}$,
c. the mean and variance of the posterior predictive distribution of a future observation $\tilde{Y}$,
d. a 95% prediction interval for a future observation $\tilde{Y}$.

6. A particular measuring device has normally distributed error with mean zero and unknown variance σ^2. In an experiment to estimate σ^2, n independent evaluations of this error are obtained.

    (i) If the prior distribution for σ^2 is inverse gamma (see A.1) with parameters m and β, show that the posterior distribution is also inverse gamma, with parameters m^* and β^* and derive expressions for m^* and β^*.

    (ii) Show that the predictive distribution for the error, Z, of a further observation made by this device has p.d.f.

    $$f(z) \ \propto \ \left(1 + \frac{z^2}{2\beta^*}\right)^{-m^* - \frac{1}{2}} \qquad z \in \mathbb{R}.$$

7. Assume $Y_1, Y_2, \ldots, Y_n$ are independent observations which have the distribution $Y_i \sim N(\beta x_i, \sigma^2)$, $i = 1, 2, \ldots, n$, where the x_is and σ^2 are known constants, and β is an unknown parameter, which has a normal prior distribution with mean β_0 and variance σ_0^2, where β_0 and σ_0^2 are known constants.

    (i)Derive the posterior distribution of β.

    (ii)Show that the mean of the posterior distribution is a weighted average of the prior mean β_0, and the maximum likelihood estimator of β.

    (iii)Find the limit of the posterior distribution as $\sigma_0^2 \to \infty$, and discuss the result.

    (iv)How would you predict a future observation from the population $N(\beta x_{n+1}, \sigma^2)$, where x_{n+1} is known?

8. Let $Y_1, Y_2, \ldots, Y_n$ be a sequence of independent, identically distributed random variables with the exponential distribution with parameter λ where λ is positive but unknown. Suppose that λ has a gamma(m, β) prior distribution.

    (i)Show that the posterior distribution of λ given $Y_1 = y_1, Y_2 = y_2, \ldots, Y_n = y_n$ is gamma$(n + m, \beta + t)$ where $t = \sum\limits_{i=1}^{n} y_i$.

    (ii)Show that the (predictive) density of Y_{n+1} given the n observations $Y_1 = y_1, Y_2 = y_2, \ldots, Y_n = y_n$ is

    $$\pi(y_{n+1} | y_1, \ldots, y_n) \ = \ \frac{(n+m)(\beta + t)^{n+m}}{(y_{n+1} + \beta + t)^{n+m+1}}.$$

    (iii)Find the joint (predictive) density of Y_{n+1} and Y_{n+2} given $Y_1 = y_1, Y_2 = y_2, \ldots, Y_n = y_n$.

9. **Taste Test** Let $Y_1, Y_2, \ldots, Y_6$ be a sequence of independent, identically distributed Bernoulli random variables with parameter θ, and suppose that $y_1 = y_2 = y_3 = y_4 = y_5 = 1$ and $y_6 = 0$.

   Derive the posterior model probabilities for Model $0 : \theta = \frac{1}{2}$ and Model $1 : \theta > \frac{1}{2}$, assuming the following prior distributions:

   (i) $P(M_0) = 0.5$, $P(M_1) = 0.5$, $\pi_1(\theta) = 2$; $\theta \in \left(\frac{1}{2}, 1 \right)$.

   (ii) $P(M_0) = 0.8$, $P(M_1) = 0.2$, $\pi_1(\theta) = 8(1 - \theta)$; $\theta \in \left(\frac{1}{2}, 1 \right)$.

   (iii) $P(M_0) = 0.2$, $P(M_1) = 0.8$, $\pi_1(\theta) = 48 \left(\theta - \frac{1}{2} \right) (1 - \theta)$; $\theta \in \left(\frac{1}{2}, 1 \right)$.

10. Suppose that:

$$M_0 : Y_1, Y_2, \ldots, Y_n | \theta_0 \sim f_0(x|\theta_1) = \theta_0(1 - \theta_0)^x, \quad x = 0, 1, \ldots.$$

$$M_1 : Y_1, Y_2, \ldots, Y_n | \theta_1 \sim f_1(x|\theta_1) = e^{-\theta_1} \theta_1^x / x!, \quad x = 0, 1, \ldots.$$

Suppose that θ_0 and θ_1 are both unknown. Assume that $\pi_0(\theta_0)$ is the beta distribution with parameters α_0 and β_0 and $\pi_1(\theta_1)$ is the Gamma distribution with parameters α_1 and β_1. Compute the (prior) predictive means under the two models. Obtain the Bayes factor. Hence study the dependence of the Bayes factor on prior data combinations. Calculate numerical values for $n = 2$ and for two data sets $y_1 = y_2 = 0$ and $y_1 = y_2 = 2$ and two sets of prior parameters $\alpha_0 = 1, \beta_0 = 2, \alpha_1 = 2, \beta_1 = 1$ and $\alpha_0 = 30, \beta_0 = 60$, $\alpha_1 = 60, \beta_1 = 30$. Write a R program to calculate the Bayes factor for given values of $y_1, \ldots, y_n$ and the parameters.

5

Bayesian computation methods

5.1 Introduction

The Bayesian machinery grinds to a halt at the point of making inference where it requires us to solve mainly integrals for realistic practical modeling problems. The integrals are non-standard for all but few problems where conjugate prior distributions are assumed for likelihoods which are based on standard statistical distributions. The nature of non-standardness of the integrals limits the scope of pure mathematical tools for integration which can be applied in general for all the problems. Hence a natural conclusion here is to look for numerical approximations.

Numerical approximations for solving integration are around and do present themselves as candidate numerical quadrature methods for Bayesian computation. Textbook numerical methods such as the trapezoidal rule, Simpson's rules, Newton-Coates formula are available and will work well for low dimensional problems. Hence these rules are to be used when there are only a few parameters in the statistical model and the posterior distribution is relatively well behaved. Indeed, one of our examples below will illustrate the use of a standard numerical integration technique. However, these integration techniques do not work well for moderately high dimensional problems. There is no universal benchmark regarding a threshold dimension but these methods will be unreliable for any dimension higher than a handful, e.g. four or five.

Faced with the failure and unsuitability of numerical integration methods for general purpose Bayesian modeling, many researchers proposed stochastic integration techniques based on the Laplace (normal) approximation of the posterior distribution. For large samples the posterior distribution can be approximated by a multivariate normal distribution and then inference can be allowed to flow through using the known properties of normal distributions. These approximation methods will work for large sample sizes and for low dimensional parameter space. However, these will not provide satisfactory approximations for the full posterior distributions in practical, especially spatio-temporal, modeling problems for large dimensional parameter spaces.

Monte Carlo integration techniques, with the advent of cheap computing power, have taken the center stage in Bayesian computation methods. The basic idea behind Monte Carlo integration is the same basic idea in statistics: that unknown population characteristics can be estimated by obtaining a large

DOI: 10.1201/9780429318443-5

enough random sample from the population. Statistical theorems, known as the laws of large numbers and the Central Limit Theorem, guarantee that the estimation will be as accurate as desired. Moreover, the errors in such estimation can be quantified providing a sense of security while using these Monte Carlo methods. This chapter will introduce and discuss many such Monte Carlo integration methods.

The Markov chain based Monte Carlo methods are the most widely used in Bayesian computation, and hence deserve an introduction. Markov chains in probability start with an initial value and determines its next value in the chain by using a probability distribution called a transition distribution. In so doing the Markov chain maintains its Markovian property – the one which states that the next value depends only on the current value and not on its distant past. A primary requirement on the transition distribution is that it is easy to draw samples from. A Markov chain so constructed with some knowledge and requirements regarding the target posterior distribution can be shown to converge to the target distribution itself. Hence simulation of the Markov chain will provide samples from the posterior distribution of interest. However, there is a problem that these samples are not going to be random as is the usual requirement of the statistical limit theorems, viz. laws of large numbers and the CLT. However, these limit theorems have been shown to hold even when the samples are drawn using a purposefully constructed Markov chain. Such nicely behaved Markov chains are easy to code as computer algorithms without learning the deep theories of converging Markov chains. This has contributed to the popularity of the Markov chain Monte Carlo (MCMC) algorithms for Bayesian computation.

5.2   Two motivating examples for Bayesian computation

♡ **Example** 5.1. Suppose that $Y_1, \ldots, Y_n \sim N(\theta, 1)$ independently and the prior distribution is the standard Cauchy distribution $\pi(\theta) = \frac{1}{\pi} \frac{1}{1+\theta^2}, -\infty < \theta < \infty$ instead of the conjugate normal prior distribution we assumed earlier. The non-conjugate Cauchy prior distribution results in a non-standard posterior distribution which cannot be analyses analytically. As a result numerical integration techniques will be required and this example will illustrate many of the proposed Bayesian computation methods.

Here the posterior distribution is given by

$$
\begin{aligned}
\pi(\theta|\mathbf{y}) &\propto \left[\prod_{i=1}^{n} \frac{1}{\sqrt{2\pi}} \exp\left[-\frac{1}{2}(y_i - \theta)^2\right]\right] \frac{1}{\pi} \frac{1}{1+\theta^2} \\
&\propto \frac{1}{1+\theta^2} \exp\left[-\frac{1}{2} \sum_{i=1}^{n}(y_i - \theta)^2\right], \quad -\infty < \theta < \infty.
\end{aligned}
$$

The identity $\sum_{i=1}^{n}(y_i - \theta)^2 = \sum_{i=1}^{n}(y_i - \bar{y})^2 + n(\theta - \bar{y})^2$ is used to further simplify the posterior distribution to

$$\pi(\theta|\mathbf{y}) \propto \frac{1}{1+\theta^2} \exp\left[-\frac{n}{2}(\theta - \bar{y})^2\right], \quad -\infty < \theta < \infty.$$

by noting that the term $\exp\left[-\frac{1}{2\sigma^2} \sum_{i=1}^{n} (y_i - \bar{y})^2\right]$ is absorbed in the constant of proportionality. Let

$$I_1 = \int_{-\infty}^{\infty} \frac{1}{1+\theta^2} \exp\left[-\frac{n}{2}(\theta - \bar{y})^2\right] d\theta$$

be that constant. Hence,

$$\pi(\theta|\mathbf{y}) = \frac{1}{I_1} \frac{1}{1+\theta^2} \exp\left[-\frac{n}{2}(\theta - \bar{y})^2\right], \quad -\infty < \theta < \infty.$$

The posterior mean is given by

$$E(\theta|\mathbf{y}) = \int_{-\infty}^{\infty} \theta\, \pi(\theta|\mathbf{y}) d\theta = \frac{1}{I_1} \int_{-\infty}^{\infty} \frac{\theta}{1+\theta^2} \exp\left[-\frac{n}{2}(\theta - \bar{y})^2\right] d\theta,$$

which is seen to be ratio of two integrals, I_2 and I_1 where

$$I_2 = \int_{-\infty}^{\infty} \frac{\theta}{1+\theta^2} \exp\left[-\frac{n}{2}(\theta - \bar{y})^2\right] d\theta.$$

These two integrals are analytically intractable, i.e., there is no exact solution for either of them. However, numerical integration techniques can be applied to evaluate these.

For numerical illustration purposes we simulate $n = 25$ observations from the $N(\theta = 1, 1)$ distribution in R. With the random number seed fixed at 44, the resulting samples give $\bar{y} = 0.847$. By applying the built in `integrate` function in R we obtain $E(\theta|\mathbf{y}) = \mathbf{0.809}$ with $I_1 = 0.2959$ and $I_2 = 0.2394$. Thus, with $n = 25$, $\bar{y} = 0.847$ the target for Bayesian computation methods will be to estimate the value 0.809 of $E(\theta|\mathbf{y})$. The numbers reported here can be verified with the `bmstdr` command, `BCauchy(true.theta=1, n=25)`.

□

♡ **Example** 5.2. For the second example we return to the normal distribution example in Section 4.14 where both parameters are unknown. The joint posterior distribution of θ and $\lambda^2(= 1/\sigma^2)$ has been given in (4.14). To find the constant of proportionality one has to perform two double integrations with respect to θ and λ^2. However, by exploiting conjugacy Section 4.14 obtains the marginal posterior distributions $\theta|\mathbf{y}$ and $\sigma^2|\mathbf{y}$ and also the posterior predictive distribution $Y_0|\mathbf{y}$. However, in general it is difficult to estimate other non-standard parametric functions exactly analytically such as the coefficient of variation $\frac{\sigma}{\theta}$. Here we may want to find $E\left(\frac{\sigma}{\theta}|\mathbf{y}\right)$ and $\mathrm{Var}\left(\frac{\sigma}{\theta}|\mathbf{y}\right)$. These tasks require non-standard bivariate integrations for which there are no exact analytical solutions. □

5.3 Monte Carlo integration

Suppose that we can draw independent samples $\boldsymbol{\theta}^{(1)}, \boldsymbol{\theta}^{(2)}, \ldots, \boldsymbol{\theta}^{(N)}$ from $\pi(\boldsymbol{\theta}|\mathbf{y})$. Then for any one-to-one function $b(\boldsymbol{\theta})$ we can estimate

$$E[b(\boldsymbol{\theta})|\mathbf{y}] \approx \bar{b}_N = \frac{1}{N} \sum_{j=1}^{N} b\left(\boldsymbol{\theta}^{(j)}\right). \tag{5.1}$$

We can use the laws of large numbers to show that $\bar{b}_N$ approaches $E_\pi[b(\boldsymbol{\theta})|\mathbf{y}]$ as $N \to \infty$. This is called *Monte Carlo integration*, and it uses the following basic idea in statistics:

> features of an unknown distribution can be discovered once a large enough random sample from that distribution has been obtained.

The superscript (j), e.g. in $\boldsymbol{\theta}^{(j)}$, will be used to denote the jth Monte Carlo sample through out, with the parenthesis to protect the replication number.

As an example of Monte Carlo, suppose we have random samples $\theta^{(1)}, \lambda^{2(1)}, \theta^{(2)}, \lambda^{2(2)} \ldots, \theta^{(N)}, \lambda^{2(N)}$ from the joint posterior distribution $\pi(\theta, \lambda^2|\mathbf{y})$ given in (4.14). We now obtain the transformed values

$$\sigma^{(j)} = \frac{1}{\lambda^{(j)}}$$

for $j = 1, \ldots, N$. Then we estimate the coefficient of variation $\frac{\sigma}{\theta}$ $(= b(\boldsymbol{\theta})$ in the above notation since $\boldsymbol{\theta} = (\theta, \lambda^2)$ here) by the posterior mean

$$E\left(\frac{\sigma}{\theta}\Big|\mathbf{y}\right) \approx \frac{1}{N} \sum_{j=1}^{N} \frac{\sigma^{(j)}}{\theta^{(j)}} \equiv \frac{1}{N} \sum_{j=1}^{N} \left(\lambda^{(j)}\theta^{(j)}\right)^{-1}.$$

Note that the sampled $\sigma^{(j)}$'s are obtained by using the above transformation after completion of sampling from the posterior distribution. No other adjustments, e.g. Jacobian calculation for transformation, are necessary. Indeed, if we are interested in finding estimate of any other parametric function then we simply form sample average for that function. For example, to estimate $E(\theta^2|\mathbf{y})$ we simply form the average $\frac{1}{N}\sum_{j=1}^{N}\theta^{2(j)}$ by defining $b(\theta) = \theta^2$. Moreover, the sample 2.5% and 97.5% quantiles of the N sample values $\frac{\sigma^{(j)}}{\theta^{(j)}}, j = 1, \ldots, N$ provide a 95% credible interval for the parameter $\frac{\sigma}{\theta}$. There is no need to do any further work (analytical or sampling) for assessing uncertainty of the posterior estimates. However, we still need to assess the Monte Carlo error of the posterior estimates. This will be discussed later in Section 5.11.3.

Note that the data set $\mathbf{y}$ does not play any part in the Monte Carlo integration, although it does influence the shape of the posterior distribution.

The data values (and their summaries) are treated as fixed parameters of the posterior distribution. Hence this notation may be suppressed for convenience.

The Monte Carlo integration method detailed above makes a very strong assumption that direct random sampling is possible from the population distribution $\pi(\boldsymbol{\theta}|\mathbf{y})$. However, for most Bayesian model fitting problems direct sampling is not yet possible theoretically, although there are exceptions. This limitation fueled the growth of literature in drawing approximate samples from the target posterior distribution $\pi(\boldsymbol{\theta}|\mathbf{y})$. The sections below describe some of these Monte Carlo sampling techniques used widely in Bayesian computation.

5.4 Importance sampling

Importance sampling is one of the oldest attempt at performing Monte Carlo integration when direct sampling from the target distribution is not possible but it is possible to sample easily from a surrogate distribution which looks like the target distribution. Such a surrogate distribution is called an importance sampling distribution. Let $g(\boldsymbol{\theta})$ denote the importance sampling density, which is easy to sample from. Implicitly we also assume that $g(\boldsymbol{\theta})$ and $\pi(\boldsymbol{\theta}|\mathbf{y})$ have the same support, i.e. they are non-negative for the same set of values of $\boldsymbol{\theta}$. Moreover, it should be ensured that there does not exist a value of θ for which $g(\boldsymbol{\theta}) = 0$ but $\pi(\theta|\mathbf{y}) \neq 0$. If there exists such a set of values, then the importance sampling density $g(\boldsymbol{\theta})$ will never draw a sample from such a set since $g(\boldsymbol{\theta}) = 0$ for such a set. Hence the posterior probability behavior in that region of the parameter space will never be explored. This would lead to bias in the posterior parameter estimates. Figure 5.1 illustrates various scenarios for the importance and target densities.

As is clear now that in most cases (at least for the example in Section 5.2) we only can evaluate the non-normalized posterior density, $h(\boldsymbol{\theta})$ say, where the full posterior density is given by:

$$\pi(\boldsymbol{\theta}|\mathbf{y}) = \frac{h(\boldsymbol{\theta})}{\int_{-\infty}^{\infty} h(\boldsymbol{\theta})d\boldsymbol{\theta}}.$$

Note that in $h(\boldsymbol{\theta})$ we suppress its dependence on conditioning of the observed data values $\mathbf{y}$. Once these observed values are plugged into the posterior distribution (likelihood × the prior), those are suppressed for ease of notation. Now we rewrite the posterior expectation $E(b(\boldsymbol{\theta})|\mathbf{y})$ as a ratio of two expectations with respect to the importance sampling distribution having density function $g(\boldsymbol{\theta})$. Thus,

$$
\begin{aligned}
E_\pi\left(b(\boldsymbol{\theta})|\mathbf{y}\right) &= \int_{-\infty}^{\infty} b(\boldsymbol{\theta})\,\pi(\boldsymbol{\theta}|\mathbf{y})d\boldsymbol{\theta} \\
&= \int_{-\infty}^{\infty} b(\boldsymbol{\theta})\frac{h(\boldsymbol{\theta})}{\int_{-\infty}^{\infty} h(\boldsymbol{\theta})d\boldsymbol{\theta}}d\boldsymbol{\theta} \\
&= \left[\int_{-\infty}^{\infty} h(\boldsymbol{\theta})d\boldsymbol{\theta}\right]^{-1}\int_{-\infty}^{\infty} b(\boldsymbol{\theta})\,h(\boldsymbol{\theta})d\boldsymbol{\theta} \\
&= \left[\int_{-\infty}^{\infty} \frac{h(\boldsymbol{\theta})}{g(\boldsymbol{\theta})}g(\boldsymbol{\theta})d\boldsymbol{\theta}\right]^{-1}\int_{-\infty}^{\infty} b(\boldsymbol{\theta})\,\frac{h(\boldsymbol{\theta})}{g(\boldsymbol{\theta})}g(\boldsymbol{\theta})d\boldsymbol{\theta} \\
&= \left[\int_{-\infty}^{\infty} w(\boldsymbol{\theta})g(\boldsymbol{\theta})d\boldsymbol{\theta}\right]^{-1}\int_{-\infty}^{\infty} b(\boldsymbol{\theta})\,w(\boldsymbol{\theta})g(\boldsymbol{\theta})d\boldsymbol{\theta}, \\
&= \left[E_g(w(\boldsymbol{\theta}))\right]^{-1} E_g\left[b(\boldsymbol{\theta})\,w(\boldsymbol{\theta})\right]
\end{aligned}
$$

where

$$
w(\boldsymbol{\theta}) = \frac{h(\boldsymbol{\theta})}{g(\boldsymbol{\theta})} \tag{5.2}
$$

is defined to be the importance sampling weight function. Note that the expectation symbol E has been given the suffix π or g to denote the averaging density, i.e. the density with respect to which the expectation is taken. Now we will have to estimate expectations of two functions $w(\boldsymbol{\theta})$ and $b(\boldsymbol{\theta})\,w(\boldsymbol{\theta})$ under the importance sampling density $g(\theta)$, which is easy to sample from. Let $\boldsymbol{\theta}^{(1)}, \boldsymbol{\theta}^{(2)}, \ldots, \boldsymbol{\theta}^{(N)}$ be a random sample from $g(\boldsymbol{\theta})$. Then

$$
\bar{b}_N^{(\mathrm{is})} = \frac{\sum_{j=1}^{N} b(\boldsymbol{\theta}^{(j)})w^{(j)}}{\sum_{j=1}^{N} w^{(j)}} \quad \text{where} \quad w^{(j)} = \frac{h(\boldsymbol{\theta}^{(j)})}{g(\boldsymbol{\theta}^{(j)})},
$$

provide a Monte Carlo integration for $E_\pi\left(b(\boldsymbol{\theta})|\mathbf{y}\right)$. Notice that in effect we have done two Monte Carlo integrations: one for the numerator and the other for the denominator.

The importance sampling weight function $w(\boldsymbol{\theta}) = \frac{h(\boldsymbol{\theta})}{g(\boldsymbol{\theta})}$ does not depend on the normalizing constant for the posterior distribution and hence is easily computable. Moreover, in $\bar{b}_N^{(\mathrm{is})}$ the normalizing constant in the importance sampling density $g(\boldsymbol{\theta})$ does not need to be calculated since that cancels in the ratio of two integrals. In addition, the weight function $w(\boldsymbol{\theta})$ becomes the likelihood function if $g(\boldsymbol{\theta})$ is chosen to be the prior density $\pi(\boldsymbol{\theta})$ since the un-normalized posterior density $h(\boldsymbol{\theta})$ is of the form likelihood × prior.

For $\bar{b}_N^{(\mathrm{is})}$ to be a good estimate, we need $w^{(j)}$'s to be well behaved, i.e., should be roughly equal. This is very hard to achieve when the dimension of $\boldsymbol{\theta}$ is high. In the next section we describe a rejection sampling method which improves on the importance sampling methodology.

♡ **Example 5.3. Cauchy prior** We shall illustrate with the example described in Section 5.2. We take the importance distribution to be the standard Cauchy distribution, which is the prior distribution here. Thus, in this case

$$
g(\theta) \propto \frac{1}{1+\theta^2}, \quad h(\theta) = \frac{1}{1+\theta^2}\exp\left[-\frac{n}{2}(\theta-\bar{y})^2\right]
$$

and

$$w(\theta) = \frac{h(\theta)}{g(\theta)} = \exp\left[-\frac{n}{2}(\theta - \bar{y})^2\right]$$

which is the likelihood function of θ.

We draw $N = 10,000$ samples from the standard Cauchy distribution and then calculate the weights

$$w\left(\theta^{(j)}\right) = \exp\left[-\frac{n}{2}(\theta^{(j)} - \bar{y})^2\right]$$

for $j = 1, \ldots, N$. To estimate the posterior mean $E(\theta|\mathbf{y})$ we take $b(\theta) = \theta$ and obtain

$$\bar{b}_N^{(\text{is})} = \frac{\sum_{j=1}^{N} \theta^{(j)} w^{(j)}}{\sum_{j=1}^{N} w^{(j)}} = 0.811$$

which is close to the value 0.809 obtained using the numerical approximation methods in Example 5.2. Here N can be increased to achieve better accuracy. For example, $N = 20,000$ provides the better estimate 0.808. The numbers reported can be verified with the `bmstdr` command:

```
BCauchy(method="importance", true.theta = 1, n=25, N=10000)
```

□

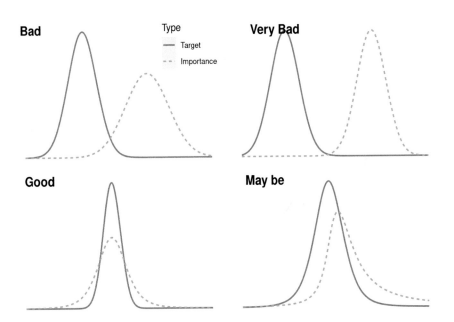

FIGURE 5.1: How different choices of g influences the computation.

5.5   Rejection sampling

Rejection sampling is a method for generating independent samples from the
target posterior distribution $\pi(\boldsymbol{\theta}|\mathbf{y})$ once we can generate independent samples
from an importance sampling density $g(\boldsymbol{\theta})$. Suppose that,

$$M = \text{supremum}_{\boldsymbol{\theta}} \frac{h(\boldsymbol{\theta})}{g(\boldsymbol{\theta})}$$

is available. For practical purposes the supremum in the above definition can
be replaced by maximum. The supremum does not need to look very intimi-
dating. For example, if the prior distribution $\pi(\boldsymbol{\theta})$ is taken as the importance
sampling distribution $g(\boldsymbol{\theta})$ then M can be calculated by finding the maximum
likelihood estimate of $\boldsymbol{\theta}$ since

$$M = \sup_{\boldsymbol{\theta}} \frac{h(\boldsymbol{\theta})}{g(\boldsymbol{\theta})} = \sup \frac{\text{Likelihood} \times \text{Prior}}{\text{Prior}} = \max_{\boldsymbol{\theta}} \text{Likelihood}.$$

This strategy will work for low dimensional problems where it is easy to find
the maximum likelihood estimates. The rejection sampling method has the
following steps. Draw a random sample $\boldsymbol{\theta} \sim g(\boldsymbol{\theta})$ and a random sample U
following the uniform distribution in $(0, 1)$ independently. Retain the drawn
sample $\boldsymbol{\theta}$ as a sample from $\pi(\boldsymbol{\theta}|\mathbf{y})$ if $u \leq \frac{h(\boldsymbol{\theta})}{M g(\boldsymbol{\theta})}$ otherwise generate another
sample from $g(\boldsymbol{\theta})$ and repeat. The quantity:

$$\alpha(\boldsymbol{\theta}) = \frac{h(\boldsymbol{\theta})}{M g(\boldsymbol{\theta})}$$

is called the acceptance probability of a candidate sample $\boldsymbol{\theta}$ generated from the
importance sampling distribution. Note also that in order to implement the
method we do *not* need the normalizing constant in $\pi(\boldsymbol{\theta}|\mathbf{y})$. Now we show that
a sample drawn using the rejection sampling method has the exact distribution
$\pi(\boldsymbol{\theta}|\mathbf{y})$. This proof can be avoided in the first reading. First, see that:

$$P\left(\boldsymbol{\theta} \text{ retained} \mid \boldsymbol{\theta} \sim g\right) = \frac{h(\boldsymbol{\theta})}{Mg(\boldsymbol{\theta})}.$$

Hence

$$P\left(\boldsymbol{\theta} \text{ retained }\right) = \int_{-\infty}^{\infty} \frac{h(\boldsymbol{\theta})}{Mg(\boldsymbol{\theta})} g(\boldsymbol{\theta}) d\boldsymbol{\theta} = \frac{1}{M} \int_{-\infty}^{\infty} h(\boldsymbol{\theta}) d\boldsymbol{\theta}.$$

Our aim is to prove that $\boldsymbol{\theta}$ retained through the rejection method has the distribution $\pi(\boldsymbol{\theta}|\mathbf{y})$. We have

$$
\begin{aligned}
P\left(\boldsymbol{\theta} < c | \boldsymbol{\theta} \text{ retained }\right) &= \frac{P(\boldsymbol{\theta} < c \text{ and } \boldsymbol{\theta} \text{ retained })}{P(\boldsymbol{\theta} \text{ retained })}\\
&= \left(\tfrac{1}{M}\int_{-\infty}^{\infty} h(\boldsymbol{\theta})d\boldsymbol{\theta}\right)^{-1}\int_{\boldsymbol{\theta}<c}\int_{0}^{\alpha(\boldsymbol{\theta})} g(\boldsymbol{\theta})du\, d\boldsymbol{\theta}\\
&= \left(\tfrac{1}{M}\int_{-\infty}^{\infty} h(\boldsymbol{\theta})d\boldsymbol{\theta}\right)^{-1}\int_{\boldsymbol{\theta}<c}\frac{h(\boldsymbol{\theta})}{M\,g(\boldsymbol{\theta})}g(\boldsymbol{\theta})d\boldsymbol{\theta}\\
&= \left(\int_{-\infty}^{\infty} h(\boldsymbol{\theta})d\boldsymbol{\theta}\right)^{-1}\int_{\boldsymbol{\theta}<c} h(\boldsymbol{\theta})d\boldsymbol{\theta}\\
&= \int_{\boldsymbol{\theta}<c} \pi(\boldsymbol{\theta}|\mathbf{y})\, d\boldsymbol{\theta}.
\end{aligned}
$$

This derivation shows that the retained $\boldsymbol{\theta}$ has the exact distribution $\pi(\boldsymbol{\theta}|\mathbf{y})$ as was claimed. $\qquad\square$

♡ **Example 5.4. Cauchy prior example revisited** We continue with the Cauchy prior example and choose the prior distribution as the importance sampling density. The likelihood function $\exp\left[-\frac{n}{2}(\theta - \bar{y})^2\right]$ is maximized at $\theta = \bar{y}$ and consequently $M = 1$. As before, we draw $N = 10,000$ samples from the standard Cauchy distribution and then calculate the weights

$$
w\left(\theta^{(j)}\right) = \exp\left[-\frac{n}{2}(\theta^{(j)} - \bar{y})^2\right]
$$

for $j = 1,\ldots,N$. We also draw N independent samples, $u^{(j)}$ for $j = 1,\ldots,N$ from the $U(0,1)$ distribution and keep the sample $\theta^{(j)}$ if $u^{(j)} < w\left(\theta^{(j)}\right)$. The retained samples are simply averaged to estimate $E(\theta|\mathbf{y})$. With the fixed seed as before, 937 samples are retained from a draw of $N = 10,000$ samples from the prior distribution. Using the command,

```
BCauchy(method="rejection", true.theta = 1, n=25, N=10000)
```

we obtain the estimate 0.816 for $E(\theta|\mathbf{y})$ with a 95% credible interval given by $(0.434, 1.222)$

$\qquad\square$

5.6 Notions of Markov chains for understanding MCMC

As models become more complex in high dimension the posterior distributions become analytically intractable. Simpler computations methods fail to produce accurate estimates of posterior expectations. MCMC methods provide a solution that works for typical problems in general purpose Bayesian modeling. Here a Markov chain is simulated whose limiting (or stationary) distribution is the target posterior distribution, $\pi(\boldsymbol{\theta}|\mathbf{y})$. Features of the posterior distribution $\pi(\boldsymbol{\theta}|\mathbf{y})$ are discovered (accurately) by forming averages (also called the *ergodic averages*) as in (5.1). It turns out that $\bar{b}_N$ still accurately estimates $E(b(\boldsymbol{\theta})|\mathbf{y})$ if we generate samples using a Markov chain. Theorems like the CLT and laws of large numbers can be proven as discussed below.

A Markov chain is generated by sampling

$$\boldsymbol{\theta}^{(t+1)} \sim p(\boldsymbol{\theta}|\boldsymbol{\theta}^{(t)})$$

at time point $t > 0$ (also called iteration) starting with an initial value $\boldsymbol{\theta}^{(0)}$. The distribution $p(\cdot|\cdot)$ is called the *transition kernel* of the Markov chain. Note that the transition kernel, and hence $\boldsymbol{\theta}^{(t+1)}$, depends only on $\boldsymbol{\theta}^{(t)}$, not on its more distance past $\boldsymbol{\theta}^{(t-1)}, \ldots, \boldsymbol{\theta}^{(0)}$. This is a requirement for the chain $\boldsymbol{\theta}^{(t)}$ to be first order Markov. For example,

$$\boldsymbol{\theta}^{(t+1)} \sim N(\boldsymbol{\theta}^{(t)}, 1)$$

defines a Markov chain. We now discuss four most fundamental concepts in the theory of Markov chains which together guarantee the desired theoretical properties.

- **Stationarity** As $t \to \infty$, the Markov chain converges in distribution to its *stationary* distribution, $\pi(\boldsymbol{\theta}|\mathbf{y})$ say. This is also called its invariant distribution. This concept of convergence is the same as the mode of convergence we learned in the Central Limit Theorem. That is,

$$\lim_{t \to \infty} P\left(\boldsymbol{\theta}^{(t)} < \mathbf{a}\right) = \int_{-\infty}^{\mathbf{a}} \pi(\boldsymbol{\theta}|\mathbf{y})d\boldsymbol{\theta},$$

  for any vector of real values $\mathbf{a}$.

- **Irreducibility:** *Irreducible* means any set of values (more technically called states) for each of which $\pi(\boldsymbol{\theta}|\mathbf{y}) > 0$ can be reached from any other state in a finite number of transitions. This is a connectedness condition that rules out the possibility of getting trapped in any island state having positive probability under the target distribution $\pi(\boldsymbol{\theta}|\mathbf{y})$.

- **Aperiodicity:** A Markov chain taking only a finite number of values is *aperiodic* if the greatest common divisor (g.c.d.) of return times to any particular state i say, is 1. Think of recording the number of steps taken to return to the state 1. The g.c.d. of those numbers should be 1. If the g.c.d. is bigger than 1, 2 say, then the chain will return in cycles of 2, 4, 6, ... number of steps. This is not allowed for aperiodicity. This definition can be extended to general state space cases where the Markov chain can take any value in a continuous interval, see e.g. Tierney (1996).

- **Ergodicity:** Suppose that we have an aperiodic and irreducible Markov chain with stationary distribution $\pi(\boldsymbol{\theta}|\mathbf{y})$. Then we have an *ergodic* theorem:

$$\begin{aligned} \bar{b}_N &= \tfrac{1}{N}\sum_{j=1}^{N} b\left(\boldsymbol{\theta}^{(j)}\right) \\ &\to E_\pi[b(\boldsymbol{\theta})|\mathbf{y}] \text{ as } N \to \infty. \end{aligned}$$

Under certain mild conditions this convergence can be expected to happen geometrically fast, see e.g. Smith and Roberts (1993). If the convergence is geometric then a version of the CLT has also been proved. The CLT states that: provided $\mathrm{Var}(b(\boldsymbol{\theta})|\mathbf{y}) < \infty$, as $N \to \infty$:

$\bar{b}_N$ follows the normal distribution with mean $E[b(\boldsymbol{\theta})|\mathbf{y}]$ and a finite variance.

The expression for the limiting variance is complicated because of the dependence present in the chain $\left\{ b\left(\boldsymbol{\theta}^{(j)}\right)\right\}$. Several authors have provided expressions for the variance and its approximation, see e.g. Roberts (1996). For geometrically convergent Markov chains this variance decreases as N increases, and hence we can make the variance of $\bar{b}_N$ as small as we like by increasing N. We use the method of batching in Section 5.11.3 to estimate the variance.

5.7   Metropolis-Hastings algorithm

The theory of convergence sketched in the previous section invites real life applications where the theory can be used. The fundamental problem then is how can we construct a Markov chain which has the nice properties (stationarity, irreducibility, aperiodicity and ergodicity) and whose stationary distribution is the target posterior distribution $\pi(\boldsymbol{\theta}|\mathbf{y})$. Metropolis et al. (1953) were first to construct one such Markov chain and Hastings (1970) generalized their methods. The combined theoretical construction method is known as the Metropolis-Hastings algorithm described below. Let $q(\boldsymbol{\phi}|\boldsymbol{\theta})$ be easy to sample from probability density function of $\boldsymbol{\phi}$ given the parameter value $\boldsymbol{\theta}$. The notation $\boldsymbol{\phi}$ is of the same dimension and structure as the parameter notation $\boldsymbol{\theta}$.

1. Start the Markov chain at any value within the parameter space, $\boldsymbol{\theta}^{(0)}$ and say we have $\boldsymbol{\theta}^{(t)} = \boldsymbol{\theta}$ at the tth iteration.

2. Generate $\boldsymbol{\phi}$ from $q(\boldsymbol{\phi}|\boldsymbol{\theta})$. The sample value $\boldsymbol{\phi}$ is called a *candidate point* and $q(\cdot|\cdot)$ is called a *proposal distribution*.

3. Calculate the acceptance probability

$$\alpha(\boldsymbol{\theta}, \boldsymbol{\phi}) = \min\left\{ 1, \frac{\pi(\boldsymbol{\phi}|\mathbf{y})q(\boldsymbol{\theta}|\boldsymbol{\phi})}{\pi(\boldsymbol{\theta}|\mathbf{y})q(\boldsymbol{\phi}|\boldsymbol{\theta})} \right\}.$$

4. Independently draw a value u from the uniform $(0,1)$ random variable.

5. Let

$$\boldsymbol{\theta}^{(t+1)} = \begin{cases} \boldsymbol{\phi} & \text{if } u \leq \alpha(\boldsymbol{\theta}, \boldsymbol{\phi}) \\ \boldsymbol{\theta} & \text{otherwise.} \end{cases}$$

Several remarks and special cases are given below to understand the algorithm.

- There is no need to know the normalizing constant in the posterior distribution to implement this algorithm since the target density $\pi(\boldsymbol{\theta}|\mathbf{y})$ only enters through the ratio $\frac{\pi(\boldsymbol{\phi}|\mathbf{y})}{\pi(\boldsymbol{\theta}|\mathbf{y})}$.

- Suppose $q(\boldsymbol{\phi}|\boldsymbol{\theta}) = g(\boldsymbol{\phi})$ for a suitable importance sampling density $g(\boldsymbol{\phi})$. In this case, the proposal distribution does not depend on the current value $\boldsymbol{\theta}\left(=\boldsymbol{\theta}^{(t)}\right)$ and is fixed throughout the MCMC algorithm. In this case,

$$\alpha(\boldsymbol{\theta}, \boldsymbol{\phi}) = \min\left\{1, \frac{\pi(\boldsymbol{\phi}|\mathbf{y})q(\boldsymbol{\theta})}{\pi(\boldsymbol{\theta}|\mathbf{y})q(\boldsymbol{\phi})}\right\} = \min\left\{1, \frac{w(\boldsymbol{\phi}|\mathbf{y})}{w(\boldsymbol{\theta}|\mathbf{y})}\right\}, \qquad (5.3)$$

where $w(\boldsymbol{\theta}|\mathbf{y}) = \pi(\boldsymbol{\theta}|\mathbf{y})/q(\boldsymbol{\theta})$ is the importance sampling weight function defined above in (5.2). (Note that $\pi(\boldsymbol{\theta}|\mathbf{y}) \propto h(\boldsymbol{\theta})$ in our notation.) This version of the Metropolis-Hastings algorithm is called the *Independence Sampler*. In practice, the independence samplers are either very good or very bad depending on the level of agreement between the proposal and target densities. The importance sampling weight function must be bounded to achieve geometric convergence, i.e. the tail of the importance sampling density must dominate the tail of the target posterior density.

- Continue to assume that the proposal distribution $q(\boldsymbol{\phi}|\boldsymbol{\theta})$ does not depend on $\boldsymbol{\theta}$, the current point. In addition, assume that it is the posterior distribution itself, i.e. $q(\boldsymbol{\phi}|\boldsymbol{\theta}) = \pi(\boldsymbol{\phi}|\mathbf{y})$. This choice only helps us understand the algorithm better, and it is unrealistic as MCMC algorithms will not be required if we can draw samples from the posterior distribution itself in the first place.

  - In this case $\alpha(\boldsymbol{\theta}, \boldsymbol{\phi}) = 1$ and as a consequence the algorithm will accept all candidate points.

  - This makes sense since the proposal distribution itself is the target distribution. Hence the algorithm will not waste any valid sample drawn from the target distribution.

  - In fact, here the samples will be independent and identically distributed.

  - This choice for $\alpha(\boldsymbol{\theta}, \boldsymbol{\phi}) = 1$ helps to remember the above density ratio. If $q(\cdot) = \pi(\cdot)$ then the arguments, $\boldsymbol{\theta}$ and $\boldsymbol{\phi}$ are placed in such a way that the density ratio is 1.

- The original Metropolis algorithm is a very popular special case when the proposal distribution $q(\boldsymbol{\phi}|\boldsymbol{\theta})$ is symmetric in its arguments, i.e. $q(\boldsymbol{\phi}|\boldsymbol{\theta}) = q(\boldsymbol{\theta}|\boldsymbol{\phi})$. In this case, the proposal density ratio drops out of the acceptance probability and consequently,

$$\alpha(\boldsymbol{\theta}, \boldsymbol{\phi}) = \min\left\{1, \frac{\pi(\boldsymbol{\phi}|\mathbf{y})}{\pi(\boldsymbol{\theta}|\mathbf{y})}\right\}. \qquad (5.4)$$

- Continue to assume that the proposal distribution $q(\phi|\theta)$ is symmetric in its arguments. In addition, assume that $q(\phi|\theta) = q(|\phi - \theta|)$, i.e. the proposal distribution only draws an additive increment over and above the current value θ. This version of the algorithm is called the *Random-Walk* Metropolis algorithm. Here the proposal value is a random increment away from the current value.

- The acceptance rate, as measured by the proportion of accepted candidate points, impacts the speed of convergence of the underlying Markov chain to its stationary distribution. By tuning the scale (variance) of the proposal distribution we are able to control the proportion of candidate values that are accepted. If the scaling parameter is too small then we are more likely to propose small jumps. These will be accepted with high probability. However, many iterations will be needed to explore the entire parameter space. On the other hand, if the scale is too large then many of the proposed jumps will be to areas of low posterior density and will be rejected. This again will lead to slow exploration of the space.

- For the original Metropolis algorithm, if $\pi(\theta|\mathbf{y})$ can be factorized into iid components we have the asymptotic result that as $p \to \infty$ the optimal acceptance rate is 0.234 (Roberts et al., 1997; Gelman et al., 1996). For $p = 1$ Gelman et al. (1996) showed that the optimal acceptance rate is 0.44, and so for univariate components-wise updating algorithms we look to tune the proposal distribution to achieve this rate.

♡ **Example** 5.5. **Cauchy prior continued** For the independence sampler with the standard Cauchy distribution as the proposal distribution the acceptance probability 5.3 is calculated by obtaining the ratio of the likelihood function at the proposal sample and the current value. The independence sampler is run using the command:

```
BCauchy(method="independence", true.theta = 1, n=25, N=10000)
```

We obtain the estimated value 0.806 with a 95% credible interval given by (0.419, 1.226). Although this is a reasonable result, the acceptance rate is only about 12%, which is not very good for the independence sampler algorithm. This shows that we need to improve the proposal distribution. We, however, do not investigate any further and instead go on to illustrate the Metropolis algorithm.

To implement the random walk Metropolis algorithm we calculate the probability ratio in the acceptance probability (5.4) by first writing a routine to calculate the log of the posterior density at any value θ. The probability ratio is then calculated by exponentiating the difference of the log densities at the proposal value and at the current iteration value θ. The proposal value itself is generated from $N(\theta, \tau^2)$ where τ^2 is the tuning parameter of the algorithm. The command

> BCauchy(method="randomwalk", true.theta = 1, n=25, tuning.sd =1)

illustrates the method. With $N = 10,000$ iterations and $\tau^2 = 1$ we obtain 24.2% acceptance rate with the estimates 0.801 and (0.411, 1.193) which compare well with previous results. With $\tau^2 = 0.25$ we have 43.71% acceptance rate and the estimates are 0.802 and (0.414, 1.189). The point estimate here is the closest so far to the truth 0.809.

□

5.8 The Gibbs sampler

The Gibbs sampler is perhaps the most popular and widely used MCMC algorithm. It can be viewed as a special case of the general Metropolis-Hastings algorithm where the proposal distribution is taken as what is known as the full (or complete) conditional distribution. Let the parameter vector $\boldsymbol{\theta} = (\theta_1, \ldots, \theta_p)$ be $p > 1$ dimensional. The full conditional distribution of a component θ_k for any k is the conditional distribution of θ_k given the particular values of all other θ's: θ_i for $i = 1, \ldots, k-1, k+1, \ldots p$. Suppressing dependence of the posterior distribution on observed data $\mathbf{y}$, we write $\pi(\boldsymbol{\theta})$ to be the posterior distribution (not the prior distribution). The full conditional distribution of θ_k is written as

$$\pi(\theta_k | \theta_1, \ldots, \theta_{k-1}, \theta_{k+1}, \ldots, \theta_p), \ k = 1, \ldots, p.$$

These distributions are similarly defined if θ_k is a block of more than one parameter. The Gibbs sampler simulates from these full conditional distributions, in turn, to complete one iteration of the algorithm. Thus, it simplifies sampling from a lower dimensional full conditional distribution of a block of parameters. The Gibbs sampler accepts all the samples from the full conditional distributions as it can be shown to be a special case of the original Metropolis-Hastings algorithm with acceptance probability one.

To draw samples from a target posterior distribution $\pi(\boldsymbol{\theta})$ the Gibbs sampler makes a Markov transition from $\boldsymbol{\theta}^{(t)}$ to $\boldsymbol{\theta}^{(t+1)}$ as follows.

$$
\begin{aligned}
\theta_1^{(t+1)} &\sim \pi(\theta_1 | \theta_2^{(t)}, \theta_3^{(t)}, \cdots, \theta_p^{(t)}) \\
\theta_2^{(t+1)} &\sim \pi(\theta_2 | \theta_1^{(t+1)}, \theta_3^{(t)}, \cdots, \theta_p^{(t)}) \\
&\vdots \quad \vdots \quad \vdots \\
\theta_p^{(t+1)} &\sim \pi(\theta_p | \theta_1^{(t+1)}, \theta_2^{(t+1)}, \cdots, \theta_{p-1}^{(t+1)}).
\end{aligned}
$$

Note that always the most up-to-date version of $\boldsymbol{\theta}$ is used, i.e. to sample $\theta_2^{(t+1)}$

the already sampled value $\theta_1^{(t+1)}$ is used (and not $\theta_1^{(t)}$). Figure 5.2 illustrates how the Gibbs sampler works for a two-dimensional parameter space.

♡ **Example 5.6.** We now return to the normal distribution example with both parameters unknown in Section 4.14. The joint posterior distribution of θ and λ^2 has been given in (4.14). From the un-normalized joint posterior distribution we have already identified the full conditional distribution of $\lambda^2|\theta, \mathbf{y}$ in (4.15) and $\theta|\lambda^2, \mathbf{y}$ in (4.18).

Starting with a reasonable value of $\lambda^2 > 0$, the Gibbs sampler for this example samples θ from the normal distribution (4.18) by plugging in the current value of λ^2 and then it draws a sample value for λ^2 from the gamma distribution (4.15) by plugging in the most recent sampled value of θ. Thus, we gather the samples $\theta^{(1)}, \sigma^{2(1)}, \theta^{(2)}, \sigma^{2(2)} \dots, \theta^{(N)}, \sigma^{2(N)}$ for a large value of N where $\sigma^{2(j)} = 1/\lambda^{2(j)}$ for $j = 1, \dots, N$.

For the New York air pollution data example after running $N = 10,000$ iterations we obtain the parameter estimates in Table 5.1 corresponding to the exact values we obtained in Example 4.14. Notice that we have estimated the coefficient of variation $E((\frac{\sigma}{\theta}|\mathbf{y})$ easily for which we did not have the exact results. Table 5.1 can be reproduced by issuing the commands provided in the help file for the `bmstdr` command `Bnormal`. Here it is assumed that $k = 1$, $M = 1$ and the prior for $1/\sigma^2$ is the gamma distribution $G(2, 1)$.

Parameter	$E(\theta|\mathbf{y})$	$\text{Var}(\theta|\mathbf{y})$	$E(\sigma^2|\mathbf{y})$	$\text{Var}(\sigma^2|\mathbf{y})$	$E\left(\frac{\sigma}{\theta}|\mathbf{y}\right)$
Exact	47.906	0.686	19.901	28.289	
Estimate	47.909	0.688	19.936	27.937	0.092
95% (lower)	46.290	–	12.099	–	0.072
95% (upper)	49.551	–	32.747	–	0.120

TABLE 5.1: Gibbs sampler estimates for the $N(\theta, \sigma^2)$ model for the New York air pollution data. The hyper parameter values are $k = 1, m = 1, a = 2$, and $b = 1$.

□

In the above example the full conditional distributions are easy to sample from because they are standard. Alternative strategies are used when some of the complete conditionals are non-standard. When they are *log-concave*, i.e., second derivative of the log density is strictly decreasing, the *adaptive rejection sampling* developed by Gilks and Wild (1992) can be used. Otherwise, a Metropolis-Hastings step can be used to sample from any full conditional, which is non-standard.

There are many other issues regarding the implementation of the Gibbs sampler for practical problems as studied by Roberts and Sahu (1997). The order of sampling has some influence on the convergence depending on the cross-correlation structure of the components of $\boldsymbol{\theta}$. Blocking, i.e. grouping parameters together and then sampling of the multivariate full conditional distributions hastens convergence. Moreover, parameterisations of the Bayesian model has a large influence. Faster convergence is achieved for models for which the joint posterior distribution has less cross-correlation, see e.g. Gelfand et al. (1995a).

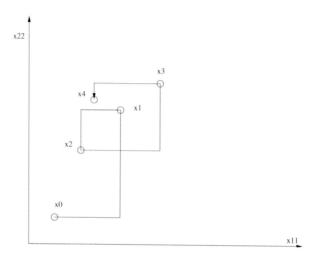

FIGURE 5.2: Diagram showing how the Gibbs sampler works.

5.9 Hamiltonian Monte Carlo

The Bayesian computation methods described so far run into problems for exploring complex problems because those methods are too general in nature. Those methods do not usually exploit any features of the posterior distribution that may help guide them into targeted learning about the peaks, troughs and steepness of the posterior distribution. Hamiltonian Monte Carlo (HMC) (Neal, 2011) uses local posterior geometry to explore high posterior probability areas but at the same time uses a Metropolis-Hastings step to correct the over sampling of the peaks ignoring low probability areas completely. The algorithm does so by general random directions for proposed moves. There are many excellent texts available to understand HMC, see e.g. Chapter 15 of Lambert (2018) where a very non-technical discussion covers a lot of aspects of HMC. In the remainder of this section we simply describe the algorithm to aid a quick understanding of HMC. This discussion is not sufficient to code the method from scratch – a task which we do not require to perform since the method has been coded in the software package STAN (Stan Development Team, 2015) that we will invoke using the R-language interface to fit Bayesian models.

The key term to understand in HMC is the Hamiltonian, borrowed from Physics, which stands for total energy in a physical system. The total energy in a system is the simple sum of potential energy at a particular point $\boldsymbol{\theta}$, and its kinetic energy which depends on the momentum m. (This m is not to be confused with the prior sample size m for the specific normal-normal example discussed so far. We only require to use the momentum m in this Section only for discussing theory.)

Assume that $\boldsymbol{\theta}$ is one-dimensional, so that we simply write θ instead of $\boldsymbol{\theta}$ for simplicity in the following discussion. The method seamlessly generalize to higher dimensional cases. In HMC the potential energy is the negative log-posterior density, which is a function of the parameter θ. The kinetic energy is proportional to the square of the momentum m – an everyday fact in Physics. Here we take the kinetic energy to be $\frac{1}{2}m^2$ bearing in mind the log of the standard normal density $\left(-\frac{1}{2}\log(2\pi) - \frac{1}{2}m^2\right)$ from where the proposal values will come from in a random walk type Metropolis algorithm. Thus, the Hamiltonian in HMC is:

$$H(\theta, m) = -\log\left[\pi\left(\theta|\mathbf{y}\right)\right] + \frac{m^2}{2}.$$

HMC samples from the joint distribution of θ and m having the pdf

$$\pi(\theta, m|\mathbf{y}) \propto \exp\left[-H(\theta, m)\right] \tag{5.5}$$

which is known as the canonical distribution corresponding temperature $T = 1$, see Neal (2011).

Each iteration of HMC is performed in two steps. In the first step a new value of the momentum m is drawn from the standard normal distribution independently of the current value of the location θ. The pair θ, m constitutes the current value of the Markov chain. In the second step a Metropolis update is performed, see discussion below (5.4), as follows. The proposal θ^*, m^* is generated by simulating a number, L, of leapfrog steps following Hamiltonian dynamics by preserving the total volume. Details regarding the number L, an associated step size (often denoted by ϵ) and the dynamics that take the current pair θ, m to θ^*, m^* are provided in Neal (2011) and omitted from here. Buried in these details is an algorithm called the no u-turn sampler (NUTS) due to Hoffman and Gelman (2014) that adapts these parameters to the peculiarities of the posterior distribution. The NUTS has been implemented in STAN and we will keep it under the hood so as not to get into more technical details. Hence, the HMC for a vector parameter $\boldsymbol{\theta}$ is performed by following the steps (Lambert, 2018):

1. Select a random starting location $\boldsymbol{\theta}^{(0)}$ from some initial proposal distribution.

2. At iteration t generate a random initial momentum from a proposal distribution (for example $\mathbf{m} \sim N(\boldsymbol{\mu}, \Sigma)$).

3. Advance the current point $\boldsymbol{\theta}, \mathbf{m}$ to $\boldsymbol{\theta}^*, \mathbf{m}^*$ by following L steps of the leapfrog algorithm as implemented by NUTS. The $\boldsymbol{\theta}^*, \mathbf{m}^*$ pair records the position of the parameter and its momentum after the L leapfrog steps.

4. Calculate the Metropolis acceptance probability $\alpha(\boldsymbol{\theta}, \boldsymbol{\phi})$ as given in (5.4), for the proposal $\boldsymbol{\phi} = (\boldsymbol{\theta}^*, \mathbf{m}^*)$ and current point $\boldsymbol{\theta} = (\boldsymbol{\theta}, \mathbf{m})$ and the target density (5.5) $\pi(\boldsymbol{\theta}, \mathbf{m}|\mathbf{y})$ where

$$\log \pi(\boldsymbol{\theta}, \mathbf{m}|\mathbf{y}) \propto \log(\pi(\boldsymbol{\theta}|\mathbf{y})) + \frac{1}{2}\mathbf{m}'\Sigma^{-1}\mathbf{m}.$$

5. Generate $u \sim U(0, 1)$ and set $\boldsymbol{\theta}^{(t+1)} = \boldsymbol{\theta}^*$ if $u < \alpha(\boldsymbol{\theta}, \boldsymbol{\phi})$ else set $\boldsymbol{\theta}^{(t+1)} = \boldsymbol{\theta}$.

In practical applications the tuning parameters L and the step size ϵ need to be adjusted for optimal performance. If L is too short the algorithm does not move far enough but a large L may be bad as well since the momentum of the sampling and U-shape curvature of the negative log-posterior density may return the sampling point where it was! Also the L leapfrog steps approximate the path of the Hamiltonian dynamics along the negative log posterior density space which may diverge from the true path in challenging posterior distributions having very steep peaks and troughs. These are implementation problems and must be tackled on a case by case basis. Otherwise, the HMC explores many practical posterior distributions much more effectively than other comparable algorithms such as the Metropolis-Hastings algorithm.

♡ **Example 5.7.** We again return to the earlier normal distribution example 5.8 applied to the New York air pollution data. The estimates from the STAN package, obtained by issuing the `bmstdr` command:

```
Bnormal(package="stan", kprior=1, prior.M =1, prior.sigma=c(2, 1))
```

are provided in Table 5.2. Notice that HMC produces near identical estimates as before.

□

Parameter	$E(\theta|\mathbf{y})$	$\text{Var}(\theta|\mathbf{y})$	$E(\sigma^2|\mathbf{y})$	$\text{Var}(\sigma^2|\mathbf{y})$	$E\left(\frac{\sigma}{\theta}|\mathbf{y}\right)$
Exact	47.906	0.686	19.901	28.289	
Estimate	47.915	0.678	19.779	27.726	0.092
95% (lower)	46.297	–	12.076		0.072
95% (upper)	49.558	–	32.447	–	0.119

TABLE 5.2: HMC estimates for the $N(\theta, \sigma^2)$ model for the New York air pollution data. The hyper parameter values are $k = 1, m = 1, a = 2$, and $b = 1$.

5.10 Integrated nested Laplace approximation (INLA)

Bayesian computation methods based on INLA are hugely popular due to the availability of user friendly software package R-INLA written in the R language. Methods based on INLA are alternative to the MCMC and provide fast computation for many modeling problems with latent Gaussian components. By being non-iterative it is faster than MCMC methods. Easy coding and fast computation led to its popularity and there are now many text books entirely based on using the R-INLA package for computation, see e.g. Blangiardo and Cameletti (2015). This section discusses the main ideas in INLA and numerical illustrations will follow in the subsequent section.

In the likelihood times prior Bayesian modeling setup the observations, Y_i for $i = 1, \ldots, n$ are given a distribution conditional on latent Gaussian random variables w_i. In other words the mean of the data Y_i is written as a function of component parameters in $\boldsymbol{\theta}$ and random effects w_i. The random effects w_i are assigned a Gaussian distribution whose parameters are included as separate components of $\boldsymbol{\theta}$. As before, joint prior distribution $\pi(\boldsymbol{\theta})$ is assumed for the parameter vector $\boldsymbol{\theta}$. The resulting Bayesian model treats the random effects w_i for $i = 1, \ldots, n$ as unknown parameters and sets out to estimate those along with the parameters $\boldsymbol{\theta}$ from the joint posterior distribution:

$$\pi(\boldsymbol{\theta}, \mathbf{w}|\mathbf{y}) \propto f(\mathbf{y}|\boldsymbol{\theta}, \mathbf{w})\,\pi(\mathbf{w}|\boldsymbol{\theta})\pi(\boldsymbol{\theta}).$$

The density terms $f(\mathbf{y}|\boldsymbol{\theta}, \mathbf{w})$ and $\pi(\mathbf{w}|\boldsymbol{\theta})$ often factorize as product of individual densities of Y_i and w_i but they do need not be for the approximation methods to work. This is often true in the case of modeling dependent spatio-temporal data. The dimensions of $\mathbf{y}$ and $\mathbf{w}$ can be high (e.g. 100 to 100,000) in moderately sized problems but often the dimension of $\boldsymbol{\theta}$ is small in low tens. INLA methods are most suited for such problems. Inferential interest lies in estimating

$$\pi(w_i|\mathbf{y}) = \int_{-\infty}^{\infty} \pi(w_i|\boldsymbol{\theta}, \mathbf{y})\pi(\boldsymbol{\theta}|\mathbf{y})d\boldsymbol{\theta}$$

and

$$\pi(\theta_k|\mathbf{y}) = \int_{-\infty}^{\infty} \pi(\boldsymbol{\theta}|\mathbf{y})d\boldsymbol{\theta}_{(-k)}$$

where $\boldsymbol{\theta}_{(-k)}$ is the $p-1$ dimensional parameter vector obtained from $\boldsymbol{\theta}$ after removing the kth component.

The nestedness in the approximations is observed in the pair of approximations:

$$\tilde{\pi}(w_i|\mathbf{y}) = \int_{-\infty}^{\infty} \tilde{\pi}(w_i|\boldsymbol{\theta}, \mathbf{y})\tilde{\pi}(\boldsymbol{\theta}|\mathbf{y})d\boldsymbol{\theta}, \quad \tilde{\pi}(\theta_k|\mathbf{y}) = \int_{-\infty}^{\infty} \tilde{\pi}(\boldsymbol{\theta}|\mathbf{y})d\boldsymbol{\theta}_{(-k)}, \quad (5.6)$$

where the generic notation $\tilde{\pi}(\cdot|\cdot)$ is used to denote an approximate conditional density of its arguments. Here $\tilde{\pi}(w_i|\boldsymbol{\theta}, \mathbf{y})$ is often a Gaussian approximation of the conditional density. This is exactly Gaussian when the random effects are assumed to be Gaussian under a top-level Gaussian model. The integrations with respect to $\boldsymbol{\theta}$ and $\boldsymbol{\theta}_{(-k)}$ are performed using numerical integration methods. The key approximation comes from:

$$\begin{aligned}
\tilde{\pi}(\boldsymbol{\theta}|\mathbf{y}) &= \frac{\pi(\mathbf{w}, \boldsymbol{\theta}|\mathbf{y})}{\pi(\mathbf{w}|\boldsymbol{\theta}, \mathbf{y})} \\
&\propto \frac{\pi(\mathbf{w}, \boldsymbol{\theta}, \mathbf{y})}{\pi(\mathbf{w}|\boldsymbol{\theta}, \mathbf{y})} \\
&\approx \frac{\pi(\mathbf{w}, \boldsymbol{\theta}, \mathbf{y})}{\pi_G(\mathbf{w}|\boldsymbol{\theta}, \mathbf{y})}\bigg|_{\mathbf{w}=\mathbf{w}^*(\boldsymbol{\theta})}
\end{aligned}$$

where $\pi_G(\mathbf{w}|\boldsymbol{\theta}, \mathbf{y})$ is the Gaussian approximation of $\pi(\mathbf{w}|\boldsymbol{\theta}, \mathbf{y})$ and $\mathbf{w}^*(\boldsymbol{\theta})$ is the mode of the full conditional distribution for a given $\boldsymbol{\theta}$.

As noted above $\pi(\mathbf{w}|\boldsymbol{\theta}, \mathbf{y})$ is almost Gaussian since $\mathbf{x}$ is often a-priori Gaussian and data $\mathbf{y}$ are not very informative. However, $\tilde{\pi}(\boldsymbol{\theta}|\mathbf{y})$ tends to depart significantly from a Gaussian distribution. Section 3 of the main INLA article by Rue et al. (2009) suggests a number of remedies to better the approximation. They suggest to locate the mode of $\tilde{\pi}(\boldsymbol{\theta}|\mathbf{y})$ by grid search and Hessian calculation. They also recommend using a cubic spline to help approximate $\pi(w_i|\boldsymbol{\theta}, \mathbf{y})$. Here a flexible skew-normal density (Sahu et al., 2003) can be adopted instead. No further details are provided here regarding the approximation methods adopted in R-INLA. The implemented methods provide Monte Carlo samples from the approximate marginal posterior distributions. These approximate samples are used in the exact same manner as the MCMC samples for performing Monte Carlo integration for inferential interests. The methods are illustrated with a numerical example below.

♡ **Example** 5.8. We illustrate INLA with the running normal distribution example 5.8. It is not possible to fit the hierarchical prior model, $\pi(\theta|\lambda^2)\pi(\lambda^2)$ directly in INLA, although there are possible work around parameterisations which can only be implemented with very advanced knowledge and techniques. That is why, we fit the model with independent prior distributions for θ and λ^2. In particular, we assume that a-priori $\theta \sim N\left(\mu, s_y^2/m\right)$ where $\mu = \bar{y} + k\frac{s_y}{\sqrt{n}}$

for $k = 1$ and $m = 1$. Further, we assume that $\lambda^2 \sim G(2,1)$. With these choices we implement INLA and obtain the parameter estimates, reported in Table 5.3, using 1000 random draws from the marginal posterior distribution approximated by INLA. The relevant `bmstdr` command is:

```
Bnormal(package="inla", kprior=1, prior.M =1, prior.sigma=c(2, 1))
```

Notice that all the estimates in Table 5.3 are close to the corresponding ones in Table 5.2 except for the $\text{Var}(\theta|\mathbf{y})$. The INLA estimate is much smaller than the exact and HMC estimates. This difference, perhaps, can be explained by the fact that the posterior distributions are different due to the differences in the joint prior distribution for θ and λ^2.

Parameter	$E(\theta|\mathbf{y})$	$\text{Var}(\theta|\mathbf{y})$	$E(\sigma^2|\mathbf{y})$	$\text{Var}(\sigma^2|\mathbf{y})$	$E\left(\frac{\sigma}{\theta}|\mathbf{y}\right)$
Exact	47.906	0.686	19.901	28.289	
Estimate	48.700	0.043	20.665	29.313	0.093
95% (lower)	48.292	–	12.574	–	0.073
95% (upper)	49.114	–	33.528	–	0.119

TABLE 5.3: INLA estimates for the $N(\theta, \sigma^2)$ model for the New York air pollution data. The hyper parameter values are $k = 1, m = 1, a = 2$, and $b = 1$.

□

5.11 MCMC implementation issues and MCMC output processing

Theory suggests that if we run the Markov chain long enough, it will converge to the stationary distribution. However, sometimes there can be slow convergence and inclusion of early non-converged iterations which do not follow the desired target distribution bias the MCMC averages for posterior inference. To reduce bias the early iterations, called *burn-in* or *warm-up* are discarded.

After we have reached the stationary distribution, how many iterations will it take to summarize the posterior distribution? Methods for determining an optimal *burn-in* and deciding on the run-length are called convergence diagnostics. These methods also diagnose problems in MCMC convergence. Below we describe a few essential convergence diagnostics methods. However, note that even if the Markov chain gives good convergence diagnostics we cannot make the claim that we have *proved* that it has converged. Rather these diagnostics are often studied to discover potential problems in MCMC convergence.

5.11.1   Diagnostics based on visual plots and autocorrelation

Here we draw a time series plot of the different components of the Markov chain. We check to see if the simulated chain remained in a region heavily influenced by the starting distribution for many iterations. If the Markov chain has converged, the trace plot will move like a 'hairy caterpillar' around the mode of the posterior distribution. Moreover, the plots should not show any trend.

To see the effect of initial values, we rerun the Markov chain with different starting points and overlay the time series of the same parameter component. The replicate plots should criss-cross each other often. If these plots do not coalesce, we should investigate whether there is multi-modality or some other problem which may question irreducibility of the underlying Markov chain. Figure 4.7 on page 74 in the book Lunn et al. (2013) provides many examples of visual convergence and non-convergence behavior of MCMC trace-plots.

The problem with trace plots is that it may appear that we have converged, but in reality the chain may have been trapped (for a long time) in a local region (not exploring the full posterior space). Hence, we also calculate autocorrelations present in the chain for different values of the lag and see how quickly these die down as a function of the lag. For a rapidly converging (or equivalently mixing) Markov chain the autocorrelations should go down to zero at a very fast rate. When this dampening does not occur, then we have a problem and will probably want to re-parameterize the model or fit a less complex model.

Markov chains that exhibit high autocorrelation will mix more slowly and hence take longer to converge. The autocorrelation plot is sometimes used to determine the *thinning interval*. The idea is to try and achieve close to independent sample values by retaining every kth value where the autocorrelation at lag k falls below some tolerance level. Although this may seem reasonable, MacEachern and Berliner (1994) show that by throwing away information the variance of the mean of the samples can only be increased. It is far better to use the concept of an *effective sample size* or ESS (Robert and Casella, 2004, Chapter 12). The ESS is computed by dividing the number of post burn-in samples N by an estimate of the autocorrelation time κ, where

$$\kappa = 1 + 2 \sum_{k=1}^{\infty} \rho(k),$$

and $\rho(k)$ is the theoretical autocorrelation of the Markov chain at lag k. Thus

$$\text{ESS} = \frac{N}{\kappa}.$$

We can estimate κ (and hence ESS) by using the sample autocorrelations of the chain and truncate the infinite sum when the autocorrelation falls below some threshold. This may lead to a biased estimate for κ (Banerjee et al., 2015, Chapter 4).

Another method of estimating κ is to estimate the spectral density at frequency zero. If we consider scalar quantities such that $\bar{b}_N = N^{-1} \sum_{t=1}^{N} b(\boldsymbol{\theta}^{(t)})$, we have from Ripley (1987, Chapter 6) that

$$NVar(\bar{b}_N) \to v^2 \kappa = 2\pi f(0),$$

where v^2 is the variance of $b(\boldsymbol{\theta})$ and $f(0)$ is the spectral density of the chain $\{b(\boldsymbol{\theta}^{(t)})\}_{t=1}^{N}$. Hence $\kappa = 2\pi f(0)/v^2$ and

$$\text{ESS} = \frac{N}{\kappa} = \frac{Nv^2}{2\pi f(0)}$$

for large values of N. This method, and the others discussed below, have been coded in the R package CODA (Plummer et al., 2006) to estimate the ESS. The R function `effectiveSize` inside the CODA package evaluates this.

5.11.2 How many chains?

Some discussion is required to decide how many parallel chains, each starting at a different initial value, should be run for making inference. There are two schools of thought advocating the use of one long chain and several parallel chains. The first strategy argues that one long chain reaches parts of the parameter space the other scheme with shorter length chains cannot, see e.g. Geyer (1992) and Raftery and Lewis (1992). Gelman and Rubin (1992) on the other hand advocate running of several long chains. These chains give indication of convergence as they show mixing with each other.

Single chain diagnostics: Raftery and Lewis (1992) consider the problem of calculating the number of iterations necessary to estimate a posterior quantile from a single run of a Markov chain. They propose a 2-state Markov chain model fitting procedure based upon pilot analysis of output from the original chain.

Suppose that we wish to estimate a particular quantile of the posterior distribution for some function $b(\boldsymbol{\theta})$ of a set of parameters $\boldsymbol{\theta}$, i.e., we wish to estimate u such that

$$P\left(b(\boldsymbol{\theta}) \leq u|\mathbf{y}\right) = q$$

for some pre-specified q and so that, given our estimate $\hat{u}$, $P\left(b(\boldsymbol{\theta}) \leq \hat{u}|\mathbf{y}\right)$ lies within $\pm r$ of the true value, say, with probability p.

Raftery and Lewis propose a method to calculate n_0, the initial number of iterations to discard (which we call the 'burn-in'), k, the thinning size, i.e., every k^{th} iterate of the Markov chain and n the number of further iterations required to estimate the above probability to within the required accuracy.

The details of the methodology can be found on the cited reference. Basically the methodology construct a discrete two state Markov chain

$$Z^{(t)} = I\left(b^{(t)} = b(\boldsymbol{\theta}^{(t)}) \leq u\right)$$

from $\boldsymbol{\theta}^{(t)}$ where $I(\cdot)$ is the indicator function of its argument taking the value 1 or 0. This two state Markov chain is analyses theoretically to find the answers. These authors also suggest looking at a quantity called the dependence factor:

$$I = \frac{n_0 + n}{n_{min}}$$

where n_{min} is the number of initial iterations performed to produce the estimates. Ideally, this factor should be close to 1. The **R** function `raftery.diag` evaluates the dependence factor I.

Multi chain diagnostics: Gelman and Rubin (1992) propose a method which assesses convergence by monitoring the ratio of two variance estimates. In particular, their method uses multiple replications and is based upon a comparison, for scalar functions of $\boldsymbol{\theta}$, of the within sample variance for each of m parallel chains, and the between sample variance of different chains. The method is based on essentially a classical analysis of variance.

The method consists of analyzing the independent sequences to form a distributional estimate for what is known about the target random variable, given the observations simulated so far. This distributional estimate, based upon the *Student's t* distribution, is somewhere between the starting and target distributions, and provides a basis for an estimate of how close the process is to convergence and, in particular, how much we might expect the estimate to improve with more simulations.

The method proceeds as follows. We begin by independently simulating $m \geq 2$ sequences of length N, each beginning at a different starting point which are over dispersed with respect to the stationary distribution. Let $\boldsymbol{\theta}_i^{(j)}$ for $i = 1, \ldots, m$ and $j = 1, \ldots, N$ denote these simulations. From these we calculate $b_{ij} = b\left(\boldsymbol{\theta}_i^{(j)}\right)$ for each scalar functional of interest $b(\boldsymbol{\theta})$. Using the b_{ij}'s, we calculate V/N, the variance between the m sequence means, which we denote by $\bar{b}_{i\cdot}$. Thus, we define:

$$V = \frac{N}{m-1} \sum_{i=1}^{m} (\bar{b}_{i\cdot} - \bar{b}_{\cdot\cdot})^2,$$

where

$$\bar{b}_{i\cdot} = \frac{1}{N} \sum_{j=1}^{N} b_{ij}, \quad \bar{b}_{\cdot\cdot} = \frac{1}{m} \sum_{i=1}^{m} \bar{b}_{i\cdot}.$$

We then calculate W, the mean of the m within-sequence variances, s_i^2. Thus, W is given by

$$W = \frac{1}{m} \sum_{i=1}^{m} s_i^2,$$

where

$$s_i^2 = \frac{1}{N-1} \sum_{j=1}^{N} (b_{ij} - \bar{b}_{i\cdot})^2.$$

The between sequence variance V contains a factor of N because it is based on the variance of the within-sequence means, $\bar{b}_{i.}$, each of which is an average of N values b_{ij}.

We can then estimate the target posterior variance $\sigma_b^2 = \text{Var}(b(\boldsymbol{\theta})|\mathbf{y})$ by a weighted average of V and W, given by

$$\hat{\sigma}_b^2 = \frac{N-1}{N}W + \frac{1}{N}V,$$

which overestimates the true value under the assumption that the starting points for the m chains are over-dispersed. On the other hand, for any finite N, the within-sequence variance W should underestimate σ_b^2 because the individual sequences have not had time to explore the whole of the target distribution and, as a result will have less variability. In the limit as $N \to \infty$ both $\hat{\sigma}_b^2$ and W approach σ_b^2, but from opposite directions.

Convergence is now monitored by estimating the factor by which the larger estimate $\hat{\sigma}_b^2$ might be reduced by calculating the potential scale reduction factor

$$\hat{R} = \frac{\hat{\sigma}_b^2}{W},$$

which is the ratio of the estimated upper and lower bounds for σ_b^2. As $N \to \infty$ the Markov chains converge and $\hat{R}$ should shrink to 1. Values of $\hat{R}$ much greater than 1 indicate a failure to converge, with values less than 1.1 or 1.2 considered low enough to be satisfied that convergence has been achieved, see e.g. (Gilks et al., 1996, Chapter 8). There are other methods and justifications for calculating $\hat{R}$, see e.g. Section 5.3.4 of Banerjee et al. (2015). The R function `gelman.diag` evaluates this diagnostics.

5.11.3 Method of batching

We now return to the problem of estimating the Monte Carlo error of the MCMC averages discussed in Section 5.6. As before let $b(\boldsymbol{\theta})$ be the scalar parameter of interest. For example, $b(\boldsymbol{\theta})$ can be taken as the coefficient of variation $\frac{\sigma}{\mu}$ for the second motivating example of this chapter. Suppose that we have decided to discard first M burn-in iterations and would like to estimate $E(b(\boldsymbol{\theta})|\mathbf{y})$ by the delayed average

$$\bar{b}_{MN} = \frac{1}{N-M} \sum_{j=M+1}^{N} b\left(\boldsymbol{\theta}^{(j)}\right).$$

instead of the $\bar{b}_N$ introduced before.

The problem of auto-correlation in the sequence $b\left(\boldsymbol{\theta}^{(j)}\right)$ is overcome by dividing the sequence

$$b\left(\boldsymbol{\theta}^{(M+1)}\right), \ldots, b\left(\boldsymbol{\theta}^{(N)}\right)$$

into k equal length batches each of size $r = \frac{N-M}{k}$, assuming that $N - M$ is divisible by k (if not we can slightly adjust N and or M), where k is a suitable number greater than or equal to 20. We then calculate the batch mean,

$$b_i = \frac{1}{r} \sum_{j=1}^{r} b \left(\boldsymbol{\theta}^{(M+r(i-1)+j)} \right), \ i = 1, \ldots, k.$$

We perform diagnostics checks on the sequence $b_1, \ldots, b_k$ to see if these are approximately uncorrelated. We estimate lag-1 autocorrelation of the sequence $\{b_i\}$. If this auto-correlation is high then a longer run of the Markov chain should be used, giving larger batches. Now the Monte Carlo error (also called the numerical standard error, nse) is estimated by:

$$\widehat{\mathrm{nse}} \left(\bar{b}_{MN} \right) = \sqrt{ \frac{1}{k(k-1)} \sum_{i=1}^{k} (b_i - \bar{b})^2 }.$$

The CODA package has implemented this in their function `batchSE`.

5.12 Computing Bayesian model choice criteria

The Bayesian model choice criteria, namely DIC, WAIC and PMCC, discussed in Section 4.16 are hard to evaluate analytically except for simple models. In this section we discuss computation of each of those using samples $\boldsymbol{\theta}^{(j)}$ for $j = 1, \ldots, J$ from the posterior distribution $\pi(\boldsymbol{\theta}|\mathbf{y})$. Note that the parameters $\boldsymbol{\theta}$ here also include any spatio-temporal (or independent) random effects that may have been sampled along with all other fixed (not varying with data indices) parameters in the model.

5.12.1 Computing DIC

The DIC, as defined in (4.22), is

$$\mathrm{DIC} = -2 \log \left\{ f \left(\mathbf{y} | \hat{\boldsymbol{\theta}}_{\mathrm{Bayes}} \right) \right\} + 2\, p_{\mathrm{DIC}}.$$

where

$$p_{\mathrm{DIC}} = 2 \left[\log \left\{ f \left(\mathbf{y} | \hat{\boldsymbol{\theta}}_{\mathrm{Bayes}} \right) \right\} - E_{\boldsymbol{\theta}|\mathbf{y}} \left\{ \log \left(f(\mathbf{y}|\boldsymbol{\theta}) \right) \right\} \right].$$

To compute the DIC we first obtain

$$\hat{\boldsymbol{\theta}}_{\mathrm{Bayes}} = \frac{1}{B} \sum_{j=1}^{N} \boldsymbol{\theta}^{(j)}$$

as the estimate of the mean of the posterior distribution $\pi(\boldsymbol{\theta}|\mathbf{y})$. Similar sample averages are formed to estimate the term. We first calculate

$$b\left(\boldsymbol{\theta}^{(j)}\right) = \log\left(f(\mathbf{y}|\boldsymbol{\theta}^{(j)})\right)$$

for $j = 1, \ldots, N$ where $f(\mathbf{y}|\boldsymbol{\theta})$ is the log-likelihood function. That is, we simply evaluate the log-likelihood function at each sampled parameter point $\boldsymbol{\theta}^{(j)}$ and calculate the average

$$\widehat{E}_{\theta|\mathbf{y}}\left\{\log\left(f(\mathbf{y}|\boldsymbol{\theta})\right)\right\} = \bar{b} \equiv \frac{1}{N}\sum_{j=1}^{N} b\left(\boldsymbol{\theta}^{(j)}\right).$$

The alternative penalty $p_{\text{DIC alt}}$, as suggested by Gelman et al. (2014),

$$p_{\text{DIC alt}} = 2\,\text{Var}_{\theta|\mathbf{y}}\left[\log\{f(\mathbf{y}|\boldsymbol{\theta})\}\right]$$

is estimated by taking the twice of the sample variance of $b\left(\boldsymbol{\theta}^{(j)}\right), j = 1, \ldots, N$. That is,

$$\hat{p}_{\text{DIC alt}} = 2\,\frac{1}{N-1}\sum_{j=1}^{N}\left\{b\left(\boldsymbol{\theta}^{(j)}\right) - \bar{b}\right\}^2,$$

where $\bar{b}$ is given above.

5.12.2 Computing WAIC

The WAIC has been defined in Section 4.16.2 as

$$\text{WAIC} = -2\sum_{i=1}^{n}\log\left\{f(y_i|\mathbf{y})\right\} + 2\,p_{\text{waic}}$$

with two forms of p_{waic}:

$$\begin{array}{rl} p_{\text{waic 1}} &= 2\sum_{i=1}^{n}\left[\log\left\{f(y_i|\mathbf{y})\right\} - E_{\theta|\mathbf{y}}\left\{\log\left(f(y_i|\boldsymbol{\theta})\right)\right\}\right], \\ p_{\text{waic 2}} &= \sum_{i=1}^{n}\text{Var}_{\theta|\mathbf{y}}\left[\log\{f(y_i|\boldsymbol{\theta})\}\right]. \end{array}$$

Using the posterior samples $\boldsymbol{\theta}^{(j)}$ for $j = 1, \ldots, N$ we first estimate

$$\widehat{f}(y_i|\mathbf{y}) = \frac{1}{N}\sum_{j=1}^{N} f(y_i|\boldsymbol{\theta}^{(j)}, \mathbf{y}), i = 1, \ldots, n$$

to perform a Monte Carlo integration of (4.8). Now let

$$b_i\left(\boldsymbol{\theta}^{(j)}\right) = \log\{f(y_i|\boldsymbol{\theta}^{(j)})\}.$$

We now estimate

$$\widehat{E}_{\theta|\mathbf{y}}\left\{\log\left(f(y_i|\boldsymbol{\theta})\right)\right\} \equiv \bar{b}_i = \frac{1}{N}\sum_{j=1}^{N} b_i\left(\boldsymbol{\theta}^{(j)}\right)$$

and

$$\widehat{\text{Var}}_{\boldsymbol{\theta}|\mathbf{y}}\left[\log\{f(y_i|\boldsymbol{\theta})\}\right] = \frac{1}{N-1}\sum_{j=1}^{N}\left\{b_i\left(\boldsymbol{\theta}^{(j)}\right) - \bar{b}_i\right\}^2.$$

Note that in WAIC calculations we work with $E_{\theta|\mathbf{y}}\left\{\log\left(f(y_i|\boldsymbol{\theta})\right)\right\}$ and $\text{Var}_{\boldsymbol{\theta}|\mathbf{y}}\left[\log\{f(y_i|\boldsymbol{\theta})\}\right]$ individually for each i and then sum them, which is in contrast to finding similar quantities $E_{\theta|\mathbf{y}}\left\{\log\left(f(\mathbf{y}|\boldsymbol{\theta})\right)\right\}$ and $\text{Var}_{\boldsymbol{\theta}|\mathbf{y}}\left[\log\{f(\mathbf{y}|\boldsymbol{\theta})\}\right]$ for the full log-likelihood. The two methods may produce different results and different values of the DIC and the WAIC.

5.12.3   Computing PMCC

The PMCC described in Section 4.16.3 is computed by what is known as composition sampling as follows. Compositional sampling is used to perform to draw samples from the posterior predictive distribution $f(y_0|y_1,\ldots,y_n)$ defined in (4.8). To perform a Monte Carlo integration of (4.8) first we first draw a sample from the posterior distribution $\pi(\boldsymbol{\theta}|\mathbf{y})$ and then draw a sample y_0 from the distribution $f(y_0|\boldsymbol{\theta},y_1,\ldots,y_n)$. Note that this is the conditional distribution of a new observation given the parameters $\boldsymbol{\theta}$ and the observations $y_1,\ldots,y_n$. If there is no dependence between the sample observations and the hypothetical future replicate Y_0 then this distribution is simply the top level distribution $f(y_0|\boldsymbol{\theta})$ as assumed in the model. If instead there is spatial or temporal dependence between the observations then the samples must be drawn from the full conditional distribution $f(y_0|\boldsymbol{\theta},y_1,\ldots,y_n)$.

For computing the PMCC we draw a sample $y_{i0}^{(j)}$ from $f\left(y_{i0}|\boldsymbol{\theta}^{(j)},y_1,\ldots,y_n\right)$ corresponding to each posterior sample $\boldsymbol{\theta}^{(j)}$ for each $i=1,\ldots,n$. Note the observation index i as a suffix in the argument y_{i0} of the density $f\left(y_{i0}|\boldsymbol{\theta}^{(j)},y_1,\ldots,y_n\right)$. The observation index i is there to emphasize the need to replicate hypothetical data corresponding to the ith observation y_i in the data set. Thus, any particular peculiarity of the ith observation, e.g. covariate values, must be taken care of when drawing samples from $f\left(y_{i0}|\boldsymbol{\theta}^{(j)},y_1,\ldots,y_n\right)$. This compositional sampling procedure provides us samples $y_{i0}^{(j)}$, for $i=1,\ldots,n$ and $j=1,\ldots,N$. Now we form the averages:

$$\hat{E}(Y_{i0}|\mathbf{y}) \equiv \bar{y}_{i0} = \frac{1}{N}y_{i0}^{(j)} \quad \text{and} \quad \widehat{\text{Var}}(Y_{i0}|\mathbf{y}) = \frac{1}{N-1}\sum_{j=1}^{N}\left(y_{i0}^{(j)} - \bar{y}_{i0}\right)^2$$

for $i = 1, \ldots, n$. The PMCC criterion (4.27) is now estimated as

$$\widehat{\text{PMCC}} = \sum_{i=1}^{n} \left\{ y_i - \hat{E}(Y_{i0}|\mathbf{y}) \right\}^2 + \sum_{i=1}^{n} \widehat{\text{Var}}\,(Y_{i0}|\mathbf{y}).$$

Thus, estimation of the PMCC criteria requires us to sample from the sampling distribution of the data itself given the parameter values. This additional data generation step conditional on the parameter values drawn from the posterior distribution adds desirable further knowledge regarding the properties of the assumed model. This becomes useful when we perform Bayesian model checking by noting that if the assumed model is correct then the observed y_i is a particular realization of the distribution from which we have a large sample $y_{i0}^{(j)}$. The model assumption may deem to be satisfactory if y_i is somewhere in the middle of the underlying distribution of the drawn sample $y_{i0}^{(j)}$. We return to this model checking issue later on.

5.12.4 Computing the model choice criteria for the New York air pollution data

We now illustrate computation of the DIC, WAIC and PMCC. We first assume that σ^2 is known. We continue to use: $\mu = \bar{y} + k\frac{\sigma}{\sqrt{n}}$ and $\tau^2 = \frac{\sigma^2}{m}$. For $k = 1$ and $m = 1$ the posterior distribution $\pi(\theta|\mathbf{y})$ is $N(\mu_p = 47.91, \sigma_p^2 = 0.58)$. To illustrate the computation of the three model choice criteria: DIC, WAIC and PMCC we simply simulate $N = 10,000$ samples from the posterior distribution $N(\mu_p, \sigma_p^2)$. Let $\theta^{(j)}$, $j = 1, \ldots, N$ denote these samples. For PMCC calculation we sample $y_{i0}^{(j)}$ from $N(\theta^{(j)}, \sigma^2)$ for $j = 1, \ldots, N$ and $i = 1, \ldots, n = 28$.

For the both parameter unknown case we continue to assume $k = 1$, $m = 1$ and the prior mean is $\mu = \bar{y} + k\frac{s_y}{\sqrt{n}}$. We draw N samples from the joint posterior distribution using the code for the Gibbs sampler in Example 5.8 in Section 5.8. Table 5.4 provides the exact values of the model choice criteria obtained in Example 4.16.3 and the computed estimates under both the known and unknown σ^2 case. The table entrees are reproduced by issuing the commands:

```
Bmchoice(case="Exact.sigma2.known")
Bmchoice(case="MC.sigma2.known")
Bmchoice(case="MC.sigma2.unknown")
```

5.13   Conclusion

This chapter introduces Bayesian computation methods which are essential for conducting Bayesian inference from hierarchical modeling. Solving mul-

| | θ unknown | | θ & σ^2 unknown |
	Exact	Estimate	Estimate
p_{DIC}	0.97	0.96	2.07
$p_{\text{DIC alt}}$	0.93	0.95	13.58
DIC	166.95	166.94	169.20
DIC alt	166.89	166.93	192.22
$p_{\text{waic 1}}$	0.92	0.91	1.82
$p_{\text{waic 2}}$	0.96	0.94	2.52
WAIC 1	166.91	166.90	168.95
WAIC 2	167.00	166.96	170.35
gof	594.30	593.29	593.54
penalty	637.24	634.58	575.80
PMCC	1231.54	1227.87	1169.34

TABLE 5.4: Different model choice criteria.

tidimensional integrals is the biggest problem in applying Bayesian inference methods to real data modeling problems. Numerical approximations and Monte Carlo integration techniques are at the heart of the engine in the machine to solve these problems. This chapter has discussed a suit of basic concepts we require to understand those techniques for integration. Understanding this step ladder from the hard integration surface to the real world of numerical Bayesian computation is crucial to achieving success in modeling. A few exercises are provided below to enhance the understanding of the reader.

5.14   Exercises

1. Suppose that $y_1, \ldots, y_n$ are i.i.d. observations from a Bernoulli distribution with mean θ. A logistic normal prior distribution is proposed for θ (a normal distribution for $\log \frac{\theta}{1-\theta}$). Show that if the prior mean and variance for $\log \frac{\theta}{1-\theta}$ are 0 and 1 respectively then the prior density function for θ is

$$\pi(\theta) \;=\; \frac{1}{\sqrt{2\pi}\theta(1-\theta)} \exp\left(-\tfrac{1}{2}\left(\log \tfrac{\theta}{1-\theta}\right)^2\right)$$

As this prior distribution is not conjugate, the Bayes estimator $E(\theta|y_1, \ldots, y_n)$ is not directly available. It is proposed to estimate it using a Monte Carlo sample generated by the Metropolis-Hastings method. One possible algorithm involves generating proposals from the prior distribution, independently of the current observation. Derive the acceptance probability of this algorithm.

2. Assume that $Y_1, Y_2, \ldots, Y_n$ are independent identically distributed $N(\theta, 1)$ observations. Suppose that the prior distribution for θ is Cauchy with density

$$\pi(\theta) = \frac{1}{\pi} \frac{1}{1 + \theta^2} \quad -\infty < \theta < \infty.$$

Derive, upto a constant of proportionality, the posterior density of θ. Suppose that the importance sampling distribution is the prior distributions given above. Obtain the acceptance probability for the rejection method and the Metropolis-Hastings independence sampler.

3. Assume that $Y_1, Y_2, \ldots, Y_n$ are independent and identically distributed $N(\theta, \sigma^2)$ observations. Suppose that the joint prior distribution for θ and σ^2 is

$$\pi(\theta, \sigma^2) = \frac{1}{\sigma^2}.$$

(a) Derive, upto a constant of proportionality, the joint posterior density of θ and σ^2.

(b) Derive the conditional posterior distributions of θ given σ^2 and σ^2 given θ.

(c) Derive the marginal posterior density of θ.

4. Suppose that $y_1, \ldots, y_n$ are random observations from a model $f(y|\theta)$ and a prior distribution $\pi(\theta)$ is assumed for $\boldsymbol{\theta}$. Denote $\tilde{Y}_i$ to be the future observation corresponding to y_i and its posterior predictive distribution is given by $f(\tilde{y}_i|\mathbf{y})$ where $\mathbf{y} = (y_1, \ldots, y_n)$.

(a) Assume the loss function

$$L(\tilde{\mathbf{Y}}, \mathbf{y}) = \sum_{i=1}^{n} (\tilde{Y}_i - y_i)^2$$

where $\tilde{\mathbf{Y}} = \left(\tilde{Y}_1, \ldots, \tilde{Y}_n \right)$. Show that:

$$E\left[L(\tilde{\mathbf{Y}}, \mathbf{y}) | \mathbf{y} \right] = \sum_{i=1}^{n} E\left[\tilde{Y}_i - E(\tilde{Y}_i|\mathbf{y}) | \mathbf{y} \right]^2 + \sum_{i=1}^{n} \left[y_i - E(\tilde{Y}_i|\mathbf{y}) \right]^2.$$

Provide an interpretation for each of the two terms on the right hand side of the above equation and discuss how the terms can be estimated when Monte Carlo samples are available from $f(\tilde{y}_i|\mathbf{y})$. State how you would use the above to compare models.

(b) Define the deviance information criterion (DIC) for a model. State how DIC can be used to compare models.

6

Bayesian modeling for point referenced spatial data

6.1 Introduction

In order to obtain a high degree of accuracy in analysis and predictions of a response variable, such as the amount of air pollution, statistical models are employed which explicitly include the underlying uncertainty in the data. Such models are statistical in nature and, if appropriately chosen, allow accurate forecasting in future time periods and interpolation over the entire spatial region of interest. In addition, they allow us to estimate the size of possible errors associated with our out of sample predictions.

Statistical models are constructed to capture the full scale of variability, which may come from many different sources, present in the data. The primary task in statistical modeling is to separate out variation that can be explained by deterministic non-random factors and other random variation that cannot be explained by as yet known causes and explanatory factors. Statistical models lay down the basic rules and regulations for individual members of a population of interest. For example, for a simple linear regression model all observations are commanded to lie on a straight line except for the added random residual errors which must be tolerated until a suitable explanation can be found through, e.g. additional data collection with better precision or by including spatial and/or temporal effects not yet present in the model. It may also be the case that we may never be able to account for *all* sources of variation and the statistical model, using all the ground level physical knowledge, is the best approximation for the true model for the data.

Modeling real life phenomena for which we have recorded random observations involves a series of complex tasks. First of all, the modeler must find a set of plausible models that does not violate any of the constraints observed by the data. For example, a standard Poisson distribution is not a correct model for data which are constrained to be structurally positive. Covariate selection and stipulation of parameterisation for the set of models are also among the important tasks at this stage. Exploratory data analysis tools help in performing these tasks. For example, by drawing simple scatter plots and or calculating interesting statistics (which may have already been proposed for similar problems) one may decide to select a subset of the covariates for

DOI: 10.1201/9780429318443-6

153

exploration by explicit modeling. This type of screening is demanded by our aims to fit parsimonious models. Of course, left out data can be examined in a re-modeling stage later on, especially if the postulated models fail to achieve the stated modeling objectives.

Model fitting and parameter estimation come as natural next steps in the modeling exercise. These estimation tasks involve mathematical integrations and optimizations in non-standard forms for which closed form solutions may not exist. Hence computer intensive methods are often needed. The Bayesian computation techniques, described in the previous Chapter 5 can be employed for this task. The resulting parameter estimates are to be interpreted in the context of the data to check for coherency and practical validity. Moreover, the posterior predictive checks, described in Section 4.17, should be applied to diagnose the model fit. If there is more than one competing models all of which fit the data well then the model choice criteria described in Section 4.16 must be applied to choose a particular model with which to make inference regarding the parameters of the model.

Statistical models are always based on probability distributions which distribute the total probability among the possible values of the random quantity being modeled. Having accounted for the total probability the statistical models allow us to compute probabilities of any current and future events for members of that population. For example, a probabilistic model will let us calculate the probability that a future individual in the same population will be above or below a threshold value or a tipping point. In the context of an air pollution modeling example we may be interested estimating the probability that a particular site or region has exceeded the legislated air pollution standard. Besides the probabilities, a statistical model may also be used to perform out of sample predictions. But in that case the associated uncertainties must also be estimated and reported.

The modeling steps and exercise altogether, in order to be successful, must address various difficult questions. Is the model adequate for the data? Is their any model uncertainty? Can a better model be found? Can data values that are outliers relative to the model be identified and explained away? How accurate are the predictions? How scalable is the whole modeling exercise? Will the model be able to cope with future, possibly more complex, data sets? Can the model be deployed in real time situations demanding immediate answers once data have been collected? How would the modeler balance the need for complexity to take care of all sources of variation and parsimony to interpret and deliver results using easy to use, preferably publicly available, computing tools? Failure to address and answer these questions early on may lead to total project failure which must be avoided.

6.2 Model versus procedure based methods

One of the popular alternatives to modeling is what is often called procedure based methods. In these methods one often applies a sequence of techniques to the data to come to the final conclusion. For example, a transformation may be applied to stabilize variability. Then separate aggregated statistics may be calculated for data classified by separate grouping factors such as site types, e.g. rural and urban in air pollution studies. Then if it is a time series data, a linear trend equation may be fitted and residuals calculated separately for each group. Residuals so obtained may then be 'kriged' to introduce spatial statistical procedures. The kriged residuals may then be converted back by adding the estimated trend equation and then group specific summaries may be predicted for studying spatio-temporal trend in unobserved sites. Such procedure based methods often produce realistic outcomes and predictions. These predictions are then validated by calculating an overall error rate, e.g. the root mean-square prediction error. There are many other procedure based methods abundant in the literature. How would one compare the corresponding results based on rigorous application of statistical models?

Explicit statistical model based methods are preferable to purely procedure based methods. There are several reasons for this.

- **Uncertainty evaluation** Explicit modeling allows us to evaluate uncertainty for any individual inference statement we make. For example, we may evaluate uncertainty based on posterior distribution of model parameters. If we use the fitted model for prediction then we evaluate uncertainty for every individual prediction we make based on the predictive distribution (4.8). Such automatic evaluation of individual level uncertainty is more statistically rigorous and attractive than stating overall measures such as root mean-square errors of a number of predictions. Inference methods based on Bayesian posterior and predictive distributions enable automatic evaluation without any extra effort such as derivative calculation for Hessian matrix evaluation in likelihood based theory.

- **Uncertainty integration** Modeling based on hierarchical Bayesian methods enables us to keep track of uncertainty throughout the data analysis process. For example, modeling components, e.g. regression models, spatial effects, temporal factors are all integrated in a hierarchical model so that the overall uncertainty in prediction is integrated from uncertainty in estimation of all these effects. Such integrated uncertainty measures are lost when one applies several procedures sequentially, e.g. transforming and de-trending the data, fitting linear regression, performing kriging on the residuals.

- **Explicit assumption and their verification** Assumptions regarding the nature and distribution of the data are made explicit at the beginning of a modeling process. In addition, the model formula itself (e.g. linear regression) states some of the underlying assumptions. By fitting the theoretical model to the data and by performing posterior predictive model checking on the model fit we automatically perform some aspects of assumption verification. It is only rarely that a modeler will recommend a badly fitted model where the poor fitting may automatically be judged to have come from implausible model assumptions.

- **Accounting for rather than adjusting for effects** Explicit modeling allows us to account for effects from all sources and carry those effects into prediction and inference. This is in contrast to procedure based methods where de-trending and de-seasonalization, for example, may occur outside of the model fitting exercise. Modeling of such effects allows us to account for such variation in prediction along with explicit accounting for uncertainty that may arise due to such effects.

- **Benefit of learning by pooling all observations** A unified hierarchical model for all the observations allows us to obtain the best estimates of uncertainty for any individual level inference. For example, by fitting a joint model for all groups (e.g. male/female) of observations we are more likely to get a better estimate of overall variation than if we were to fit separate models completely. We admit that it may so happen that a separate model may be justified by the nature of data collection and the data itself may support separate modeling. However, a unified hierarchical model may still allow us to provide better understanding of all sources of variation in the data.

- **Scientific benefit from application of coherent methods** Modeling, in this book, is an umbrella term that encompasses model statement, fitting, checking, model based prediction and re-modeling. By using the mathematically coherent Bayesian methods at each of these modeling sub-tasks we guarantee scientific validity of the conclusions we make. Biases which may arise from personal choices regarding model implementations are eliminated by applying the scientifically sound Bayesian methods. Once a model has been written down the Bayesian machinery takes over for making any inference we may wish to.

We end this section with a cautionary note regarding the usefulness of modeling. There is a popular saying, "All models are wrong but some are useful." Indeed, the whole idea of statistics to recreate nature based on observations alone is a daunting task that is not guaranteed to produce the correct result with 100% certainty. Hence statistical modeling may fail to produce the most accurate result in a given situation. There may be several possible explanations for this. For example, is the model fitted best possible model for the data keeping in mind the see-saw relationship between goodness-of-fit and predictive capabilities of a model? Is the statistical model using the same set of information as the other superior method? How about uncertainty evaluation for the predictions made? Lastly, are the results robust for possible variations in validation data set and so on. Thus, an unfavorable result for a statistical modeling exercise should not automatically mean the end of statistical modeling per se.

6.3 Formulating linear models

Linear statistical models are often starting points in many practical modeling situations in spatial and spatio-temporal modeling. Some arguments regarding the failure of standard linear models must be made before attempting more complex modeling methods. This is our primary motivation for introducing the linear regression models in this section and more broadly in this chapter. In so doing we aim to achieve three fundamental goals:

1. Generalization of the earlier Bayesian theory results as far as practicable. The theoretical developments are included so that the Bayesian computation methods of the previous chapter can be checked using exact analytical results where possible. However, the theory can be skipped if required.

2. Introduction to Bayesian model fitting software packages using the theoretical examples and thereby verifying both the software packages and the theoretical calculations. This is undertaken in Section 6.6 below.

3. Introduction to various modes of cross-validation methods. Using the simplicity of linear models we illustrate the new ideas so that these can be applied to more complex spatio-temporal modeling in the later chapters. This is undertaken in Section 6.8 below.

This gentle and slow introduction to modeling is designed to convince the reader of the scientific soundness of the Bayesian and computational methods. At the same time the familiarity of the linear models is exploited to learn new computation methods and model validation methods.

6.3.1 Data set preparation

As detailed in the exploratory data analysis chapter a great deal of data cleaning and preparation is required and is hereby reminded before we undertake the task of modeling. In this stage the response variable should be identified and the continuously explanatory variables should be normalized (mean subtracted and divided by the standard deviation) to reduce correlations and thereby to enhance computational stability. These investigations often provide the much needed sanctity checks required for successful modeling. Attention should also be paid to understand the structure of the data, e.g. how the data are organized in space and time. Investigations should be carried out regarding any missing observations and covariate values. The experimenter should consult the data provider to gain insight regarding any patterns in missingness and then decide how best to resolve those. Most Bayesian model fitting methods treat the missing observations as unknown parameters and estimate those along with other model parameters. In fact, this is a general trick that is often used to perform prediction using Bayesian models. This trick is also employed in the computations performed in Chapter 7. However, in order to reduce uncertainty in inference, it is always better to avoid having missing data if at all possible.

It is recommended that a simple linear regression model is fitted with the `lm` command in R. The output of the `lm` commanded should be recorded for comparison with the output of the Bayesian modeling of this chapter. Subsequently these output will also be compared with the advanced spatio-temporal modeling results that will follow this chapter. In this stage the researcher may also try out possible data transformation to achieve better model fitting results. These transformations can be applied in the Bayesian modeling setup as well. The experimentation here will enable us to answer 'what-if' questions. If the response is categorical then the model fitting should be performed by using the `glm` command in R instead of the `lm` command. However, unless otherwise mentioned we assume that we are fitting normal linear models.

At the end of this pre-modeling stage the researcher should have a very good idea of the data set at hand and how to model it using simple methods. The researcher should also note any sort of anomaly that may arise because of model fitting using the simple methods. For example, is there any fitted or predicted value which lies outside the range of the data? Is their any systematic lack of fit which may possibly be explained by a spatial or temporal effect parameter? Identifying such problems early on is a very good idea before investing the time to fit more advanced models with advanced technology.

6.3.2 Writing down the model formula

The next task after data preparation is to write down the model equation using statistical symbols. The formula used in the `lm` command serves up the most important clue here. The response variable in the left hand side of the `lm` command is denoted as Y_i and the explanatory variables in the right hand side are denoted as x_{ij} for $j = 1, \ldots, p$. The subscript i denotes the ith data point which may actually be classified by further categories such as race, gender, site types, spatial location and or time. For the sake of model development and theoretical calculations below we assume that a single identifier i is sufficient to denote rows in the data spreadsheet. The p explanatory columns may contain a constant column of 1's for the intercept term and any factor variable, e.g. site types taking two or more values, e.g. rural, urban and suburban are coded using dummy variables. The R extractor command `model.matrix` provides the explanatory variable columns, x_{ij} for $j = 1, \ldots, p$. Now the general linear regression model is written as:

$$Y_i = \beta_1 x_{i1} + \ldots + \beta_p x_{ip} + \epsilon_i, i = 1, \ldots, n \qquad (6.1)$$

where $\beta_1, \ldots, \beta_p$ are unknown regression coefficients and ϵ_i is the error term that we assume to follow the normal distribution with mean zero and variance σ^2. The usual linear model assumes the errors ϵ_i to be independent for $i = 1, \ldots, n$. In the theory discussed below we assume a known correlation structure between the errors but we get the usual linear model as a special case when there is zero correlation between the ϵ_i's. This allows us to generalize the linear model to be suitable for dependent spatio-temporal data. The independent error model is a generalization of the normal distribution based examples in the Bayesian inference Chapter 4 with $p = 1$ and $\beta_1 = \theta$.

The regression model is also written as

$$Y_i \sim N(\mathbf{x}'_i \boldsymbol{\beta}, \ \sigma^2)$$

where $\mathbf{x}_i = (x_{i1}, \ldots, x_{ip})$ is a column vector of the covariate (explanatory variable) values for the ith data point for $i = 1, \ldots, n$; and $\boldsymbol{\beta} = (\beta_1, \ldots, \beta_p)$.

The above model is more succinctly written using a common vector notation as described below. We write $\mathbf{Y} = (Y_1, \ldots, Y_n)$, and $\boldsymbol{\epsilon} = (\epsilon_1, \ldots, \epsilon_n)$ as $n \times 1$ column vectors. The p explanatory variables for n data points are collected in a $n \times p$ matrix X. The matrix X, obtained as a stacking of the row vectors $\mathbf{x}'_1, \ldots, \mathbf{x}'_n$ each of which is of length p, is known as the design matrix of the model. Then the model (6.1) is written as

$$\mathbf{Y} \sim N\left(X\boldsymbol{\beta}, \sigma^2 H\right) \qquad (6.2)$$

where H is a known correlation matrix of $\boldsymbol{\epsilon}$, i.e. $H_{ij} = \mathrm{Cor}(\epsilon_i, \epsilon_j)$ for $i, j = 1, \ldots, n$. The multivariate normal distribution $N\left(X\boldsymbol{\beta}, \sigma^2 H\right)$, see (A.24) has the n dimensional mean vector $X\boldsymbol{\beta}$ and covariance matrix $\sigma^2 H$. The unknown parameters in the model are p components of $\boldsymbol{\beta}$ and σ^2. The assumption of known correlation parameters will be relaxed in Chapter 7 and later.

For convenience we work with $\lambda^2 = 1/\sigma^2$, the error precision as previously in Section 4.13 in the Bayesian inference Chapter 4. The likelihood function for $\boldsymbol{\beta}$ and λ^2 is now given by:

$$f(\mathbf{y}|\boldsymbol{\beta}, \sigma^2) = \left(\frac{\lambda^2}{2\pi}\right)^{\frac{n}{2}} |H|^{-\frac{1}{2}} \exp\left[-\frac{\lambda^2}{2}(\mathbf{y} - X\boldsymbol{\beta})'H^{-1}(\mathbf{y} - X\boldsymbol{\beta})\right].$$

As before we specify a hierarchical prior distribution $\pi(\boldsymbol{\beta}|\lambda^2)$ and $\pi(\lambda^2)$. We continue to assume that $\pi(\lambda^2)$ is a gamma distribution $G(a, b)$ for given values of $a > 0$ and $b > 0$. The prior distribution for $\boldsymbol{\beta}$ given λ^2 is assumed to be the multivariate normal distribution with mean $\boldsymbol{\beta}_0$ and covariance matrix $\sigma^2 M^{-1} \left(\equiv \frac{M^{-1}}{\lambda^2}\right)$. This is a generalization of the prior distribution for

θ in Section 4.7. This conjugate prior distribution allows us to get standard distributions for making posterior and predictive inference. In this way we generalize the results presented in Section 4.13 for the linear regression model case of this chapter. The joint prior density for $\boldsymbol{\beta}$ and λ^2 is now given by:

$$\pi(\boldsymbol{\beta}, \lambda^2) \propto \left(\lambda^2\right)^{\frac{p}{2}+a-1} \exp\left[-\frac{\lambda^2}{2}\left\{2b + (\boldsymbol{\beta} - \boldsymbol{\beta}_0)' M(\boldsymbol{\beta} - \boldsymbol{\beta}_0)\right\}\right].$$

The joint posterior distribution of $\boldsymbol{\beta}$ and λ^2, $\pi\left(\boldsymbol{\beta}, \lambda^2 | \mathbf{y}\right)$, is:

$$\propto \left(\lambda^2\right)^{\frac{n+p}{2}+a-1} \exp\left[-\frac{\lambda^2}{2}\left\{2b + (\mathbf{y} - X\boldsymbol{\beta})' H^{-1}(\mathbf{y} - X\boldsymbol{\beta}) + (\boldsymbol{\beta} - \boldsymbol{\beta}_0)' M(\boldsymbol{\beta} - \boldsymbol{\beta}_0)\right\}\right].$$

$$(6.3)$$

Now we use the matrix identity:

$$2b + (\mathbf{y} - X\boldsymbol{\beta})' H^{-1}(\mathbf{y} - X\boldsymbol{\beta}) + (\boldsymbol{\beta} - \boldsymbol{\beta}_0)' M(\boldsymbol{\beta} - \boldsymbol{\beta}_0) = 2b^* + (\boldsymbol{\beta} - \boldsymbol{\beta}^*)' M^*(\boldsymbol{\beta} - \boldsymbol{\beta}^*)$$

where

$$M^* = M + X'H^{-1}X, \quad \boldsymbol{\beta}^* = (M^*)^{-1}\left(M\boldsymbol{\beta}_0 + X'H^{-1}\mathbf{y}\right) \qquad (6.4)$$

and

$$2b^* = 2b + \boldsymbol{\beta}_0' M\boldsymbol{\beta}_0 + \mathbf{y}'H^{-1}\mathbf{y} - (\boldsymbol{\beta}^*)' M^*(\boldsymbol{\beta}^*).$$

Hence the joint posterior distribution is given by:

$$\pi\left(\boldsymbol{\beta}, \lambda^2 | \mathbf{y}\right) \propto \left(\lambda^2\right)^{\frac{n+p}{2}+a-1} \exp\left[-\frac{\lambda^2}{2}\left\{2b^* + (\boldsymbol{\beta} - \boldsymbol{\beta}^*)' M^*(\boldsymbol{\beta} - \boldsymbol{\beta}^*)\right\}\right].$$

$$(6.5)$$

Now the full conditional posterior distributions are given by:

$$\boldsymbol{\beta} | \lambda^2, \mathbf{y} \sim N\left(\boldsymbol{\beta}^*, \frac{1}{\lambda^2}(M^*)^{-1}\right) \qquad (6.6)$$

$$\lambda^2 | \boldsymbol{\beta}, \mathbf{y} \sim G\left(\frac{n+p}{2} + a, \; b^* + \frac{1}{2}(\boldsymbol{\beta} - \boldsymbol{\beta}^*)' M^*(\boldsymbol{\beta} - \boldsymbol{\beta}^*)\right).$$

By direct integration the marginal posterior distributions are obtained as follows:

$$\boldsymbol{\beta} | \mathbf{y} \sim t_p\left(\boldsymbol{\beta}^*, \frac{2b^*}{n+2a}(M^*)^{-1}, n+2a\right), \quad \lambda^2 | \mathbf{y} \sim G\left(\frac{n}{2} + a, b^*\right). \qquad (6.7)$$

We use these marginal posterior distributions to make inference. Here $\pi(\boldsymbol{\beta}|\mathbf{y})$ has the multivariate t-distribution (A.28). As a result

$$E(\boldsymbol{\beta}|\mathbf{y}) = \boldsymbol{\beta}^* \text{ and } \text{Var}(\boldsymbol{\beta}|\mathbf{y}) = \frac{2b^*}{n+2a-2}(M^*)^{-1}. \qquad (6.8)$$

Thus, $\boldsymbol{\beta}^*$ provides a point estimate for the parameter $\boldsymbol{\beta}$. We obtain a credible interval for the kth component, β_k, for $k = 1, \ldots, p$ by using its marginal posterior distribution, which is a t-distribution with $n + 2a$ degrees of freedom having mean β_k^* and scale parameter γ_k^2 where $\gamma_k^2 = \frac{2b^*}{n+2a} (M^*)_{kk}^{-1}$ where $(M^*)_{kk}^{-1}$ is the kth diagonal entry of $(M^*)^{-1}$. Now it is straightforward to see that an equal-tailed $(1 - \alpha)100\%$ credible interval for β_k is given by

$$\beta_k^* \pm \gamma_k \, t_{\alpha/2;n+2a}$$

where $P(X > t_{\alpha/2;n+2a}) = \alpha/2$ when X follows the standard t-distribution, $t(0, 1, n + 2a)$.

Similarly we estimate σ^2 by the posterior expectation

$$E\left(\sigma^2 | \mathbf{y}\right) = E\left(\frac{1}{\lambda^2} | \mathbf{y}\right) = \frac{2b^*}{n + 2a - 2}$$

which follows from the properties of the Gamma distribution (A.18). Moreover,

$$\text{Var}\left(\sigma^2 | \mathbf{y}\right) = 2\frac{(2b^*)^2}{(n + 2a - 2)^2(n + 2a - 4)}.$$

Here also we can find an equal tailed credible interval for σ^2 by using the probability identity

$$P\left(g_{\alpha/2;n/2+a,\gamma} \leq \frac{1}{\sigma^2} \leq g_{1-\alpha/2;n/2+a,\gamma}\right) = 1 - \alpha$$

where $g_{\alpha;\nu,\gamma}$ is such that $P(Z < g_{\alpha;\nu,\gamma}) = \alpha$ for any $0 < \alpha < 1$ when Z follows $G(\nu, \gamma)$.

6.3.3 Predictive distributions

We can predict a future observation using the above linear model as follows. The theoretical development here is slightly more general than the usual independent error linear model because of the possibility of spatio-temporal correlation in the observations. Of course, the general theory will render the results for the independent error model as special cases .

Let $\mathbf{x}_0$ denote the p-dimensional vector of values of the regression variables for the new observation Y_0. Because of the dependence between the Y_0 and the observed random variables $\mathbf{Y}$, we will have to use (4.8) to obtain the posterior predictive distribution of Y_0 given the n observations $\mathbf{y}$. In order to do this, we first construct the joint distribution:

$$\begin{pmatrix} Y_0 \\ \mathbf{Y} \end{pmatrix} \sim N\left\{\begin{pmatrix} \mathbf{x}_0'\boldsymbol{\beta} \\ X\boldsymbol{\beta} \end{pmatrix}, \sigma^2 \begin{pmatrix} 1 & \Sigma_{12} \\ \Sigma_{21} & H \end{pmatrix}\right\}, \quad (6.9)$$

where $\Sigma_{21} = \Sigma_{12}'$ and Σ_{12} is the n dimensional vector with elements given by $\text{Cor}(Y_0, Y_i)$. Now we obtain the conditional distribution of $Y_0 | \mathbf{y}, \boldsymbol{\beta}, \sigma^2$ as

$$N\left\{\mathbf{x}_0'\boldsymbol{\beta} + \Sigma_{12}H^{-1}\left(\mathbf{y} - X\boldsymbol{\beta}\right), \sigma^2\left(1 - \Sigma_{12}H^{-1}\Sigma_{21}\right)\right\}.$$

Now we write $\lambda^2 = 1/\sigma^2$ as before and let

$$\delta^2 = 1 - \Sigma_{12}H^{-1}\Sigma_{21}$$

be the multiplication factor in the conditional variance. We need to integrate out $\boldsymbol{\beta}$ and $\lambda^2(= 1/\sigma^2)$ from the above distribution to obtain the required predictive distribution. Let

$$\mu_0 = \mathbf{x}_0'\boldsymbol{\beta} + \Sigma_{12}H^{-1}\left(\mathbf{y} - X\boldsymbol{\beta}\right)$$

denote the conditional mean of $Y_0|\mathbf{y}, \boldsymbol{\beta}, \lambda^2$. Now

$$
\begin{aligned}
y_0 - \mu_0 &= y_0 - \mathbf{x}_0'\boldsymbol{\beta} - \Sigma_{12}H^{-1}\mathbf{y} + \Sigma_{12}H^{-1}X\boldsymbol{\beta} \\
&= y_0 - \Sigma_{12}H^{-1}\mathbf{y} - \left\{\mathbf{x}_0' - \Sigma_{12}H^{-1}X\right\}\boldsymbol{\beta} \\
&= \tilde{y}_0 - \mathbf{g}'\boldsymbol{\beta}, \quad \text{say}
\end{aligned}
$$

where

$$\tilde{y}_0 = y_0 - \Sigma_{12}H^{-1}\mathbf{y} \quad \text{and} \quad \mathbf{g}' = \mathbf{x}_0' - \Sigma_{12}H^{-1}X.$$

Now we first find the posterior predictive distribution of

$$\tilde{Y}_0 = Y_0 - \Sigma_{12}H^{-1}\mathbf{y}$$

and then the distribution of Y_0 is obtained by adding $\Sigma_{12}H^{-1}\mathbf{y}$ to the mean of the distribution of $\tilde{Y}_0$. We now have

$$\pi(\tilde{Y}_0|\mathbf{y}, \boldsymbol{\beta}, \lambda^2) \quad \propto (\lambda^2)^{\frac{1}{2}} \exp\left[-\frac{\lambda^2}{2\delta^2}\left\{\tilde{y}_0 - \mathbf{g}'\boldsymbol{\beta}\right\}^2\right]$$

This shows that

$$\tilde{Y}_0|\mathbf{y}, \boldsymbol{\beta}, \lambda^2 \sim N\left(\mathbf{g}'\boldsymbol{\beta}, \frac{\delta^2}{\lambda^2}\right).$$

To obtain the required predictive density we must integrate the above normal density with respect to the joint posterior distribution $\pi(\boldsymbol{\beta}, \lambda^2|\mathbf{y})$ in (6.5) according to the general definition of posterior predictive density (4.8). This integration is performed in two steps by using the hierarchical representation:

$$\pi(\boldsymbol{\beta}, \lambda^2|\mathbf{y}) = \pi(\boldsymbol{\beta}|\lambda^2, \mathbf{y})\,\pi(\lambda^2|\mathbf{y}).$$

The regression coefficients $\boldsymbol{\beta}$ are first integrated out by keeping λ^2 fixed and using the conditional posterior distribution $\pi(\boldsymbol{\beta}|\lambda^2, \mathbf{y})$ given in (6.6). The result is then integrated out using the marginal posterior distribution $\pi(\lambda^2|\mathbf{y})$.

    To perform the first step integration with respect to $\boldsymbol{\beta}$ we simply obtain the mean variance of $\tilde{Y}_0|\lambda^2, \mathbf{y}$ using the results discussed in the two remarks made just below the normal posterior predictive distribution (4.10). Namely,

$$
\begin{aligned}
E(\tilde{Y}_0|\lambda^2, \mathbf{y}) &= E_{\boldsymbol{\beta}|\lambda^2, \mathbf{y}}E(\tilde{Y}_0|\boldsymbol{\beta}, \lambda^2, \mathbf{y}) \\
&= E_{\boldsymbol{\beta}|\lambda^2, \mathbf{y}}(\mathbf{g}'\boldsymbol{\beta}) \\
&= \mathbf{g}'\boldsymbol{\beta}^*
\end{aligned}
$$

by using (6.6) to obtain the last result. Now

$$
\begin{aligned}
\text{Var}(\tilde{Y}_0|\lambda^2, \mathbf{y}) &= E_{\boldsymbol{\beta}|\lambda^2, \mathbf{y}} \text{Var}\left[\tilde{Y}_0|\boldsymbol{\beta}, \lambda^2, \mathbf{y}\right] + \text{Var}_{\boldsymbol{\beta}|\lambda^2, \mathbf{y}}\left[E\left(\tilde{Y}_0|\boldsymbol{\beta}, \lambda^2, \mathbf{y}\right)\right] \\
&= E_{\boldsymbol{\beta}|\lambda^2, \mathbf{y}}\left(\frac{\delta^2}{\lambda^2}\right) + \text{Var}_{\boldsymbol{\beta}|\lambda^2, \mathbf{y}}(\mathbf{g}'\boldsymbol{\beta}) \\
&= \frac{\delta^2}{\lambda^2} + \frac{1}{\lambda^2}\mathbf{g}'\left(M^*\right)^{-1}\mathbf{g} \\
&= \frac{1}{\lambda^2}\left(\delta^2 + \mathbf{g}'\left(M^*\right)^{-1}\mathbf{g}\right).
\end{aligned}
$$

Hence, the first step of integration results in

$$
\tilde{Y}_0|\mathbf{y}, \lambda^2 \sim N\left(\mathbf{g}'\boldsymbol{\beta}^*, \frac{1}{\lambda^2}\left(\delta^2 + \mathbf{g}'\left(M^*\right)^{-1}\mathbf{g}\right)\right)
$$

where normality is guaranteed by the underlying hierarchical specification. By integrating this with respect to the marginal posterior distribution of λ^2 in (6.7), we obtain the posterior predictive distribution of $\tilde{Y}_0$ given $\mathbf{y}$ as:

$$
\tilde{Y}_0|\mathbf{y} \sim t\left(\mathbf{g}'\boldsymbol{\beta}^*, \frac{2b^*}{n+2a}\left(\delta^2 + \mathbf{g}'\left(M^*\right)^{-1}\mathbf{g}\right), n+2a\right).
$$

Now the mean of $Y_0|\mathbf{y}$ is

$$
\begin{aligned}
\mathbf{g}'\boldsymbol{\beta}^* + \Sigma_{12}H^{-1}\mathbf{y} &= \left(\mathbf{x}_0' - \Sigma_{12}H^{-1}X\right)\boldsymbol{\beta}^* + \Sigma_{12}H^{-1}\mathbf{y} \\
&= \mathbf{x}_0'\boldsymbol{\beta}^* + \Sigma_{12}H^{-1}(X\boldsymbol{\beta}^* - \mathbf{y}).
\end{aligned}
$$

Therefore, the posterior predictive distribution of Y_0 given $\mathbf{y}$ is:

$$
t\left(\mathbf{x}_0'\boldsymbol{\beta}^* + \Sigma_{12}H^{-1}(\mathbf{y} - X\boldsymbol{\beta}^*), \frac{2b^*}{n+2a}\left(\delta^2 + \mathbf{g}'\left(M^*\right)^{-1}\mathbf{g}\right), n+2a\right). \quad (6.10)
$$

The variance of this posterior predictive distribution is:

$$
\text{Var}(Y_0|\mathbf{y}) = \frac{2b^*}{n+2a-2}\left(\delta^2 + \mathbf{g}'\left(M^*\right)^{-1}\mathbf{g}\right)
$$

where expressions for b^*, δ^2, $\mathbf{g}$ and M^* are given above. For the independent error regression model and for the associated prediction of a new independent observation Y_0 the above formulae simplify considerably. For the independent model, H is the identity matrix, Σ_{12} is a null vector and consequently

$$
\delta^2 = 1 - \Sigma_{12}H^{-1}\Sigma_{21} = 1 \quad \text{and} \quad \mathbf{g}' = \mathbf{x}_0'.
$$

Then the distribution (6.10) reduces to

$$
t\left(\mathbf{x}_0'\boldsymbol{\beta}^*, \frac{2b^*}{n+2a}\left(1 + \mathbf{x}_0'\left(M^*\right)^{-1}\mathbf{x}_0\right), n+2a\right) \quad (6.11)
$$

with the simplified expressions

$$
M^* = M + X'X, \quad \boldsymbol{\beta}^* = \left(M^*\right)^{-1}\left(M\boldsymbol{\beta}_0 + X'\mathbf{y}\right)
$$

and
$$2b^* = 2b + \boldsymbol{\beta}_0' M \boldsymbol{\beta}_0 + \mathbf{y}' H^{-1} \mathbf{y} - (\boldsymbol{\beta}^*)' M^* (\boldsymbol{\beta}^*).$$

Now we return to the example in Section 4.13. The example there is a special case of the regression formulation here with $p = 1$, $\beta_1 = 0$, $\beta_0 = \mu$, X is a n-dimensional column vector of 1's, H is the identity matrix, and $\mathbf{x}_0 = 1$ is a singleton. With these choices the posterior distributions and the posterior predictive distribution for that example match with the respective ones obtained here. We use the following algebra to verify this claim. Now $X'X = n$, $X'\mathbf{y} = \sum_{i=1}^n y_i = n\bar{y}$, $\mathbf{y}'\mathbf{y} = \sum_{i=1}^n y_i^2$. We write the scalar notation m instead of the matrix notation M and thus $m^* = n + m$.

$$
\begin{aligned}
\delta^2 + \mathbf{g}' (M^*)^{-1} \mathbf{g} &= 1 + x_0 (m^*)^{-1} x_0 \\
&= 1 + \frac{1}{n+m} \\
&= \frac{n+m+1}{n+m},
\end{aligned}
$$

$$
\begin{aligned}
\beta^* &= (M^*)^{-1} \left(M \boldsymbol{\beta}_0 + X' H^{-1} \mathbf{y} \right) \\
&= (n + m)^{-1} (m\mu + n\bar{y}) \\
&= \frac{m\mu + n\bar{y}}{n+m} \\
&= \mu_p,
\end{aligned}
$$

$$
\begin{aligned}
2b^* &= 2b + \boldsymbol{\beta}_0' M \boldsymbol{\beta}_0 + \mathbf{y}' H^{-1} \mathbf{y} - (\boldsymbol{\beta}^*)' M^* (\boldsymbol{\beta}^*) \\
&= 2b + \mu^2 m + \sum_{i=1}^n y_i^2 - (n + m) \left(\frac{m\mu + n\bar{y}}{n+m} \right)^2 \\
&= 2b + (n - 1)s_y^2 + \frac{mn}{n+m} (\mu - \bar{y})^2,
\end{aligned}
$$

after simplification.

6.4   Linear model for spatial data

The general theory developed in the previous Section 6.3 is now being applied to model spatially varying point referenced data. Let $Y(\mathbf{s}_i)$ denote the spatial random variable at location $\mathbf{s}_i$ for $i = 1, \ldots, n$. As a result of this explicit spatial reference we let $\mathbf{Y} = (Y(\mathbf{s}_1), \ldots, Y(\mathbf{s}_n))$ and define the observed values $\mathbf{y}$ accordingly. Similarly we let X be the $n \times p$ matrix of covariates and the notation $\mathbf{x}_i$ is changed to $\mathbf{x}(\mathbf{s}_i)$ to emphasis the explicit spatial reference. With these changes in notation we still assume the linear model (6.2) $\mathbf{Y} \sim N \left(X\boldsymbol{\beta}, \sigma^2 H \right)$. What remains to be discussed is how do we form the matrix H. We use the Matèrn covariance function $C(d|\boldsymbol{\psi})$ in (2.1) for this purpose. Thus, we write the i, jth element of $\sigma^2 H$ as

$$\sigma^2 h_{ij} = C(d_{ij}|\boldsymbol{\psi})$$

where $d_{ij} = ||\mathbf{s}_i - \mathbf{s}_j||$ denotes the distance between the two locations $\mathbf{s}_i$ and $\mathbf{s}_j$ for $i, j = 1, \ldots, n$. The Matèrn covariance function depends on the additional parameter ψ. For simplicity, we shall assume the special case of exponential covariance function whereby we write

$$h_{ij} = \exp\left(-\phi d_{ij}\right)$$

and treat the decay parameter ϕ as known. For example, following on from earlier discussion in Section 2.6 we may choose a ϕ for which the effective range d_0 is a suitable multiple of the maximum possible distance between the n observation location. Thus, we may set $\phi = 3/d_0$. We may also choose ϕ using cross-validation approaches; see practical examples later in this chapter.

The marginal posterior distributions (6.7) are now readily computed for making inference. Out of sample prediction at a new location s_0 is performed by evaluating the posterior predictive distribution (6.10) of $Y_0 = Y(s_0)$ given $\mathbf{y}$. In order to calculate this posterior predictive distribution we need to evaluate two key quantities $\Sigma_{12}H^{-1}$ and $\delta^2 = 1 - \Sigma_{12}H^{-1}\Sigma_{21}$. The n dimensional cross-correlation vector Σ_{12} has the ith element given by $\exp\left(-\phi d_{i0}\right)$ where d_{i0} is the distance between the locations s_i and s_0 for $i = 1, \ldots, n$.

We repeat the above prediction procedure for predicting at, m say, multiple locations. The predictions obtained this way are from the marginal distributions of the m-dimensional joint predictive distribution. This is what is often done in practice and the software packages we will discuss later automatically render marginal predictions. Calculation of the multivariate joint predictive distribution requires further theoretical development starting by writing (6.9) for a vector valued $\mathbf{Y}_0$ instead of the scalar random variable Y_0. This has not been persuaded at all in this book as we do not require the multivariate joint predictive distribution unless we want to learn joint behavior of predictions at many different locations.

The general theory in Section 6.3 has so far enabled us to perform two most important tasks, viz. estimation and prediction. How do we calculate the three Bayesian model choice criteria, DIC, WAIC and PMCC? For PMCC, the posterior predictive distribution of $Y(s_i)|\mathbf{y}$ will simply be the distribution of the fitted values in the usual linear model theory. Thus, estimation of PMCC requires calculation of the mean and variance of the fitted values $X\beta$ given the observations $\mathbf{y}$. The required mean and variance are calculated using (6.8) and the results stated under the discussion of the multivariate normal distribution in Section A.1 in Appendix A.

$$E(X\beta|\mathbf{y}) = X\beta^* \text{ and } \mathrm{Var}(X\beta|\mathbf{y}) = \frac{2b^*}{n+2a-2}X\left(M^*\right)^{-1}X'.$$

Thus in the PMCC defining equation (4.27), we take $E(Y_{i*}|\mathbf{y})$ as the ith component of $X\beta^*$ and $\mathrm{Var}\left(Y_{i*}|\mathbf{y}\right)$ is the ith diagonal element of $\frac{2b^*}{n+2a-2}X\left(M^*\right)^{-1}X'$ for $i = 1, \ldots, n$.

The other two model choice criteria, DIC and WAIC are not easily available in closed form. Hence we will use the sampling based methodology detailed in Section 5.12 to estimate those. Samples from the two marginal posterior distributions $\beta|\mathbf{y}$ and $\lambda^2|\mathbf{y}$, see (6.7), are easily generated since they are standard distributions. An alternative sampling approach first draws samples from the marginal posterior distribution (6.7) of $\lambda^2|\mathbf{y}$ and then draws β from the full conditional posterior distribution (6.6) of $\beta|\lambda^2, \mathbf{y}$. With either of the two approaches we do not need to use any MCMC technique. Thus, this exact method is able to provide an important sanity check for the Bayesian computations using MCMC methods. Numerical illustrations of model choice criteria are provided in Section 6.7 of this chapter.

6.4.1   Spatial model fitting using `bmstdr`

We return to the New York air pollution data example. We are now able to include the three covariates, maximum temperature, wind speed and relative humidity. In addition, we assume that the error distribution is spatially correlated with a given value of effective range. Using the Bayesian model choice criteria we will now be able to compare the two models:

M1 Standard Bayesian linear regression model with the three covariates.

M2 Spatial linear model of this section.

For the regression coefficients β we provide a normal prior distribution with mean zero $(= \beta_0)$ and variance $10^4 (= M^{-1})$ for all components. We illustrate with a gamma prior distribution $G(a = 2, b = 1)$ for the precision parameter $\lambda^2 = \frac{1}{\sigma^2}$. We assume the exponential covariance function and take $\phi = 0.4$ which implies an effective range of 7.5 kilometers. The choice of this value of ϕ is discussed later.

The `bmstdr` package includes the function `Bspatial` for fitting regression models to point referenced spatial data. The arguments to this function has been documented in the help file which can be viewed by issuing the `R` command `?Bspatial`. The options **model="lm"** and **model="spat"** are used for fitting and analysis using models M1 and M2 respectively. Two required arguments for these functions are a formula and a data frame just like the similar ones for a classical linear model fitting using the `lm` function. For the built-in data `nyspatial` data set in `bmstdr` we can use the formula yo3 $\sim$ xmaxtemp + xwdsp + xrh.

Default values of the arguments `prior.beta0`, `prior.M` and `prior.sigma2` defining the prior distributions for β and $\lambda^2 = 1/\sigma^2$ are provided. An optional vector argument `validrows` providing the row numbers of the supplied data frame for model validation can also be given. This is discussed further in the model validation Section 6.8. The argument `scale.transform` can take one of three possible values: `NONE, SQRT` and `LOG` which defines the on the fly transformation for the response variable which appears on the left hand side of the formula. The model choice statistics are calculated on the opted scale but model validations and their uncertainties are calculated on the original scale of the response for ease of interpretation. This strategy of a possible transformed modeling scale but predictions on the original scale is adopted throughout this book.

Three additional arguments, `coordtype`, `coords` and `phi` are used to fit the spatial model M2 but not M1. The `coords` argument provides the coordinates of the data locations. The type of these co-ordinates, specified by the `coordtype` argument, taking one of three possible values: `utm`, `lonlat` and `plain` determines various aspects of distance calculation and hence model fitting. The default for this argument is utm when it is expected that the coordinates are supplied in units of meter. The `coords` argument provides the actual coordinate values and this argument can be supplied as a vector of size two identifying the two column numbers of the data frame to take as coordinates. Or this argument can be given as a matrix of number of sites by 2 providing the coordinates of all the data locations.

The parameter `phi` determines the rate of decay of the spatial correlation for the assumed exponential covariance function. The default value, if not provided, is taken to be 3 over the maximum distance between the data locations so that the effective range is the maximum distance, see Section 2.6.

There are other arguments for `Bspatial`, e.g. `verbose`, which control various aspects of model fitting and return values. Some of the other arguments are only relevant for specifying prior distributions and performing specific tasks as we will see throughout the remainder of this chapter. The return value is a list of class `bmstdr` providing parameter estimates, model choice statistics if requested and validation predictions and statistics as discussed in the following sections. The S3 methods **print**, **plot**, **summary**, **fitted**, and **residuals** have been implemented for objects of the `bmstdr` class.

The function calls

```
M1 <- Bspatial(formula=yo3 ~ xmaxtemp+xwdsp+xrh, data=nyspatial)
```

and

```
M2 <- Bspatial(model = "spat", formula=yo3 ~ xmaxtemp+xwdsp+xrh,
data=nyspatial, coordtype="utm", coords=4:5, phi = 0.4)
```

with the default arguments respectively fit the independent linear regression model and the spatial model exactly using the posterior and predictive distributions obtained in Section 6.3 for the running New York air pollution data example without any data transformation. The three commonly used S3 methods, **print**, **plot** and **summary** respectively, prints, plots and summarizes the model fitted object. For example, summary(M2) will print a summary of the model fit and plot(M2) will draw a fitted versus residual plot as in the case of model fitting using the lm function. Figure 6.1 provides the fitted versus residual plot and a plot of the residuals against observation numbers to detect serial correlation for M2. Such plots will be more useful in spatio-temporal modeling in Chapter 7. As noted above, the additional option `scale.transform="SQRT"` will fit the above models on the square-root scale for the response.

The `Bspatial` command also calculates the model choice statistics, DIC and WAIC and the PMCC. We, however, postpone our discussion on model choice until Section 6.7 so that we can compare the similar model fits using other **R** packages. The parameter estimates from the two models M1 and M2 are provided in Table 6.1. The estimates are very similar and what is remarkable is that the β_2 parameter for wind speed remains significant in the spatial model.

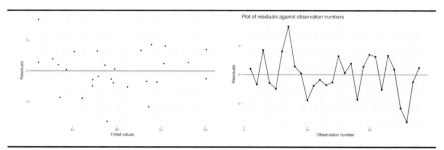

FIGURE 6.1: Fitted versus residual plots for M2 (left) and the residuals against observation number (right) to detect serial correlation.

| | Model M1 | | | | Model M2 | | | |
Param	Est	sd	2.5%	97.5%	Est	sd	2.5%	97.5%
β_0	−52.78	52.91	−157.13	51.57	−58.17	53.84	−164.35	48.01
β_1	1.53	0.99	−0.42	3.48	1.65	1.00	−0.33	3.62
β_2	3.25	1.19	0.91	5.59	3.28	1.19	0.94	5.62
β_3	11.24	8.46	−5.45	27.93	11.74	8.61	−5.25	28.73
σ^2	11.00	2.94	6.67	18.04	10.90	2.91	6.61	17.87

TABLE 6.1: Parameter estimates from the two models M1 and M2 with $\phi = 0.4$.

6.5   A spatial model with nugget effect

The regression model with spatially colored error distribution in Section 6.3 does not incorporate one key idea in spatial statistics – the so called 'nugget effect' or micro-scale variation, see e.g. Diggle and Ribeiro (2007). The main idea here is that variation in the spatial data cannot be explained by constant spatial variance alone. In addition to the spatial variance σ^2, it is thought, there is another source of small scale variation which may be also be attributed to measurement error.

The modeling incorporating a nugget effect is performed by splitting the error in the regression model (6.1) into two parts one accounting for spatial variation and the other accounting for the nugget effect. Thus, the model equation (6.1) is modified to:

$$Y(\mathbf{s}_i) = \mathbf{x}'(\mathbf{s}_i)\boldsymbol{\beta} + w(\mathbf{s}_i) + \epsilon(\mathbf{s}_i) \tag{6.12}$$

for all $i = 1, \ldots, n$. In the above equation, the pure error term $\epsilon(\mathbf{s}_i)$ is assumed to follow the independent zero mean normal distribution with variance σ_ϵ^2 for all $i = 1 \ldots, n$. The stochastic process $w(\mathbf{s})$ is assumed to follow a zero mean GP with the Matèrn covariance function (2.1) with unknown parameters $\boldsymbol{v} > 0$.

Using vectorised notation we write this model hierarchically as

$$\mathbf{Y}|\mathbf{w} \sim N(X\boldsymbol{\beta} + \mathbf{w}, \sigma_\epsilon^2 I) \tag{6.13}$$
$$\mathbf{w} \sim N(\mathbf{0}, \sigma_w^2 S_w) \tag{6.14}$$

where I is the identity matrix of order n, S_w is the $n \times n$ correlation matrix whose elements are formed using the Matèrn correlation function given above. Note that the matrix S_w does not contain the variance parameter σ_w^2 but it is a complex non-linear function of the parameters $\boldsymbol{v}$ describing the correlation function of the GP.

A full Bayesian model specification requires specification of the prior distributions for all the unknown parameters $\boldsymbol{\theta} = (\boldsymbol{\beta}, \sigma_\epsilon^2, \sigma_w^2, \boldsymbol{v})$. We can continue to assume a conjugate multivariate normal prior distribution for $\boldsymbol{\beta} \sim N(\boldsymbol{\beta}_0, M^{-1})$ without the scaling factor σ_ϵ^2 in the variance of $\boldsymbol{\beta}$. The scaling factor will not matter here because of two reasons: (i) we will use mostly non-informative prior distributions and (ii) we no longer aim to get exact posterior and predictive distributions. Instead, we shall use available software packages which do not require us to place a scaling factor up-front.

For non-informative prior specification, M will be taken as a diagonal matrix with very small values for the diagonal entries. We continue to work with precisions $\lambda_w^2 = 1/\sigma_w^2$ and $\lambda_\epsilon^2 = 1/\sigma_\epsilon^2$ and assign independent $G(a_w, b_w)$ and $G(a_\epsilon, b_\epsilon)$ prior distributions. Other type of prior distributions such as half-Cauchy for σ_ϵ and σ_w are also possible. Let $\pi(\boldsymbol{\theta})$ denote the combined joint prior distribution for all the parameters. The log-posterior density from the above model specifications and prior distributions is now written as:

$$\log \pi(\boldsymbol{\theta}, \mathbf{w}|\mathbf{y}) \quad \propto \quad -\frac{n}{2}\log(\sigma_\epsilon^2) - \frac{1}{2\sigma_\epsilon^2}(\mathbf{y} - X\boldsymbol{\beta} - \mathbf{w})'(\mathbf{y} - X\boldsymbol{\beta} - \mathbf{w})$$
$$-\frac{1}{2}\log|\sigma_w^2 S_w| - \frac{1}{2\sigma_w^2}\mathbf{w}'S_w^{-1}\mathbf{w} + \log \pi(\boldsymbol{\theta}). \tag{6.15}$$

This joint posterior distribution can be computed in several ways as described in the following subsections. Below we also discuss how to predict at an unobserved location $\mathbf{s}_0$.

6.5.1 Marginal model implementation

The spatial random effect $\mathbf{w}$ is easily integrated out from the hierarchical specifications (6.13) and (6.14). Thus, we have the marginal model for the data vector $\mathbf{Y}$:

$$\mathbf{Y} \sim N\left(X\boldsymbol{\beta}, \sigma_\epsilon^2\, I + \sigma_w^2 S_w\right). \tag{6.16}$$

Let $H = \sigma_\epsilon^2\, I + \sigma_w^2 S_w$ denote the covariance matrix of the marginal distribution of $\mathbf{Y}$. The log-posterior density function of $\boldsymbol{\theta}$ is now written as:

$$\log \pi(\boldsymbol{\theta}|\mathbf{y}) \propto -\frac{1}{2}\log\left(|H|\right) - \frac{1}{2}(\mathbf{y} - X\boldsymbol{\beta})'H^{-1}(\mathbf{y} - X\boldsymbol{\beta}) + \log \pi(\boldsymbol{\theta}). \tag{6.17}$$

The above log-posterior density function is free of the random effects $\mathbf{w}$ as intended and as a result MCMC implementation of this model avoids having to sample them. Because of this reduced sampling demand and $\pi(\boldsymbol{\theta}|\mathbf{y})$ is a density on a lower dimensional space, the marginal model may lead to faster MCMC convergence. However, conjugacy is lost for sampling the variance components σ_ϵ^2 and σ_w^2 since the joint posterior distribution $\pi(\boldsymbol{\theta}|\mathbf{y})$ in (6.17) is a complex non-linear function of the variance components. Hence sampling of the variance components will require tuning if implemented using a Metropolis-Hastings sampler. This model has been successfully implemented using the spLM function in the spBayes software package which we will illustrate with the New York pollution data example.

Prediction of $Y(\mathbf{s}_0)$ given $\mathbf{y}$ can proceed using the marginal model as follows. Note that, marginally

$$Y(\mathbf{s}_0)|\boldsymbol{\theta} \sim N(\mathbf{x}'(\mathbf{s}_0)\boldsymbol{\beta}, \sigma_\epsilon^2 + \sigma_w^2).$$

But to calculate the posterior predictive distribution $f(y(\mathbf{s}_0)|\mathbf{y})$ we require the conditional distribution of $Y(\mathbf{s}_0)|\boldsymbol{\theta}, \mathbf{y}$. This can be obtained by noting that, given $\boldsymbol{\theta}$:

$$\begin{pmatrix} Y(\mathbf{s}_0) \\ \mathbf{Y} \end{pmatrix} \sim N\left[\begin{pmatrix} \mathbf{x}'(\mathbf{s}_0)\boldsymbol{\beta} \\ X\boldsymbol{\beta} \end{pmatrix}, \begin{pmatrix} \sigma_\epsilon^2 + \sigma_w^2 & \sigma_w^2 S_{w,12} \\ \sigma_w^2 S_{w,21} & H \end{pmatrix} \right]$$

where $S_{w,12}$ and $S_{w,21}$ are similarly defined as in (6.9). Using the distribution theory for the multivariate normal distribution in Section A.1, we conclude that:

$$Y(\mathbf{s}_0)|\boldsymbol{\theta}, \mathbf{y} \sim N\left(\mu_y,\ \sigma_y^2\right) \tag{6.18}$$

where

$$\begin{aligned} \mu_y &= \mathbf{x}'(\mathbf{s}_0)\boldsymbol{\beta} + \sigma_w^2 S_{w,12} H^{-1}(\mathbf{y} - X\boldsymbol{\beta}), \\ \sigma_y^2 &= \sigma_\epsilon^2 + \sigma_w^2 - \sigma_w^4 S_{w,12} H^{-1} S_{w,21}. \end{aligned}$$

Now we re-write the posterior predictive density

$$f(y(\mathbf{s}_0)|\mathbf{y}) = \int_{-\infty}^{\infty} f\left(y(\mathbf{s}_0)|\boldsymbol{\theta}, \mathbf{y}\right) \pi(\boldsymbol{\theta}|\mathbf{y})\, d\boldsymbol{\theta}, \tag{6.19}$$

where $f(y(s_0)|\boldsymbol{\theta}, \mathbf{y})$ is the normal density in (6.18). As before, samples $\boldsymbol{\theta}^{(j)}$ from the posterior distribution can be used to estimate (6.19). The extra step required is to generate $y^{(j)}(s_0)$ from (6.18) by plugging in $\boldsymbol{\theta}^{(j)}$.

MCMC sampling using the marginal model does not automatically generate the spatial random effects $\mathbf{w}$. However, these may be required in various investigations to explore spatial variation and out of sampling predictions using the (6.18) if desired instead of the marginal (6.19). After sampling $\boldsymbol{\theta}^{(j)}$ from the marginal model we can generate $\mathbf{w}^{(j)}$'s using the full conditional distribution:

$$\mathbf{w}|\boldsymbol{\theta}, \mathbf{y} \sim N\left\{\Sigma_{[\mathbf{w}|\boldsymbol{\theta},\mathbf{y}]}\left(\frac{1}{\sigma_\epsilon^2}\mathbf{y} + \frac{S_w^{-1}}{\sigma_w^2}X\boldsymbol{\beta}\right), \ \Sigma_{[\mathbf{w}|\boldsymbol{\theta},\mathbf{y}]}\right\}$$

where

$$\Sigma_{[\mathbf{w}|\boldsymbol{\theta},\mathbf{y}]} = \left(\frac{I}{\sigma_\epsilon^2} + \frac{S_w^{-1}}{\sigma_w^2}\right)^{-1}.$$

6.6 Model fitting using software packages

Software packages are now introduced to fit and predict the data using the more advanced models with nugget effect. Exact posterior sampling by brute force coding, as has been done so far in the numerical examples, is difficult because of the complexity of the models with nugget effect that has introduced more than one variance component. In addition, we aim to learn several software packages so that we can tackle various data complexities such as those arising from having missing observations. We take a gentle and gradual approach to introducing the software packages and return back to fitting simpler models so that we can welcome new ideas and contrast those with what has been learned so far. Below we introduce the three software packages: (i) spBayes, (ii) R-Stan and (iii) R-inla.

6.6.1 spBayes

The spBayes package is very versatile as it is able to fit univariate and multivariate point referenced data including spatio-temporal data. This package was introduced to overcome limitations of generic all purpose Bayesian computing package such as BUGS which performs poorly in matrix computations especially for large spatial data sets. Moreover, BUGS and other packages such as geoR (Diggle and Ribeiro, 2007) are difficult to work with for fitting more advanced and multivariate models. The geoR package does not allow Bayesian computation using the full flexibility of the MCMC methods. spBayes, on the other hand, uses highly efficient and publicly available LAPACK and BLAS libraries to perform all matrix computations. As a result it is much faster and is thus able to provide the user with much greater flexibility in modeling. `spBayes` offers greater control to the user in designing the MCMC updates using Gibbs and Metropolis steps. This control is sometimes crucial in obtaining better mixing and faster convergence. Here are the key `spBayes` functions we need to learn.

- `spLM` and `spMvLM` are the main functions for fitting univariate and multivariate Bayesian spatial regression models respectively. The package allows the users to fit the marginal model (6.16) easily. Using the `spLM` template users can also code up their own covariance structures. This allows fitting of sophisticated multivariate spatial models, space-varying regression models, and even dynamic some spatio-temporal models.

- `spPredict` acts on `spLM` or `spMvLM` output object and performs predictions (univariate and joint) on sites provided by the user.

- `spDiag` acts on `spLM` or `spMvLM` output object to compute the DIC and PMCC.

- `prior` is a function that is used to specify prior distributions.

Arguments for `spLM`

- formula: specifies the fixed effect regression

- data: data-frame to look for Y, X and other data variables.

- `run.control`: specifies number of MCMC iterations.

- `n.samples`: specifies number of MCMC iterations.

- `beta.update.control`: prior for beta and how beta should be updated.

- `priors`: prior for model parameters using list argument.

- `tuning`: a list of tuning parameters used for the Metropolis-Hastings update.

- `var.update.control`: identifies parameters to be updated in Σ, specifies their priors and how they should be updated. Priors are specified through here.

- `starting`: a list of initial values for starting the MCMC chain.

- `beta.update.control` takes a method argument that can be "gibbs" for a Gibbs update for β (i.e. a normal draw) or "mh" for a Metropolis-Hastings update (with a multivariate normal proposal).

- `spRecover` is used to recover the model parameters and spatial random effects from the `spLM` output.

The `spPredict` function is used for spatial prediction at arbitrary sites. It requires the following inputs:

- `sp.obj`: output from a `spLM` model fitting.

- `pred.coords`: an $r \times 2$ matrix of r prediction point coordinates in two-dimensional space (e.g., easting and northing).

- `pred.covars`: An $r \times p$ matrix or data frame containing the regressors associated with `pred.coords`.

Following on from Section 6.4.1 we now define model M3 to be the marginal model (6.16) with the same three covariates. We adopt the exponential correlation function and hence we have three parameters $\sigma_\epsilon^2, \sigma_w^2$ and ϕ describing the covariance structure. For each of the four regression parameters we adopt independent normal prior distribution with mean 0 and variance 1000. For σ_w^2 we adopt the inverse gamma distribution with both shape and scale parameter value fixed at 2 and for the nugget effect parameter σ_ϵ^2 we adopt the inverse gamma distribution with shape parameter 2 and scale parameter 0.1. The shape parameter value at 2 guarantees that the prior distribution has finite mean but the distribution is sufficiently vague that it does not have finite variance. The adopted scale parameter values expresses the preference that the prior mean of σ_ϵ^2 is less than that of σ_w^2. This is justified because we expect the nugget (or measurement error, σ_ϵ^2) to be much smaller than the data variance, σ_w^2 here.

For the correlation decay parameter ϕ we adopt the uniform prior distribution $U(0.005, 2)$ which corresponds to an effective range between 1.5 to 600 kilometers. Recall that model M2 was fitted with a fixed value of $\phi = 1$ which correspond to an effective range of 3 kilometers. It is important to note that the results we report below are sensitive to the choice of the prior distributions for ϕ, σ_ϵ^2 and σ_w^2. Ideally, a prior sensitivity study should be conducted to explore the effect of the adopted prior distributions. This is omitted in the interest of brevity and we continue to study the evolution of the models by adding complexity gradually.

With reasonable starting and tuning values we run the MCMC chains for 5000 iterations and to make inference we use samples from the last 4000 iterations after discarding samples from the first 1000 iterations. We have performed MCMC convergence checks using the CODA package which we do not report here. The parameter estimates are reported in Table 6.2. These new parameter estimates are similar in magnitude to the ones we obtained for M2 in Table 6.1. The parameter β_2 for wind speed remains significant under all three models and the estimate of σ^2 under M2 tallies with the estimate of the total variance $\sigma_w^2 + \sigma_\epsilon^2$ under model M3. Models M2 and M3 achieve this with the help of the similar fixed and estimated values of ϕ under the two models. We defer a discussion on model choice until later in this chapter.

The `bmstdr` model fitting function `Bspatial` admits the option `package="spBayes"` for analysis and model fitting using `spBayes`. The required arguments are similar to the previous function call to `Bspatial`. The additional arguments required here includes: `prior.tau2` and `prior.phi` which are both two dimensional vectors. The `prior.tau2` provides the parameters of the inverse gamma distribution for the prior on the nugget effect, σ_ϵ^2. In this case the `prior.sigma2` provides the parameters of the inverse gamma distribution for the prior on spatial variance, σ_w^2. The two values in `prior.phi` are the lower and upper limits of the uniform prior distribution for ϕ the spatial decay parameter. If this is not specified the default values are chosen so that the effective range is uniformly distributed between 25% and 100% of the maximum distance between data locations. For our illustration below we set `prior.phi=c(0.005, 2)` which corresponds to an effective range between 1.5 and 600 kilometers. The maximum distance between the data locations in New York is about 591 kilometers. The `Bspatial` command for model fitting using the `package="spBayes"` option is given by:

```
M3 <- Bspatial(formula=yo3 ~ xmaxtemp+xwdsp+xrh, data=nyspatial,
    coordtype="utm", coords=4:5, prior.phi=c(0.005, 2), package="
    spBayes")
```

The parameter estimates from the fitted model M3 are provided in Table 6.2. These estimates are comparable to those in Table 6.1 for models M1 and M2. The last row for ϕ estimates an effective range of about 2.74 kilometers.

	mean	sd	2.5%	97.5%
β_0	-41.151	49.455	-138.316	58.492
β_1	1.320	0.941	-0.527	3.135
β_2	3.381	1.219	0.999	5.788
β_3	9.476	7.945	-6.519	24.718
σ_ϵ^2	12.518	3.654	7.458	21.306
σ_w^2	0.092	0.234	0.015	0.313
ϕ	1.094	0.617	0.060	1.962

TABLE 6.2: Parameter estimates for model M3 fitted using the `spBayes` software package.

6.6.2 R-Stan

The `Stan` software package can be used to fit the conditional model and perform predictions using the predictive distributions (6.18). We will illustrate the computations using specific numerical examples below.

Unlike `spBayes`, `Stan` is a general purpose Bayesian model fitting software package that implements HMC as introduced in Section 5.9. The package allows us to fit and predict using spatial and spatio-temporal models. In this section, we use the `rstan` package to illustrate model fitting of the full model (6.15) with exponential covariance function but keeping the decay parameter ϕ fixed at some value.

There are excellent online resources to learn the `Stan` software package and the associated language. Most notably we refer the reader to the documentation website, `https://mc-stan.org/users/documentation/` for a detailed user guide with examples. Below we discuss the main steps for getting started with `Stan` using the R-interface package `rstan`.

There are three main steps for computing using the `rstan` package. The first is a set up and data preparation stage. In this step we set up the number of chains, number of warm up iterations and number of samples, the initial values for each chain. More importantly, we also prepare a list containing the data and covariate values that we would like to pass on to the `Stan` program.

In the second step the main task is to write the `Stan` code preferably in a separate file named with a .stan extension. The stan file contains different code blocks: function, data and transformed data, parameter and transformed parameter, model and generated quantities. Each of these blocks starts with the name of the block followed by an open brace '{' and ends with a closing brace '}'. All the code blocks are optional and can be omitted if not required. Inside the blocks, comments are inserted by putting a double slash, //in front. Multiple lines of code are commented out with a /* at the start and */at the end. The `Stan` documentation website[1] states the following about the individual code blocks:

"The function-definition block contains user-defined functions. The data block declares the required data for the model. The transformed data block allows the definition of constants and transforms of the data. The parameters block declares the model's parameters – the unconstrained version of the parameters is what's sampled or optimized. The transformed parameters block allows variables to be defined in terms of data and parameters that may be used later and will be saved. The model block is where the log probability function is defined. The generated quantities block allows derived quantities based on parameters, data, and optionally (pseudo) random number generation."

In the third and final step the R function `stan` is called with all the prepared arguments in the first two steps. The `Stan` model fit object as the output of the *stan* function is summarized in the usual way to extract parameter estimates by using the drawn samples. The `Stan` output can be analyzed using the `CODA` library as well.

A few remarks are in order to help us getting started:

- Beginners may omit the function block at the start as it is not necessary to get started.

- All calculations and statements which do not need to be repeated at each MCMC iteration should be put in the data and transformed data block. This will ensure that those calculations are not repeated which may potentially slow down the computations. Any constants may be declared in the transformed data block.

---

[1] *https://mc-stan.org/docs/2_22/reference-manual/overview-of-stans-program-blocks.html*

- All parameters declared in the parameter block will be sampled and default prior distributions will be adopted unless prior distributions are explicitly provided in the model block.

- The transformed parameter block may contain declarations regarding any intermediate or auxiliary functions of parameters (and data) that may help write the model. These transformed parameters will be calculated at each iteration.

- Parameter transformation and Jacobian. This issue will arise only when a prior distribution is to be provided on a transformed scale of the original parameter. For example, `Stan` specifies the normal distribution with the standard deviation parameter σ instead of the conventional variance parameter σ^2. Hence, the parameter block should contain a declaration regarding σ, e.g. `real<lower=0> sigma;`. However, the user may decide to provide a prior distribution on σ^2, e.g. the inverse gamma distribution. In such a case the sampler will sample σ and the joint posterior density should be for σ and the other parameters. Hence, the log posterior density will need to be adjusted by the log of the absolute value of the Jacobian of the transformation from σ^2 (the old variable for which the prior distribution has been specified) to σ (the new variable which will be sampled). The Jacobian of the transformation is

$$\frac{d\,\text{old}}{d\,\text{new}} = \frac{d\sigma^2}{d\sigma} = 2\sigma.$$

Hence, the model block must contain the following statement:

```
target += log(2*sigma).
```

This Jacobian calculation can be avoided if we only use the variance parameter in all the statements and calculations. For example, the parameter declaration should be `real<lower=0> sigma2;` and the model block should contain a statement like the following:

```
y ~ normal(mu, sqrt(sigma2))
```

- Data variables and parameter declarations can be of integer, real, vector or matrix type. Also, finite ranges, if any, of each variable must be provided by using the lower and upper arguments in the declaration. Illustrations are provided in the code for the New York air pollution example.

- It may be easier to code predictions and further generations by hard coding of those in **R**, outside of `Stan`, although there are `rng` functions available in `Stan` to do this in the generated quantities block. Generating the predictions in the `Stan` requires more careful programming in `Stan` as we shall demonstrate. We provide illustrations of these methods with the help of the conditional distributions noted earlier for spatial modeling.

- Error and warning messages regarding divergent transitions are issued by Stan if there are problems with the sampler. These and other coding problems may be solved with the help of extensive documentations available online and also using community help. The examples provided in this book only helps the reader to get started.

We now define model M4 to be the full spatial model (6.15) with the same three covariates as before. We continue to adopt the exponential correlation function and have three parameters $\sigma_\epsilon^2, \sigma_w^2$ and ϕ describing the covariance structure. For each of the four regression parameters we adopt the default flat prior distributions built-in the Stan package. For σ_w^2 and σ_ϵ^2 we adopt the same inverse gamma distributions as before in the spBayes modeling Section 6.6.1. For our illustration we set $\phi = 0.4$ to get comparable results as obtained before. Stan does allow adoption of prior distributions for ϕ and we shall illustrate that for spatio-temporal modeling in the later chapters.

We save the following Stan code and save it as spatial_model.stan in the current working directory so that R can find it.

```
// data block. This must contain the same variables as data list
// that will come from R

data {
int<lower=0> n; // number of sites
int<lower=0> p; // number of covariates
vector[n] y;
matrix[n, p] X;
matrix[n, n] dist; // to hold n by n distance matrix
real<lower=0.00001> phi;
vector<lower=0>[2] priorsigma2;
vector<lower=0>[2] priortau2;
}

// These will never change during the MCMC computation.
transformed data {
  real delta=1e-5;
  vector[n] mu_0 = rep_vector(0, n);
}

// Declare all the parameters to be sampled here

parameters {
  vector[p] beta;
  real<lower=0> sigma_sq;
  vector[n] eta;
  real<lower=0> tau_sq;
}

// Model specification
```

```
model {
vector[n] xbmodel;
matrix[n, n] L;
matrix[n, n] Sigma;
real u;

// print(beta)
// print(sigma_sq)

xbmodel = X * beta;
 for (i in 1:n) {
   for (j in 1:n) {
     Sigma[i, j] = sigma_sq * exp((-1)*phi*dist[i,j]);
   }
   Sigma[i, i] = Sigma[i, i] + delta;
 }
 L = cholesky_decompose(Sigma);
 eta ~ multi_normal_cholesky(mu_0, L);
 sigma_sq ~ inv_gamma(priorsigma2[1], priorsigma2[2]);
 tau_sq ~ inv_gamma(priortau2[1], priortau2[2]);
 y ~ normal(xbmodel+eta, sqrt(tau_sq));
}
```

To run the Stan model we go through the following code in R.

```
rm(list=ls())
library(bmstdr)
head(nyspatial)
n <- nrow(nyspatial)
y <- nyspatial$yo3
X <- cbind(1, nyspatial[, 7:9])
p <- ncol(X)
head(X)
phi <- 0.4
distmat <- as.matrix(dist(as.matrix(nyspatial[, 4:5])))/1000
dim(distmat)
distmat[1:5, 1:5]
max(distmat)
datatostan <- list(n=n, p=p, y = y, X=as.matrix(X),
    priorsigma2 = c(2, 1), priortau2 = c(2, 1),
    phi=phi, dist=distmat)

M0 <- lm(formula=yo3 ~ xmaxtemp+xwdsp+xrh, data=nyspatial)
coef(M0)
initfun <- function() {
  # starting values near the lm estimates
  # variations will work as well
  list(sigma_sq = 1, tau_sq=1, beta=coef(M0))
}
library(rstan)
```

```
stanfit <- stan(data=datatostan, file = "spatial_model.stan", seed =
    44, chains = 1, iter = 1000, warmup = 500, init=initfun)

stanestimates <- rstan::summary(stanfit, pars =c("beta", "tau_sq",
    "sigma_sq"), probs = c(.025, .975))
names(stanestimates)
params <- data.frame(stanestimates$summary)
print(params)
```

The above code have been packaged in the `bmstdr` package function `Bspatial` when it is called with the option **package="stan"**. The required arguments are similar to the previous function call `Bspatial` for the linear regression model and the spatial model. The additional arguments `ad.delta`, `t.depth` are required here for controlling many aspects of the HMC algorithm underlying the `Stan` code. Suitable default values have been set for these parameters in the `Bspatial` function. See documentation for details. Now we fit the `Stan` model using the following command. The resulting parameter estimates, reported in Table 6.3 below, are similar to the ones from models M1, M2 and M3.

```
M4 <- Bspatial(package="stan", formula=yo3 ~ xmaxtemp+xwdsp+xrh,
    data=nyspatial,coordtype="utm", coords=4:5, phi=0.4)
```

	mean	sd	2.5%	97.5%
β_0	-55.127	58.871	-168.165	58.383
β_1	1.629	1.133	-0.520	3.841
β_2	3.221	1.304	0.632	5.772
β_3	11.150	9.241	-6.748	29.157
σ_ϵ^2	0.098	0.126	0.019	0.426
σ_w^2	12.432	3.702	7.213	21.064

TABLE 6.3: Parameter estimates for model M4 implemented in `Stan` with ϕ fixed at 0.4.

6.6.3   R-inla

Bayesian computations for the linear models are also performed using the INLA methodology discussed in Section 5.10. The key to fitting spatial model lies in writing down a model formula encompassing the fixed effects and the spatial random effects as follows. The R package `INLA` fits the continuous spatial model by preparing a discretized mesh of the study region. The Matèrn covariance function is used in the discretization process. We provide some details now. In the first stage we continue to assume the model (6.12) for observed data $Y(\mathbf{s}_i)$, for $i = 1, \ldots, n$. The spatial random effect $w(\mathbf{s})$ is defined using a discretized Gaussian Markov Random Field (GMRF) by:

$$w(\mathbf{s}) = \sum_{g=1}^{G} \xi_g(\mathbf{s})\tilde{w}_g \tag{6.20}$$

where G is the total number of vertices in the discretization of the spatial study region performed using a triangulation, $\{\xi_g\}$ is the set of basis functions and $\{\tilde{w}_g\}$ are zero mean Gaussian distributed random variables forming a GMRF. The basis functions are chosen to be piece wise linear on each triangle, i.e. ξ_g is 1 at vertex g and 0 elsewhere. Using the GMRF representation (6.20), the linear predictor in (6.12) is written as:

$$O(\mathbf{s}_i) = \mathbf{x}'(\mathbf{s}_i)\boldsymbol{\beta} + \sum_{g=1}^{G} \tilde{A}_{ig}\tilde{\mathbf{w}} \tag{6.21}$$

where $\tilde{A}$ is the sparse $n \times G$ matrix that maps the GMRF $\tilde{\mathbf{w}}$ from the n observation locations to the G triangulation nodes. Now the pair of specifications (6.13) and (6.14) are written as

$$\mathbf{Y} \sim N(\mathbf{O}, \sigma_\epsilon^2 I) \tag{6.22}$$

$$\mathbf{O} = \mathbf{x}'(\mathbf{s}_i)\boldsymbol{\beta} + \tilde{A}\tilde{\mathbf{w}} \tag{6.23}$$

where $\mathbf{O} = (O(\mathbf{s}_1), \ldots, O(\mathbf{s}_n))$ and $\tilde{\mathbf{w}}$, $G \times 1$ follows the GMRF as mentioned above.

The triangulation is created in R by using the helper function `inla.mesh.2d` which has arguments that include the coordinates of the data locations and many others controlling the behavior of the discretization. We omit the details of those arguments for brevity as those are provided in the help file for the function `inla.mesh.2d`. An example triangulation for our study region of the state of New York is provided in Figure 6.2. This plot has been obtained using the following commands

```
coords <- nyspatial[, c("utmx", "utmy")]/1000
max.edge <- diff(range(coords[,1]))/15
bound.outer <- diff(range(coords[,2]))/3
mesh <- inla.mesh.2d(loc = coords, max.edge = c(1,5)*max.edge,
offset = c(max.edge, bound.outer), cutoff = max.edge/5)
plot(mesh)
```

The dependence structure in the GMRF is setup by the R helper function `inla.spde2.pcmatern` which requires the created the mesh and specifications of prior distributions for the parameters in the Matèrn covariance function. The prior distributions for the spatial decay parameter ϕ and spatial variance parameter σ_w^2 are specified using what are called the penalized complexity (PC) prior distributions proposed by Simpson et al. (2017). The prior distributions are specified by two pairs of constants (ρ_0, p_ρ) and (σ_0, p_σ) such that:

$$P(\text{range} < \rho_0) = p_\rho, \quad P(\sigma_w > \sigma_0) = p_\sigma, \tag{6.24}$$

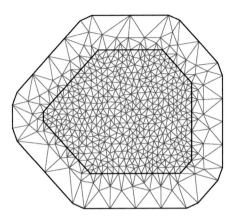

Constrained refined Delaunay triangulation

FIGURE 6.2: Triangulation of the region of the state of New York.

where range is the spatial range ($\approx 3/\phi$) for the exponential covariance function. The two constants p_ρ and p_σ may be specified as NA in which case the corresponding parameter will be fixed at the values specified by ρ_0 and σ_0. These prior distributions do not admit the inverse gamma prior distribution for σ^2 as a special case. The following **R** command provides an example of the prior PC prior distribution

```
spde <- inla.spde2.pcmatern(mesh = mesh, alpha = 1.5, prior.range =
    c(1, 0.5), prior.sigma = c(1, 0.05))
```

This example command specifies the exponential covariance function since it has set `alpha =1.5` and this value corresponds to the $\nu = 0.5$ parameter in the Matèrn covariance function. This command also implies $\rho_0 = 1$, $p_\rho = 0.5$, $\sigma_0 = 1$, and $p_\sigma = 0.05$ in (6.24).

All data and covariate values are passed to the front end **R** function `inla` in a particular way by using the `inla.stack` command. Covariate values for spatial prediction are also passed using another call to the `inla.stack` command. Let `y`, of length n, denote the dependent vector of data in **R** and let `X` denote the n by $p-1$ matrix of covariate values. This matrix has $p-1$ columns for the $p-1$ covariates without the unit vector for the intercept term. The observation weight matrix is formed using the **R** command:

```
A <- inla.spde.make.A(mesh = mesh, loc = as.matrix(coords))
```

A stack is formed by using the command

```
stack <- inla.stack(tag = "est", data = list(y = y), A = list(A, 1),
    effects = list(se = 1:spde$n.spde, Xcov = X))
```

An inverse gamma prior for the data precision parameter $1/\sigma_\epsilon^2$ is specified by using the command

```
hyper <- list(prec = list(prior = "loggamma", param = c(2,0.1)))
```

and a formula object is created using y ∼-1 + X + f(se, model = spde)

The main model fitting command is inla which takes the arguments created as above. In addition, many other optional parameters controlling different aspects of the model fitting and prediction process are passed on using the control option. An example model fitting command is:

```
ifit <- inla(formula, data = inla.stack.data(stack), family = "
    gaussian", control.family = list(hyper = hyper), control.predictor
    = list(A = inla.stack.A(stack), compute = T), control.compute = list
    (config = T, dic = T, waic = T), verbose = F)
```

On successful execution, the above command saves the inla fitted model as the object ifit. Inference regarding the parameters, predictions and various model choice statistics are obtained by extracting different elements from the fitted object ifit. The marginals.fixed and marginals.hyperpar of the fitted object ifit contain the estimates of the fixed effects β and the hyper-parameters (σ_ϵ^2, σ_w^2 and ϕ) respectively. The order of the hyper-parameters in the saved object can be obtained by issuing the R command: rownames(ifit$summary.hyperpar). As in the previous chapter we generate approximate samples from the various marginal posterior distributions using inla.rmarginal function. To generate samples from the marginal distribution of β_j we write betaj.samp <-inla.rmarginal(N, ifit$marginals.fixed[[j]]) where N is the desired number of samples. To generate samples from $\sigma_\epsilon^2|\mathbf{y}$ we issue the commands

```
prec.samp <- inla.rmarginal(N, ifit$marginals.hyperpar[[1]])
sig2e.samp <- 1/prec.samp
```

To generate samples from $\phi|\mathbf{y}$ we use

```
range.samp <- inla.rmarginal(N, ifit$marginals.hyperpar[[2]])
phi.samp <- 3/range.samp
```

For σ_w^2 the commands are

```
sd.samp <- inla.rmarginal(N, ifit$marginals.hyperpar[[3]])
sig2w.samp <- sd.samp^2
```

The generated samples can be used to make inference on any desired parameter or any function of the parameters. Model validation by performing out of sampling predictions requires a bit of extra work involved in generating posterior samples from the spatial effects and then effectively re-constructing the models (6.23) and (6.22) for each of the drawn samples. The code lines to perform this are a bit long and are not included here. Further details regarding `inla` based model fitting are available from many online resources including the text book for fitting spatial and spatio-temporal data by Blangiardo and Cameletti (2015) and a very recent generic text book by Gómez-Rubio (2020).

As previously for spBayes and Stan , the `bmstdr` function `Bspatial` with option **package="inla"** enables model fitting, validation and calculation of model choice statistics using the INLA package. The function call also returns the INLA fitted model object which can be used for further analysis and investigation. The `bmstdr` code for fitting the running model using the INLA package is given by:

```
M5 <- Bspatial(package="inla", formula=yo3 ~ xmaxtemp+xwdsp+xrh,
   data=nyspatial, coordtype="utm", coords=4:5)
```

	mean	sd	2.5%	97.5%
β_0	−12.420	27.335	−65.774	41.775
β_1	0.830	0.603	−0.383	1.996
β_2	3.671	1.104	1.543	5.911
β_3	5.126	4.776	−4.382	14.471
ϕ	8.637	30.187	0.395	40.553
σ_ϵ^2	0.126	0.396	0.000	0.819
σ_w^2	12.582	3.420	7.349	20.708

TABLE 6.4: Parameter estimates for model M5 implemented in INLA.

6.7 Model choice

Model fitting using the `bmstdr` function `Bspatial` allows us to perform the other important tasks, viz. model choice, model checking and cross-validation. To request model choice calculation the user must set the `mchoice` flag to `TRUE` in the model fitting command. Indeed, the `bmstdr` commands for calculating the model choice criteria are same as before but with the additional option `mchoice=T`. Note that the spatial models are multivariate models and as commented at the end of Section 4.16 we use the conditional distributions to compute the model choice criteria.

The model fitting commands are given below where `<as before>` stands for the arguments:

```
formula=yo3 ~ xmaxtemp+xwdsp+xrh, data=nyspatial, coordtype="utm",
   coords=4:5
```

```
M1.c <- Bspatial(package="none", model="lm", formula=yo3 ~ xmaxtemp+
    xwdsp+xrh, data=nyspatial, mchoice=T)
M2.c <- Bspatial(package="none", model="spat", <as before>, phi=0.4,
    mchoice=T)
M3.c <- Bspatial(package="spBayes", prior.phi=c(0.005, 2),
<as before>, mchoice=T)
M4.c <- Bspatial(package="stan", phi=0.4, <as before>, mchoice=T)
M5.c <- Bspatial(package="inla", <as before>, mchoice=T)
```

For M3, fitted by using the spBayes package, the spBayes command spDiag can be used to calculate the DIC, PMCC and another score statistic, which is not discussed here. The spDiag command does not, however, calculate the WAIC statistics. The bmstdr package uses the MCMC samples from the posterior distribution to calculate WAIC.

To facilitate comparison of the intercept only model with the regression models, M1-M5, we calculate the model choice criteria values for the univariate $N(\theta, \sigma^2)$ model for the same 28 y-observations by issuing the bmstdr command:

```
M0 <- Bmchoice(case="MC.sigma2.unknown", y=ydata)
```

This M0 can also be interpreted as the intercept only base model for the regression models M1-M5.

Table 6.5 provides the different model choice criteria for the six models M0 to M5. As expected M1-M5 are much better models than M0 and the spatial models M2-M5 are better than the independent error linear regression model M1. The model choice criteria DIC and WAIC are similar for the spatial models M2-M5. However, the INLA based M5 obtains much lower PMCC penalty than the other models M2-M4. This is because the variance of the predictions are much smaller under the INLA model fit. This may seem to be a good news but this also mean that the prediction intervals are shorter and hence the model based predictions may not cover the full range of variance of the data. We shall explore this formally using a coverage criteria in Section 6.8.

6.8   Model validation methods

Statistical models are approximations for reality. But there can be very many such approximations and Bayesian model choice methods described in Chapter 4 are all relevant tools and should be used for choosing between these approximations. In addition to those tools there is a number of cross-validation criteria which are often used in practice. The common theme in all these criteria is the partition of the sample data into a training set and a testing set. The training set is used to fit the proposed models and to judge the quality of

	M0	M1	M2	M3	M4	M5
p_{DIC}	2.07	4.99	4.98	5.17	4.85	4.17
$p_{DIC\ alt}$	13.58	5.17	5.16	7.83	5.13	
DIC	169.20	158.36	158.06	158.68	157.84	157.23
DIC Alt	192.22	158.72	158.41	163.99	158.39	
$p_{waic\ 1}$	1.82	5.20	4.93	4.88	4.42	4.73
$p_{waic\ 2}$	2.52	6.32	5.91	6.77	5.29	
WAIC 1	168.95	158.57	157.51	158.70	156.99	158.46
WAIC 2	170.35	160.82	159.47	162.48	158.72	
gof	593.54	327.98	330.08	323.56	320.49	334.03
penalty	575.80	351.52	346.73	396.63	393.92	39.17
PMCC	1169.34	679.50	676.82	720.18	714.41	373.19

TABLE 6.5: Model choice criteria for various models. Note: M0 is the intercept only model and INLA does not calculate the alternative values of DIC and WAIC.

the fit. The test set, also called the validation set, is used to criticize the out of sample predictive capability of the models. A model validation statistic is a discrepancy measure between the actual observations and their model based predictions in the test set. The model with the least value of the discrepancy statistic is selected to be the best among the ones in the competition.

There are myriads of such validation statistics, each addressing a particular demand on the fitted model, and there are a large number of ways to split the sample data into training and test sets. Subsection 6.8.2 below provides some discussion on cross-validation methods. Section 9.4 introduces more methods and plots for forecast validation. However, many of those methods can be used for model validation as well.

In the interest of brevity, in this Subsection we describe only four most widely used model validation criteria especially for prediction using spatio-temporal modeling. We also discuss how to estimate those criteria using MCMC samples. We also numerically illustrate these methods with a real data example in Section 6.8.3.

6.8.1 Four most important model validation criteria

Suppose that the test set consists of r observations y_ℓ and the associated covariates x_ℓ for $\ell = 1, \ldots, r$. Assume that all these are available, i.e. there are no missing values in any of the y and x values. We pretend that we do not know the test response data y_ℓ for $\ell = 1, \ldots, r$ and use the fitted model to predict those. In the prediction procedure we assume x_ℓ values for $\ell = 1, \ldots, r$ are available. How could we predict each validation observation y_ℓ?

Bayesian theory discussed in Section 4.11 dictates that the predictions must be made by using the posterior predictive distribution $f(y_\ell|y)$ as written down in (4.9). We have seen that this distribution is only available in a few examples involving conjugate prior distributions for normally distributed errors. In general, the posterior predictive distribution $Y_\ell|\mathbf{y}$ will not be standard but we have a set of MCMC samples $y_\ell^{(j)}$, $j = 1, \ldots, J$ for a large value of J is available from the posterior distribution $f(y_\ell|\mathbf{y})$. Let $\hat{y}_\ell$ denote the chosen summary, usually either the mean or median, of the J predictive samples. Thus, $\hat{y}_\ell$ is the model based prediction for the actual observation y_ℓ for $\ell = 1, \ldots, r$.

1. **Root Mean Square Error (RMSE)** By far the most popular predictive criteria is the RMSE defined by:

$$\text{RMSE} = \sqrt{\frac{1}{r} \sum_{\ell=1}^{r} (y_\ell - \hat{y}_\ell)^2}.$$

2. **Mean Absolute Error (MAE)** is an alternative to the RMSE and is defined by:

$$\text{MAE} = \frac{1}{r} \sum_{\ell=1}^{r} |y_\ell - \hat{y}_\ell|.$$

3. **Continuous Ranked Probability Score (CRPS).** CRPS provides a better measure of discrepancy between the observations and the predictions by using the whole predictive distribution $f(y_\ell|\mathbf{y})$ rather than using a particular summary $\hat{y}_\ell$ as has been used by the RMSE and MAE. Let $F_\ell(y_\ell)$ denote the cumulative distribution function (cdf) corresponding to the predictive distribution $f(y_\ell|\mathbf{y})$, where in $F_\ell(\cdot)$ we have suppressed the conditioning on $\mathbf{y}$ for convenience. We define

$$\text{CRPS}(F_\ell, y_\ell) = E_{F_\ell}|Y_\ell - y_\ell| - \frac{1}{2} E_{F_\ell}|Y_\ell - Y_\ell^*|$$

where Y_ℓ and Y_ℓ^* are independent copies of a random variable with cdf $F_\ell(\cdot)$ having finite mean. We estimate the above measure using MCMC samples as follows:

$$\widehat{\text{CRPS}}(F_\ell, y_\ell) = \frac{1}{N} \sum_{j=1}^{N} |y_\ell^{(j)} - y_\ell| - \frac{1}{2N^2} \sum_{j=1}^{N} \sum_{k=1}^{N} |y_\ell^{(j)} - y_\ell^{(k)}|.$$

With r validation observations we calculate the overall measure, given by

$$\widehat{\text{CRPS}} = \frac{1}{r} \sum_{\ell=1}^{r} \widehat{\text{crps}}(F_\ell, y_\ell).$$

Thus, CRPS is an integrated distance between the validation predictions and the corresponding observations.

4. **Coverage (CVG)** We calculate coverage of the $100(1-\alpha)\%$, e.g. the 95% predictive intervals by using

$$\text{CVG} = 100\frac{1}{r}\sum_{\ell=1}^{r} I\left(L_\ell \leq y_\ell \leq U_\ell\right)$$

where (L_ℓ, U_ℓ) is the $100(1-\alpha)\%$ predictive interval for the predicting y_ℓ and $I(\cdot)$ is the indicator function. Often, L_ℓ and U_ℓ are simply the $2.5th$ and $97.5th$ percentile points of the MCMC samples, $y_\ell^{(j)}$ for $j = 1, \ldots, J$.

Note that CVG is not a discrepancy measure like the previous three criteria. The ideal value of CVG will be the theoretical value $100(1-\alpha)$. The model which produces a CVG value closest to $100(1-\alpha)$ is to be chosen as the best model. A point worth noting here is that very high CVG value exceeding $100(1-\alpha)$ is also not desirable as well as a very low value less than α. This is because a very high value may indicate that the prediction intervals are too wide so that those contain the true observations in a higher proportion than what would be expected. In such a case the predictions have too much uncertainty and that there is scope for reducing prediction uncertainty by re-modeling.

The reader is able to choose any number of the above four criteria for their modeling and analysis purposes. The first three of these four criteria are error rates which we seek to minimize. The last once, CVG, assesses the overall uncertainty level of the predictions and this should be near the adopted theoretical value, i.e. 95% in almost all the examples in this book.

6.8.2 K-fold cross-validation

In Bayesian K-fold cross-validation, we partition the data, preferably randomly, into K subsets for a suitable value of K, typically a number between 5 to 10 in the literature. All the data in each of these K subsets are used as test data and the data in the remaining $K-1$ subsets are used as training data. Each of the K training and test pair of data sets are used to fit and then validate models. Thus, model fitting is repeated K times and K sets of validation statistics, see the above Subsection, are obtained. In addition to the above statistics one may also evaluate the other model choice criteria such as the DIC, WAIC, and PMCC detailed in Section 4.16. This strategy of K-fold cross-validation guards against accidental choosing of a poor model due to selection of a particularly favorable validation set. Thus, this strategy adds to a desirable amount of robustness in model selection and validation.

The leave one out cross-validation (LOO-CV) is a particular case of K-fold cross-validation when $K = n$, the sample size. In LOO-CV, the training set consists of $n-1$ data points and the test is just the remaining data point.

Thus, the LOO-CV predictive density is given by

$$f(y_i|\mathbf{y}_{(-i)}) = \int_{-\infty}^{\infty} f(y_i|\theta, \mathbf{y}_{(-i)}) \, \pi(\theta|\mathbf{y}_{(-i)}) \, d\theta.$$

where $\mathbf{y}_{(-i)}$ is $n-1$ dimensional containing all observations except for y_i. For the normal-normal example in Section 4.11.1 $f(y_i|\mathbf{y}_{(-i)})$ is the normal distribution

$$N\left(\mu_p(-i), \ \sigma^2 + \sigma_p^2(n-1)\right)$$

where

$$\sigma_p^2(n-1) = \frac{1}{\frac{n-1}{\sigma^2} + \frac{1}{\tau^2}}$$

and

$$\mu_p(-i) = \sigma_p^2(n-1)\left(\frac{(n-1)\bar{y}_{-i}}{\sigma^2} + \frac{\mu}{\tau^2}\right), \quad \bar{y}_{-i} = \frac{1}{n-1}\sum_{j \neq i} y_j.$$

In general, estimation of the LOO-CV predictive density requires re-fitting the model n times which can be a computationally demanding task for moderately large values of n. However, several MCMC based approximation methods are available for this. For example, Gelfand et al. (1992) suggest the following scheme:

$$\hat{f}^{-1}(y_i|\mathbf{y}_{(-i)}) = \frac{1}{N}\sum_{j=1}^{N} \frac{1}{f(y_i|\mathbf{y}_{(-i)}, \theta^{(j)})}$$

where $\theta^{(j)}$ are the MCMC samples from the full posterior distribution $\pi(\theta|\mathbf{y})$. Gelman et al. (2014) details some other method based on importance sampling for estimating the cross-validation densities. Those methods are more suited for independently distributed data sets for which $f(y_i|\mathbf{y}_{(-i)}, \theta) = f(y_i|\theta)$. We do not consider those techniques any further in this book.

6.8.3 Illustrating the model validation statistics

The model validation statistics are calculated for the New York air pollution data example. Data from eight validation sites 8, 11, 12, 14, 18, 21, 24, and 28 are set aside and model fitting is performed using the data from the remaining 20 sites, see Figure 6.3.

> ♣ **R Code Notes 6.1. Figure 6.3** The function `geom_rect` has been used to draw the rectangle box to annotate.

The `bmstdr` command for performing validation needs an additional argument `validrows` which are the row numbers of the supplied data frame which should be used for validation. Thus, the commands for validating at the sites 8, 11, 12, 14, 18, 21, 24, and 28 are given by:

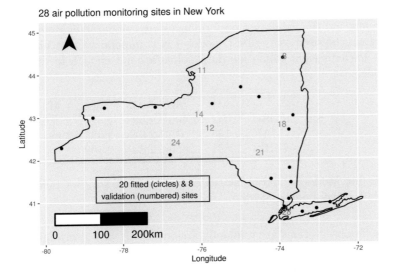

FIGURE 6.3: Eight validation sites in New York.

```
s <- c(8,11,12,14,18,21,24,28)
M1.v <- Bspatial(package="none", model="lm", formula=yo3 ~ xmaxtemp+
    xwdsp+xrh, data=nyspatial, validrows=s)
M2.v <- Bspatial(package="none", model="spat", phi=0.4, <as before>,
    validrows=s)
M3.v <- Bspatial(package="spBayes", prior.phi=c(0.005, 2),
<as before>, validrows=s)
M4.v <- Bspatial(package="stan", phi=0.4, <as before>, validrows=s)
M5.v <- Bspatial(package="inla", <as before>, validrows=s)
```

where `<as before>` stands for the arguments:

```
formula=yo3 ~ xmaxtemp+xwdsp+xrh, data=nyspatial, coordtype="utm",
    coords=4:5
```

We now return to discussing the choice of the fixed value of the spatial decay parameter ϕ in M2. Previously we have used the fixed value of 0.4 during estimation and model choice. We now use validation to find the optimal value of ϕ. We take a grid of ϕ values calculate the RMSE value for each value of ϕ in the grid. The optimal ϕ is the one that minimizes the RMSE. A simple R function can be written to perform this grid search. Indeed, the function `phichoice_sp` has been provided in the `bmstdr` package. For the New York air pollution example 0.4 turns out to be the optimal value for ϕ. All the necessary code for running this and other examples has been provided online on github[2].

---

[2]https://github.com/sujit-sahu/bookbmstdr.git

Table 6.6 present the statistics for all five models. Coverage is 100% for all five models and the validation performances are comparable. Model M2 with $\phi = 0.4$ can be used as the best model if it is imperative that one must be chosen. Perhaps eight validation data points are not enough to separate out the models.

	M1	M2	M3	M4	M5
RMSE	2.447	2.400	2.428	2.423	2.386
MAE	2.135	2.015	2.043	2.035	1.983
CRPS	2.891	2.885	2.891	2.199	1.944

TABLE 6.6: Model validation statistics for the New York air pollution data example for five models, M1-M5.

The 28 observations are randomly assigned to $K = 4$ groups of equal size for the purposes of calculating K-fold validation statistics. Table 6.7 presents the 4-fold validation statistics for M2 only. It shows a wide variability in performance with a low coverage of 57.14% for Fold 3. In this particular instance four of the seven validation observations are over-predicted, see Figure 6.4 resulting in low coverage and high RMSE. However, these statistics are based on data from seven validation sites only and as a result these may have large variability explaining the differences in the k-fold validation results.

	Fold 1	Fold 2	Fold 3	Fold 4
RMSE	2.441	5.005	5.865	2.508
MAE	1.789	3.545	5.462	2.145
CRPS	2.085	2.077	1.228	2.072
CVG	100%	85.71%	57.14%	100%

TABLE 6.7: Four-fold model validation statistics for model M2.

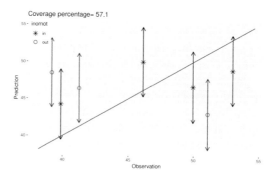

FIGURE 6.4: Observed versus prediction plot for Fold 3 using model M2.

6.9 Posterior predictive checks

Figure 6.5 provides two diagrams performing posterior predictive checks for the entire distribution of the pollution data and the maximum. These diagram show that the predictive distributions are able to capture the full scale of data variation.

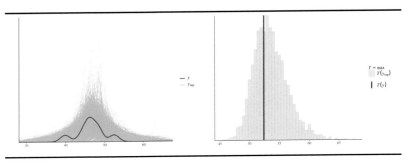

FIGURE 6.5: The left panel shows an overlay of the data density and the posterior predictive density. The right panel shows the posterior predictive distribution of the maximum of the pollution values.

6.10 Conclusion

This chapter has introduced spatial modeling for point referenced data. A hierarchy of more gradually complex models have been presented starting with a simple independent error linear regression model without any spatial effects. Model fitting has been illustrated with coding in R using the `bmstdr` package. The main model fitting command `Bspatial` resembles the lm command in R which require a data frame and a formula argument. Additional arguments providing the spatial information are required for spatial data modeling. The main objective here has been to transition the reader from fitting linear models to fitting more advanced spatial data models.

The output of `Bspatial` can be explored by the standard R commands, **plot, print, summary, fitted** and **residuals**. The **plot** command also draws the fitted versus residual plot. The familiar **predict** command, however, does not currently work with the `Bspatial` model fitted output.

Software packages `spBayes`, `Stan` and `INLA` have been introduced so that the rearder can go on experimenting those packages themselves. Glimpses of code to be written for fitting with these packages have been provided so that the reader can learn to code themselves. These packages are also invoked within the `Bspatial` command so that the reader can seamlessly use those packages without having to invest a lot of time intially before starting to use those in their practical data modeling.

By way of limitation, this chapter does not fit models for discrete data and neither does it study prior sensitivity and MCMC convergence. The first limitation can be overcome by more coding with the `bmstdr` package and the second is for the sake of brevity. The reader is urged to undertake the sensitivities studies themselves. Instead, the chapter presented the key measures and concepts in model validation, e.g. RMSE, CRPS, Coverage and k-fold cross validation methods. With this knowledge base in model fitting and validation we move to Chapter 7 for spatio-temporal modeling of point referenced data.

6.11   Exercises

1. Verify all the theoretical results reported in Section 6.3.

2. Write function in `R` to calculate the CRPS statistics based on two inputs: (i) an $m \times 1$ vector of observations – some which may be missing and (ii) an $m \times N$ matrix of MCMC samples. Study the computation time required for large values of m and N.

3. Reproduce all the tables and figures reported in this chapter. Most of the required code is provided on github.

4. For a practical data set of interest write your own code to fit models using the packages `spBayes`, `Stan` and `INLA`. You can use the `piemonte` data set, taken from Blangiardo and Cameletti (2015) and made available online through github, after aggregating it over time. See Blangiardo and Cameletti (2015) for a description of the `piemonte` data set.

5. Change the parameters `max.edge`, `offset` and `cutoff` to vary the resolution of the triangulation used in `INLA` based model fitting. Investigate the effect of such changes on the model fitting and validation. Is there an optimal set of values for these parameters?

6. Obtain cross-validation Table 6.7 but using `Stan`, , `INLA` and `spBayes`.

7. Posterior predictive checks have been illustrated in Section 6.9. Perform this for a spatial model fitted using `spBayes` and `INLA`.

7

Bayesian modeling for point referenced spatio-temporal data

7.1 Introduction

The Bayesian spatio-temporal models are represented in a hierarchical structure where we specify distributions for data, spatio-temporal process and parameters in three stages:

First $[data|process, parameter]$

Second $[process|parameter]$

Third $[parameter]$

The first stage specifies the top level distribution of the data to be modeled after suitable transformation. This specification may also be provided using a generalized linear model if the data distribution is discrete, e.g. binomial or Poisson, as has been done in Chapter 10 for areal data. The development in this chapter is for continuous data where we assume a top level Gaussian model for data. Data transformation, e.g. log or square-root or in general a Box-Cox transformation, is often necessary to stabilize the variability of the data and to encourage normality. Model fitting, checking and choice tasks are performed at the transformed scale but predictions are done at the original scale for easy interpret-ability by the practitioner and also in the interest of correctly assessing the uncertainty of the predictions at the original scale.

In the second stage of the modeling hierarchy, we specify a GP based process which can have many different forms depending on the assumptions we would like to make. For example, we may assume independent over time GP as the underlying process or we may assume a separable space-time covariance structure (Mardia and Goodall, 1993) for the spatio-temporal data. Different model structures to accommodate different types of temporal dependence may also be added. In the third stage of the hierarchy we introduce the prior distributions for the parameters and the hyper-parameters.

In the process of model extension from independent regression models to spatio-temporal regression models we extend the notation for data to include a spatial identifier $\mathbf{s}$ and a discrete temporal identifier t. Thus, in the development below we replace the earlier random variable $Y(\mathbf{s}_i)$ in Chapter 6 by

DOI: 10.1201/9780429318443-7

$Y(\mathbf{s}_i, t)$ for $i = 1, \ldots, n$ and $t = 1, \ldots, T$ to model data from n spatial locations $\mathbf{s}_1, \ldots, \mathbf{s}_n$ and T regularly spaced time points. Data observed at irregular time intervals and at moving spatial locations are discussed later in Section 8.6 in Chapter 8.

Corresponding to the response variable $Y(\mathbf{s}_i, t)$ we let $\mathbf{x}(\mathbf{s}_i, t) = (x_1(\mathbf{s}_i, t), \ldots, x_p(\mathbf{s}_i, t))$ denote the p-dimensional covariate vector. Some of these p covariates may only vary spatially, e.g. the type of location; some others may vary only temporally, e.g. seasonally and some may vary in both space and time, e.g. a continuous covariate such as temperature which varies in both space and time for modeling air pollution.

To make use of the general linear model theory we stack the nT observations in a long vector, $\mathbf{y}$, where the time index varies fastest. That is, all T observations from site 1 are put in first and then the observations from site 2 and so on. Thus,

$$\mathbf{y} = (y(\mathbf{s}_1, 1), y(\mathbf{s}_1, 2), \ldots, y(\mathbf{s}_1, T) \ldots, y(\mathbf{s}_n, 1), y(\mathbf{s}_n, 2), \ldots, y(\mathbf{s}_n, T)).$$

We follow the same convention for organizing the covariate data and we thus end up with the $nT \times p$ covariate matrix X. Note that data from all the locations and time points must be present in $\mathbf{y}$ and the covariate (design) matrix X. Missing data, if there is any, in $\mathbf{y}$ may be recorded as NA but the vector $\mathbf{y}$ should not skip any entry where there is missing data as that will break the regularity in the spatio-temporal data structure. In the modeling development in this section we do not include the possibility of any missing value of the covariates. Any missing value of the covariates must be imputed before going into modeling, although this is not totally satisfactory as this does not take care of the additional uncertainty arising from such imputation. Imputation can be avoided by explicit modeling of the covariates but that requires further bespoke code development using a general purpose Bayesian modeling software package such as BUGS. We refer the reader to Section 9.1.2 of the BUGS book by Lunn et al. (2013).

In the first stage of modeling hierarchy we assume the model

$$Y(\mathbf{s}_i, t) = \mu(\mathbf{s}_i, t) + e(\mathbf{s}_i, t), \ i = 1, \ldots, n, \ t = 1, \ldots, T, \tag{7.1}$$

where $\mu(\mathbf{s}_i, t)$ is a space-time process describing the mean structure of the data $Y(\mathbf{s}_i, t)$ and the error term $e(\mathbf{s}_i, t)$ is a zero-mean space-time process. The mean structure is modeled as a regression model using the available covariates $\mathbf{x}(\mathbf{s}_i, t)$, i.e.

$$\mu(\mathbf{s}_i, t) = \mathbf{x}'(\mathbf{s}_i, t)\boldsymbol{\beta}(\mathbf{s}_i, t) = \sum_{j=1}^{p} x_j(\mathbf{s}_i, t)\beta_j(\mathbf{s}_i, t).$$

This regression model is designed to capture high level spatial, temporal, and/or spatio-temporal trend in the data. For example, there may be a North-South divide in the data that the model should try to investigate. Or there may be a simple linear temporal trend or known seasonality that must be included in the model. The regression model is also flexible in accommodating spatio-temporally varying regression coefficients $\beta(\mathbf{s}_i, t)$. More complex models of such nature can be gradually built-upon starting from the simple model where $\beta(\mathbf{s}_i, t) = \beta$. Subsequent sections in this chapter will experiment with many different choices for the regression models. A cautionary note here is that the values of the included covariates at the desired prediction locations and time points must also be collected in order to predict at those location-time point combinations. Otherwise, the model can only learn about the effects of the covariates but will not be able to perform out-of-sample predictions.

The error term $e(\mathbf{s}_i, t)$ provides opportunities for spatio-temporal process modeling and can be thought, and hence interpreted, in many different ways. Perhaps the simplest first attempt is to treat $e(\mathbf{s}_i, t)$ as a zero mean spatio-temporal Gaussian process. This GP will be based on an assumed space-time correlation structure. A general fully flexible space-time correlation structure will require working with (e.g. inverting) the $nT \times nT$ covariance matrix of the nT dimensional error vector $\mathbf{e}$ collecting all the $e(\mathbf{s}_i, t)$ for $i = 1, \ldots, n$ and $t = 1, \ldots, T$. Gneiting (2002) provides a valid, attractive covariance function in space and time. However, model fitting under such a specification requires inversion of $nT \times nT$ matrices, which is prohibitive for moderately sized problems even for moderate values of n (e.g. 100) and T (e.g. 365).

A separable in space and time correlation structure as proposed by Mardia and Goodall (1993) avoids the above computation problem at the cost of simplifying the nature of the space-time interaction which can be learned from the data. By assuming that the joint space-time correlation function is just a product of a correlation function in space and another one in time it is possible to overcome the computational problem. Under a separable space and time correlation structure assumption, model fitting and prediction will only require working, e.g. inverting, with $n \times n$ and $T \times T$ matrices instead of the $nT \times nT$ covariance matrix for the nT dimensional vector $\mathbf{e}$. This method is further explored in Subsection 7.2 with the notation $e(\mathbf{s}_i, t)$ changed in favor of the more popular $\epsilon(\mathbf{s}_i, t)$.

The most popular modeling approach is to write the overall error term as the sum of two independent pieces $w(\mathbf{s}_i, t)$ and $\epsilon(\mathbf{s}_i, t)$ where both have mean zero, as has been done in Section 6.5 for spatial data. These two pieces serve two different purposes in modeling. The first one, $w(\mathbf{s}_i, t)$ treated as a random process, models spatio-temporal dependence while the second one is used as a pure error term. The pure error term is often assumed to be independent for all spatial locations at all time points, and it is also interpreted as the measurement error term. Another spatial statistics interpretation is to think of this as the 'nugget' effect to capture micro-scale variation. The micro-scale variation is really spatial variation on a scale smaller than the smallest distance between any two points in the sample design. In spatial only data modeling the nugget effect can be directly estimated by using repeated measurements taken at coincident locations. In our case of spatio-temporal data modeling the time series of observations at each location provide information that help in this estimation. Usually we suppose that $\epsilon(\mathbf{s}_i, t) \sim N(0, \sigma_\epsilon^2)$ independently of $w(\mathbf{s}_i, t)$ and also independently for $i = 1, \ldots, n$ and $t = 1, \ldots, T$. The nugget effect can be made temporally varying by assuming a different variance denoted by $\sigma_{\epsilon,t}^2$ and different time point t for $t = 1, \ldots, T$.

We now discuss the first term $w(\mathbf{s}_i, t)$ in the combined error term $e(\mathbf{s}_i, t)$. This term is used to model the desired spatial and temporal dependence. Chapter 11 in Banerjee et al. (2015) provides many general suggestions for modeling. For example, $w(\mathbf{s}_i, t) = \alpha(t) + w(\mathbf{s}_i)$ for suitable independent random variables $\alpha(t)$ and $w(\mathbf{s}_i)$ so that the spatio-temporal effect is just the sum of a temporal and a spatial term. The temporal term $\alpha(t)$ may provide a temporal dynamics, e.g. $\alpha(t) = \rho\alpha(t-1) + \eta(t)$ where $\eta(t)$ may be assumed to follow an independent zero mean normal distribution, i.e. a white noise process. The spatial term $w(\mathbf{s}_i)$ may be assumed to be a GP. Another possibility is to assume a separable covariance structure for $w(\mathbf{s}_i, t)$ as has been assumed for the full error term in Section 7.2. But here the nugget effect $\epsilon(\mathbf{s}_i, t)$ is retained along with the separable spatio-temporal process $w(\mathbf{s}_i, t)$. There are numerous other possible models for $w(\mathbf{s}_i, t)$. For example, $w(\mathbf{s}_i, t)$ may be assumed to be a GP independent in time. In another example of modeling, the term $w(\mathbf{s}_i, t)$ may be assigned an auto-regressive model in time,

$$w(\mathbf{s}_i, t) = \rho w(\mathbf{s}_i, t-1) + \eta(\mathbf{s}_i, t)$$

where ρ is the auto-regressive parameter and the new error term $\eta(\mathbf{s}_i, t)$ is given a temporally independent GP. Further model specification for $w(\mathbf{s}_i, t)$ may be based on process convolution which leads to non-stationary models. For example, Sahu et al. (2006) obtain a non-stationary model by writing $w(\mathbf{s}_i, t) = u(\mathbf{s}_i, t) + p(\mathbf{s}_i)v(\mathbf{s}_i, t)$ where $u(\mathbf{s}_i, t)$, $p(\mathbf{s}_i)$ and $v(\mathbf{s}_i, t)$ are independent GP's with different covariance functions.

Vectorised notations for data, processes and parameters enable us to write models and various distributions more economically. In this book we always model data at discrete time t. Hence we are allowed to use the symbol t as a suffix, e.g. $w_t(\mathbf{s})$ instead of the general $w(\mathbf{s}, t)$. The later is more general since mathematically $w(\mathbf{s}, t)$ is a function of both $\mathbf{s}$ and t and there t can be a continuous argument just like $\mathbf{s}$, but by convention we use a non-negative integer as a suffix such as z_5. In the sections below we use boldface letters to denote vectors of appropriate dimensions (as before) and the suffix t to denote time. Specifically, we let $\mathbf{Y}_t = (Y(\mathbf{s}_1, t), \ldots, Y(\mathbf{s}_n, t))$, $\boldsymbol{\epsilon}_t = (\epsilon(\mathbf{s}_1, t), \ldots, \epsilon(\mathbf{s}_n, t))$, $\boldsymbol{\mu}_t = (\mu(\mathbf{s}_1, t), \ldots, \mu(\mathbf{s}_n, t))$, $\mathbf{w}_t = (w(\mathbf{s}_1, t), \ldots, w(\mathbf{s}_n, t))$. Similarly we form the $n \times p$ matrix X_t by combining the p covariates at the n data locations. Thus, the ith row of X_t is the p-dimensional vector $\mathbf{x}(\mathbf{s}_i, t)$. Although we use the vectorised notations for convenience in expressing various quantities, we prefer the simple space-time specific notation for a single variable such as $y(\mathbf{s}_i, t)$ at the initial model formulation stage so that we encourage thinking of a model for every individual data point, such as model (7.1).

It is now evident that there is a huge number of modeling possibilities for a practical point referenced spatio-temporal data set, and it will not be prudent for any text book to prescribe one particular modeling strategy over the remaining ones. General discussions and modeling considerations are presented in Banerjee et al. (2015) and also in Cressie and Wikle (2011) among many other excellent texts now available. In the remainder of this chapter we introduce several flexible modeling strategies capturing different aspects of spatio-temporal dependence structure. In each case we describe the models, discuss fitting strategies, and obtain predictive distributions for spatio-temporal interpolation. Section 7.2 returns to the multiple regression model but with spatio-temporal error distribution without a nugget effect. The model can be fitted by exact methods as the required posterior and predictive distributions are available in closed form. The theoretical developments in this section allow the reader to gain insight into the model fitting and prediction process without being reliant on computer intensive sampling based methods such as MCMC.

7.2   Models with spatio-temporal error distribution

The very first attempt at modeling a spatio-temporal response variable should naturally consider the linear regression models discussed in Section 6.3. The iid error distribution assumed for the regression model (6.1) is no longer tenable since there is spatio-temporal dependence in the data and the model should incorporate that. Hence our first attempt at spatio-temporal modeling is to extend the model (6.1) to have a spatio-temporal error distribution.

The multiple linear regression model (6.1) is now written as

$$Y(\mathbf{s}_i, t) = \beta_1 x_1(\mathbf{s}_i, t) + \cdots + \beta_p x_p(\mathbf{s}_i, t) + \epsilon(\mathbf{s}_i, t), \tag{7.2}$$

for $i = 1, \ldots, n$, $t = 1, \ldots, T$ where $\epsilon(\mathbf{s}_i, t)$ is the spatio-temporal error term which we specify below. Note also that in our very first attempt at spatio-temporal modeling we do not allow the regression coefficients β to vary spatially or temporally which may be the case in many practical modeling situations. We write $\boldsymbol{\epsilon}$ to denote the vector of all the nT $\epsilon(s_i, t)$'s stacked in the same order as the observations $\mathbf{y}$.

The error term $\epsilon(\mathbf{s}_i, t)$ is assumed to be a zero-mean spatio-temporal Gaussian process with a separable covariance structure, given by:

$$\text{Cov}\{\epsilon(\mathbf{s}_i, t_k),\ \epsilon(\mathbf{s}_j, t_l)\} = \sigma^2\ \rho_s(|\mathbf{s}_i - \mathbf{s}_j|; \boldsymbol{v}_s)\ \rho_t(|t_k - t_l|; \boldsymbol{v}_t). \tag{7.3}$$

where $\rho_s(|\mathbf{s}_i - \mathbf{s}_j|; \boldsymbol{v}_s)$ is an isotropic correlation function of the spatial distance $|\mathbf{s}_i - \mathbf{s}_j|$ between two locations $\mathbf{s}_i$ and $\mathbf{s}_j$ and this may depend on the parameter $\boldsymbol{v}_s$ which may be more than one in number. For example, there are two parameters if the Matèrn correlation function (2.1) is assumed. Similarly, $\rho_t(|t_k - t_l|; \boldsymbol{v}_t)$ denotes the isotropic correlation function of the time difference $|t_k - t_l|$ and this function may depend on the parameters $\boldsymbol{v}_t$. The covariance function (7.3) is called a separable covariance function since it models the spatio-temporal dependence as the product of a separate dependence structure in space and another one in time (Mardia and Goodall, 1993). This model has the limitation that it does not allow learning of possible space-time interaction in the dependence structure. However, the product structure in (7.3) allows analytical investigation and may be used as the 'straw' model for model comparison purposes.

Let Σ_s denote the $n \times n$ spatial correlation matrix where Σ_s has elements $\rho_s(|\mathbf{s}_i - \mathbf{s}_j|; \boldsymbol{v}_s)$, for $i, j = 1, \ldots, n$. Similarly, let Σ_t denote the $T \times T$ temporal correlation matrix having elements $\rho_t(|t_k - t_l|; \boldsymbol{v}_t)$, for $k, l = 1, \ldots, T$. Let $\boldsymbol{v} = (\boldsymbol{v}_s, \boldsymbol{v}_t)$ denote all the correlation parameters.

Let $H = \Sigma_s \otimes \Sigma_t$ denote the $nT \times nT$ correlation matrix of $\boldsymbol{\epsilon}$ where $\otimes$ denotes the Kronecker product of two matrices. The idea of Kronecker product is to simply perform element wise multiplication of two matrices Σ_s $(n \times n)$ and Σ_t $(T \times T)$ to obtain the $nT \times nT$ matrix H. The spatio-temporal regression model (7.2) reduces to the usual regression model with independent errors when we take $H = I$, the identity matrix of order nT. This can be achieved by choosing $\rho_s(d; \boldsymbol{v}_s) = \rho_t(d; \boldsymbol{v}_t) = 1$ if $d = 0$ and 0 otherwise.

As before in Section 6.3, for convenience, we work with the precision $\lambda^2 = 1/\sigma^2$. The joint prior distribution of $\boldsymbol{\beta}, \lambda^2$ is assumed to be:

$$\pi(\boldsymbol{\beta}, \lambda^2) = N\left(\boldsymbol{\beta}_0, \frac{1}{\lambda^2}M^{-1}\right)G(a, b),$$

where $\boldsymbol{\beta}_0$, $p \times 1$, and M, $p \times p$, are suitable hyper-parameters.

7.2.1 Posterior distributions

Model (7.2) can be written as

$$\mathbf{Y} \sim N\left(X\boldsymbol{\beta}, \sigma^2 H(\boldsymbol{v})\right).$$

The joint posterior distribution of $\boldsymbol{\beta}$ and λ^2, $\pi(\boldsymbol{\beta}, \lambda^2|\mathbf{y})$, is:

$$\propto (\lambda^2)^{\frac{nT+p}{2}+a-1}$$
$$\exp\left[-\frac{\lambda^2}{2}\left\{2b + (\mathbf{y} - X\boldsymbol{\beta})'H^{-1}(\mathbf{y} - X\boldsymbol{\beta}) + (\boldsymbol{\beta} - \boldsymbol{\beta}_0)'M(\boldsymbol{\beta} - \boldsymbol{\beta}_0)\right\}\right].$$

This posterior distribution is of the same form as (6.3) in Section 6.3 with obvious changes in notation. The total number of observations n in (6.3) is nT here. Because of the similarity of the joint posterior distributions we simply state the following results without the need to carry out any further mathematical derivation.

Analogous to (6.4), we obtain:

$$M^* = M + X'H^{-1}X, \quad \beta^* = (M^*)^{-1}\left(M\beta_0 + X'H^{-1}\mathbf{y}\right)$$

and

$$2b^* = 2b + \beta_0'M\beta_0 + \mathbf{y}'H^{-1}\mathbf{y} - (\beta^*)'M^*(\beta^*).$$

The joint posterior distribution is:

$$\pi\left(\beta, \lambda^2|\mathbf{y}\right) \propto \left(\lambda^2\right)^{\frac{nT+p}{2}+a-1} \exp\left[-\frac{\lambda^2}{2}\left\{2b^* + (\beta - \beta^*)'M^*(\beta - \beta^*)\right\}\right].$$

The full conditional posterior distributions are given by:

$$\beta|\mathbf{y}, \lambda^2 \sim N\left(\beta^*, \tfrac{1}{\lambda^2}(M^*)^{-1}\right)$$
$$\lambda^2|\mathbf{y}, \beta \sim G\left(\tfrac{nT+p}{2} + a, b^* + \tfrac{1}{2}(\beta - \beta^*)'M^*(\beta - \beta^*)\right).$$

By direct integration the marginal posterior distributions are obtained as follows:

$$\beta|\mathbf{y} \sim t_p\left(\beta^*, \frac{2b^*}{nT + 2a}(M^*)^{-1}, nT + 2a\right), \quad \lambda^2|\mathbf{y} \sim G\left(\frac{nT}{2} + a, b^*\right). \quad (7.4)$$

As in Section 6.3 we use the above marginal posterior distributions (7.4) to make inference. No further details are provided here. But note the modified definitions of β^*, M^* and b^* to accommodate the correlation matrix H here. These coincide with the corresponding ones obtained in Section 6.3 when H is the nT dimensional identity matrix. Note also that a spatial only model when $T = 1$, or a temporal only model when $n = 1$, is a special case of the above development.

7.2.2 Predictive distributions

Inclusion of spatial and temporal dependence in the regression model (7.2) influences the predictive distribution of the response random variable at an unobserved location and at any time point t which may be in the future as well. Suppose the problem is to predict $Y(\mathbf{s}_0, t_0)$ at an unobserved location $\mathbf{s}_0$ at any time point t_0. This location $\mathbf{s}_0$ must be a sufficient distance away from each of the data location $\mathbf{s}_i$ $i = 1, \ldots, n$, otherwise we will encounter problems in matrix inversion due to the occurrence of singular matrices. We also note that we interpolate the spatial surface at any time point t_0 by independently predicting at a large number of locations inside the study domain. The methods below will not work for simultaneous predictions at a number new locations. That is a separate problem, which is often not interesting when the purpose is not to make joint predictive inference at multiple locations.

Let the p-dimensional vector of values of the regression variables at this new location-time combination be given by $\mathbf{x}(\mathbf{s}_0, t_0)$. Because of the spatial and temporal dependence between the $Y(\mathbf{s}_0, t_0)$ and the observed random variables $Y(\mathbf{s}_i, t)$ for $i = 1, \ldots, n$ and $t = 1, \ldots, T$, we will have to use (4.8) to obtain the posterior predictive distribution of $Y(\mathbf{s}_0, t_0)$ given the nT observations $\mathbf{y}$. In order to do this, we first construct the joint distribution:

$$\begin{pmatrix} Y(\mathbf{s}_0, t_0) \\ \mathbf{Y} \end{pmatrix} \sim N \left\{ \begin{pmatrix} \mathbf{x}'(\mathbf{s}_0, t_0)\boldsymbol{\beta} \\ X\boldsymbol{\beta} \end{pmatrix}, \sigma^2 \begin{pmatrix} 1 & \Sigma_{12} \\ \Sigma_{21} & H \end{pmatrix} \right\},$$

where $\Sigma_{21} = \Sigma'_{12}$ and Σ_{12} is the nT dimensional vector with elements given by $\sigma_s(\mathbf{s}_i - \mathbf{s}_0)\sigma_t(t - t_0)$ where $\sigma_s(\mathbf{s}_i - \mathbf{s}_0) = \rho_s(|\mathbf{s}_i - \mathbf{s}_0|; \boldsymbol{v}_s)$ and $\sigma_t(t - t_0) = \rho_t(|t - t_0|; \boldsymbol{v}_t)$. Now we obtain the conditional distribution of $Y(\mathbf{s}_0, t_0)|\mathbf{y}, \boldsymbol{\beta}, \sigma^2$ as

$$N \left\{ \mathbf{x}'(\mathbf{s}_0, t_0)\boldsymbol{\beta} + \Sigma_{12} H^{-1}(\mathbf{y} - X\boldsymbol{\beta}), \sigma^2 \left(1 - \Sigma_{12} H^{-1} \Sigma_{21} \right) \right\}.$$

Note that this conditional distribution is the Kriging step in the Bayesian analysis, and it is automatically required since we need to perform conditioning on $\mathbf{y}$ to obtain the posterior predictive distribution (4.8).

We need to integrate out $\boldsymbol{\beta}$ and $\lambda^2(= 1/\sigma^2)$ from the above distribution to obtain the required predictive distribution. Let

$$\mu(\mathbf{s}_0, t_0) = \mathbf{x}'(\mathbf{s}_0, t_0)\boldsymbol{\beta} + \Sigma_{12} H^{-1}(\mathbf{y} - X\boldsymbol{\beta})$$

denote the conditional mean of $Y(\mathbf{s}_0, t_0)|\mathbf{y}, \boldsymbol{\beta}, \sigma^2$. Now

$$\begin{aligned} y(\mathbf{s}_0, t_0) - \mu(\mathbf{s}_0, t_0) &= y(\mathbf{s}_0, t_0) - \mathbf{x}'(\mathbf{s}_0, t_0)\boldsymbol{\beta} - \Sigma_{12} H^{-1}\mathbf{y} + \Sigma_{12} H^{-1} X\boldsymbol{\beta} \\ &= y(\mathbf{s}_0, t_0) - \Sigma_{12} H^{-1}\mathbf{y} - \left\{ \mathbf{x}'(\mathbf{s}_0, t_0) - \Sigma_{12} H^{-1} X \right\} \boldsymbol{\beta} \\ &= y_*(\mathbf{s}_0, t_0) - \mathbf{g}'\boldsymbol{\beta}, \quad \text{say} \end{aligned}$$

where

$$y_*(\mathbf{s}_0, t_0) = y(\mathbf{s}_0, t_0) - \Sigma_{12} H^{-1}\mathbf{y}$$

and

$$\mathbf{g}'(\mathbf{s}_0, t_0) = \mathbf{x}'(\mathbf{s}_0, t_0) - \Sigma_{12} H^{-1} X. \tag{7.5}$$

Therefore,

$$\pi(Y(\mathbf{s}_0, t_0)|\mathbf{y}, \boldsymbol{\beta}, \sigma^2) \propto (\lambda^2)^{\frac{1}{2}} \exp\left[-\frac{\lambda^2}{2\delta^2(\mathbf{s}_0, t_0)} \left\{ y_*(\mathbf{s}_0, t_0) - \mathbf{g}'\boldsymbol{\beta} \right\}^2 \right]$$

where

$$\delta^2(\mathbf{s}_0, t_0) = 1 - \Sigma_{12} H^{-1} \Sigma_{21}. \tag{7.6}$$

This shows that

$$Y_*(\mathbf{s}_0, t_0)|\mathbf{y}, \boldsymbol{\beta}, \lambda^2 \sim N\left(\mathbf{g}'(\mathbf{s}_0, t_0)\boldsymbol{\beta}, \frac{1}{\lambda^2}\delta^2(\mathbf{s}_0, t_0) \right).$$

Hence by integrating out $\boldsymbol{\beta}$ we have

$$Y_*(\mathbf{s}_0, t_0)|\mathbf{y}, \lambda^2 \sim N\left(\mathbf{g}'(\mathbf{s}_0, t_0)\boldsymbol{\beta}^*, \frac{1}{\lambda^2} \left(\delta^2(\mathbf{s}_0, t_0) + \mathbf{g}'(\mathbf{s}_0, t_0) V^* \mathbf{g}(\mathbf{s}_0, t_0) \right) \right).$$

By integrating this with respect to the marginal posterior distribution of λ^2 in Equation (7.4), we obtain the posterior predictive distribution of $Y_*(s_0, t_0)$ given $\mathbf{y}$ as:

$$t\left(\mathbf{g}'(s_0, t_0)\boldsymbol{\beta}^*, \frac{2b^*}{nT + 2a}\left(\delta^2(s_0, t_0) + \mathbf{g}'(s_0, t_0)(M^*)^{-1}\mathbf{g}(s_0, t_0)\right), nT + 2a\right).$$
(7.7)

Now the posterior predictive distribution of

$$Y(s_0, t_0) = Y_*(s_0, t_0) + \Sigma_{12}H^{-1}\mathbf{y}$$

is the t-distribution with the same scale and degrees of freedom as in (7.7) but its mean is given by:

$$E(Y(s_0, t_0)|\mathbf{y}) = \mathbf{x}'(s_0, t_0)\boldsymbol{\beta}^* + \Sigma_{12}H^{-1}(\mathbf{y} - X\boldsymbol{\beta}^*),$$
(7.8)

where $\mathbf{g}(s_0, t_0)$ is given in (7.5) and $\delta^2(s_0, t_0)$ is given in (7.6). For large data sets calculating $\Sigma_{12}H^{-1}$ is a challenge because of the large dimensions of these matrices. However, considerable simplification can be achieved as the below derivation, taken from Sahu et al. (2011), shows.

7.2.3 Simplifying the expressions: $\Sigma_{12}H^{-1}$ and $\Sigma_{12}H^{-1}\Sigma_{21}$

Note that $H = \Sigma_s \otimes \Sigma_t$ and

$$\begin{pmatrix} 1 & \Sigma_{12} \\ \Sigma_{21} & H \end{pmatrix} = \begin{pmatrix} 1 & \Sigma_s'(\mathbf{s} - \mathbf{s}_0) \otimes \Sigma_t'(\mathbf{t} - \mathbf{t}') \\ \Sigma_s(\mathbf{s} - \mathbf{s}_0) \otimes \Sigma_t(\mathbf{t} - \mathbf{t}') & \Sigma_s \otimes \Sigma_t \end{pmatrix}$$

where $\Sigma_s(\mathbf{s} - \mathbf{s}_0)$ is an $n \times 1$ column vector with the ith entry given by $\sigma_s(\mathbf{s}_i - \mathbf{s}_0)$ and $\Sigma_t(\mathbf{t} - \mathbf{t}_0)$ is a $T \times 1$ column vector with the kth entry given by $\sigma_t(t - t_0)$. Here $H^{-1} = \Sigma_s^{-1} \otimes \Sigma_t^{-1}$. Hence the $1 \times nT$ vector $\Sigma_{12}H^{-1}$ will have elements (for $j = 1, \ldots, n$ and $k = 1, \ldots, T$)

$$\begin{aligned}
b_{jk}(s_0, t_0) &= \sum_{i=1}^{n} \sum_{m=1}^{T} \sigma_s(\mathbf{s}_i - \mathbf{s}_0)\sigma_t(m - t_0)(\Sigma_s)_{ij}^{-1}(\Sigma_t)_{mk}^{-1} \\
&= \sum_{i=1}^{n} \sigma_s(\mathbf{s}_i - \mathbf{s}_0)(\Sigma_s)_{ij}^{-1} \sum_{m=1}^{T} \sigma_t(m - t_0)(\Sigma_t)_{mk}^{-1} \\
&= b_s(j, s_0)\, b_t(k, t_0),
\end{aligned}$$

where

$$b_s(j, s_0) = \sum_{i=1}^{n} \sigma_s(\mathbf{s}_i - \mathbf{s}_0)(\Sigma_s)_{ij}^{-1}, \text{ and } b_t(k, t_0) = \sum_{m=1}^{T} \sigma_t(m - t_0)(\Sigma_t)_{mk}^{-1}.$$

The quantity $b_t(k, t_0)$ simplifies considerably by noting that it resembles the inner product of a multiple of a particular column of Σ_t and a particular row of Σ_t^{-1}. First, consider the case $t_0 \leq T$. In this case $b_t(k, t_0)$ is the inner product of the t_0th column of Σ_t and kth row of Σ_t^{-1}. Hence $b_t(k, t_0)$ will be 1 if $t_0 = k$ and 0 otherwise. Now consider the case $t_0 > T$. Suppose that we can write

$$\sigma_t(m - t_0) = \sigma_t(t_0 - T)\sigma_t(T - m) \tag{7.9}$$

for $m = 1, \ldots, T$, thus $b_t(k, t_0)$ will be $\sigma_t(t_0 - T)$ times the inner product of the Tth column of Σ_t and kth row of Σ_t^{-1}. Observe that (7.9) holds for the exponential covariance function adopted here. Thus, we have proved the following result:

$$b_t(k, t_0) = \begin{cases} \delta_{k,t_0}, & \text{if } t_0 \leq T \\ \delta_{k,T}\,\sigma_t(t_0 - T), & \text{if } t_0 > T \end{cases}$$

where $\delta_{i,j} = 1$ if $i = j$ and 0 otherwise.

Now we obtain simplified expressions for a quantity like $\Sigma_{12}H^{-1}\mathbf{a}$ where $\mathbf{a}$ is nT by 1 with elements a_{jk}, $j = 1, \ldots, n$ and $k = 1, \ldots, T$. We have:

$$\begin{aligned} \Sigma_{12}H^{-1}\mathbf{a} &= \sum_{j=1}^{n}\sum_{k=1}^{T} b_{jk}(\mathbf{s}_0, t_0)a_{jk} \\ &= \sum_{j=1}^{n}\sum_{k=1}^{T} a_{jk}b_s(j, \mathbf{s}_0)\, b_t(k, t_0) \\ &= \sum_{j=1}^{n} b_s(j, \mathbf{s}_0)\sum_{k=1}^{T} a_{jk}b_t(k, t_0). \end{aligned}$$

Now

$$\sum_{k=1}^{T} a_{jk}b_t(k, t_0) = \begin{cases} \sum_{k=1}^{T} a_{jk}\delta_{k,t_0}, & \text{if } t_0 \leq T \\ \sum_{k=1}^{T} a_{jk}\delta_{k,T}\,\sigma_t(t_0 - T), & \text{if } t_0 > T. \end{cases}$$

Thus we have,

$$\sum_{k=1}^{T} a_{jk}b_t(k, t_0) = \begin{cases} a_{jt_0}, & \text{if } t_0 \leq T \\ a_{jT}\,\sigma_t(t_0 - T), & \text{if } t_0 > T. \end{cases}$$

Finally,

$$\Sigma_{12}H^{-1}\mathbf{a} = \begin{cases} \sum_{j=1}^{n} b_s(j, \mathbf{s}_0)a_{jt_0}, & \text{if } t_0 \leq T \\ \sigma_t(t_0 - T)\sum_{j=1}^{n} b_s(j, \mathbf{s}_0)a_{jT} & \text{if } t_0 > T. \end{cases}$$

Now we simplify the expression for the conditional variance. Note that $\Sigma_{12}H^{-1}\Sigma_{21}$ is exactly equal to $\Sigma_{12}H^{-1}\mathbf{a}$ where $\mathbf{a} = \Sigma_{21}$. For this choice we have, $a_{jt} = \sigma_s(\mathbf{s}_j - \mathbf{s}_0)\sigma_t(t - t_0)$. Hence,

$$\Sigma_{12}H^{-1}\Sigma_{21} = \begin{cases} \sum_{j=1}^{n} b_s(j, \mathbf{s}_0)\sigma_s(\mathbf{s}_j - \mathbf{s}_0)\sigma_t(t_0 - t_0), & \text{if } t_0 \leq T \\ \sigma(t_0 - T)\sum_{j=1}^{n} b_s(j, \mathbf{s}_0)\sigma_s(\mathbf{s}_j - \mathbf{s}_0)\sigma_t(T - t_0), & \text{if } t_0 > T. \end{cases}$$

Let

$$a_s(\mathbf{s}_0) = \sum_{i=1}^{n}\sum_{j=1}^{n} \sigma_s(\mathbf{s}_i - \mathbf{s}_0)\left(\Sigma_s^{-1}\right)_{ij}\sigma_s(\mathbf{s}_j - \mathbf{s}_0).$$

Thus,

$$\Sigma_{12} H^{-1} \Sigma_{21} = \begin{cases} a_s(\mathbf{s}_0), & \text{if } t_0 \leq T \\ a_s(\mathbf{s}_0)\sigma_t^2(t_0 - T) & \text{if } t_0 > T. \end{cases}$$

Now

$$\delta^2(\mathbf{s}_0, t_0) = 1 - a_s(\mathbf{s}_0)\, a_t(t_0),$$

where

$$a_t(t_0) = \begin{cases} 1, & \text{if } t_0 \leq T \\ \sigma_t^2(t_0 - T) & \text{if } t_0 > T. \end{cases}$$

The expressions for $t_0 > T$ will be required for forecasting in Chapter 9.

7.2.4 Estimation of v

So far in this section we have ignored the fact that the parameters v are unknown and ideally those should be estimated within the Bayesian model as well. We postpone such estimation to more sophisticated later developments in Section 7.3. In this section we adopt cross-validation approaches (Section 6.8) sketched below to choose optimal values of v and then make inference conditional on those chosen values of v. We adopt a simple cross validation approach by splitting the data into a training and a test set. We take a grid of values of v and calculate the RMSE on the test data set. We choose the optimal value of v to be the one which produces the least RMSE.

This strategy has been adopted by several authors, see e.g. Sahu et al. (2011), Sahu et al. (2006). In effect, this is an empirical Bayes approach which uses either the full data y or a subset validation data set to estimate v, see e.g. Section 4.5 in Berger (1985).

7.2.5 Illustration of a spatio-temporal model fitting

We return to the running New York air pollution data example but now we model the full spatio-temporal data set with the missing observations imputed using a non-Bayesian linear model as described in the exploratory analysis Chapter 3. To stabilize the variance this modeling is performed in the square-root scale for response, the observed daily 8-hour maximum ozone concentration level, as done by Bakar and Sahu (2015). We avoid modeling on the original raw scale since we have encountered negative predictions of ozone levels under such modeling. We still use the three covariates: maximum temperature, wind speed and relative humidity. We assume that the spatio-temporal error distribution is Gaussian with a separable correlation structure with exponential correlation function for both space and time. Thus we take,

$$\rho_s(|\mathbf{s}_i - \mathbf{s}_j|; v_s) = \exp\{-\phi_s|\mathbf{s}_i - \mathbf{s}_j|\} \quad \text{and} \quad \rho_t(|t_k - t_l|; v_t) = \exp\{-\phi_s|t_k - t_l|\}$$

and choose specific values of the decay parameters ϕ_s and ϕ_t. As in Section 6.4.1 we compare the spatio-temporal model with the linear regression model with independent error distributions. Thus, we now compare:

M1 standard Bayesian linear regression model with the three covariates,

M2 spatio-temporal model with a separable covariance structure.

The same default $G(a = 2, b = 1)$ prior distribution is assumed for the precision parameter $\lambda^2 = \frac{1}{\sigma^2}$. For the very first illustration we take the default values for ϕ_s and ϕ_t as suggested in the bmstdr package. We perform a grid search below to obtain optimum values for these parameters.

Estimation, model choice and validation predictions are performed by the bmstdr function Bsptime with suitable options for working with different models. The options model="lm", package="none" are for fitting M1 and the options model="separable", package="none" are for the separable spatio-temporal model. Computation details by exploiting the separable covariance structure avoiding operations for $nT \times nT$ covariance matrices for the spatio-temporal model M2 are detailed in Sahu et al. (2011).

In general, the Bsptime function, documented in the package, takes a formula and a data argument as in the case of Bspatial. It is important to note that the Bsptime function **always assumes** that the data frame is first sorted by space and then time within each site in space. For performing model validation we need to send the optional argument validrows, which is a vector containing the row numbers of the model fitting data frame which should be used for validation. For the separable model the validrows argument must contain all the time points for the validation sites as the exact model fitting function is not able to handle missing values at some time points. That is, if sites $\mathbf{s}_1$ and $\mathbf{s}_2$ are to be used for validation then the row numbers of the data frame containing the observations at the $2T$ site-time pairs $(\mathbf{s}_1, 1), \ldots, (\mathbf{s}_1, T)$ and $(\mathbf{s}_2, 1), \ldots, (\mathbf{s}_2, T)$ must be sent as the validrows argument. Moreover, the exact model fitting methods for the model="lm" and model="separable" cannot handle missing data. Hence a quick fix of substituting any missing data by the grand mean of the remaining data has been adopted in the package. This is a limitation that has been allowed to be included in the package in the interest of performing exact model fitting and validation methods as detailed in the theory sections above. Subsequent model fitting options below do not suffer from this shortcoming.

Like the Bspatial function, the Bsptime function needs specification of the arguments coordtype and coords. See the documentation of the Bsptime function for further details. For fitting a **separable** model Bsptime requires specification of two decay parameters ϕ_s and ϕ_t. If these are not specified then values are chosen which correspond to the effective ranges as the maximum distance in space and the maximum length in time.

Default normal prior distributions with mean 0 and precision 10^{-4} for the regression parameters are assumed. A default $G(2, 1)$ prior distribution for the precision $(1/\sigma^2)$ parameter is assumed. Missing observations, if any, in the response are replaced by the grand mean of the data before model fitting so that the exact model fitting methods can be applied. Missing covariate values are not permitted at all.

The code lines for fitting the two models M1 and M2 are given below.

```
f2 <- y8hrmax ~ xmaxtemp+xwdsp+xrh
M1 <- Bsptime(model="lm", formula=f2, data=nysptime,
    scale.transform = "SQRT", N=5000)
M2 <- Bsptime(model="separable", formula=f2, data=nysptime,
    scale.transform = "SQRT", coordtype="utm", coords=4:5, N=5000)
```

The prior distributions are given default values and the default burn in number of iterations is 1000. The S3 methods **print**, **plot** and **summary** have been implemented for the model fitted objects using the Bsptime function. Also the fitted values and residuals can be extracted and plotted using the familiar extractor commands such as **fitted** and **residuals**.

The **summary** command has been used to obtain the parameter estimates reported in Table 7.1 for the two models, M1 and M2 when fitting has been done in the square-root scale of the response. The variance σ^2 is estimated to be slightly higher in the spatio-temporal model perhaps to compensate for the non-zero spatial correlation in the model M2. Indeed, the output of M2 contains the values of the two decay parameters ϕ_s and ϕ_t which in this case are approximately 0.005 and 0.048. These two correspond to the spatial range of 591 kilometers and the temporal range of 62 days which are the maximum possible values for the spatial and temporal domains of the data.

We see much agreement between the parameter estimates from the two models M1 and M2. Contrasting these estimates with the ones obtained in Table 6.1 we see that all three covariates are significantly different from zero since the 95% credible intervals do not include the value zero, unlike that in the spatial case. This shows that a high-resolution spatio-temporal model can be better at detecting relationship than similar models at a temporally aggregated scale. Thus, spatio-temporal analysis may be preferable to aggregated analysis either in time or space.

Table 7.4 provides the different model choice criteria for the two models. All the model choice criteria show preference for the spatio-temporal regression model. Note that the model choice criteria values are very different from the corresponding spatial only model in Table 6.5 because of several factors including the differences in scale of the modeled data and the difference due to the number of independent observations that went into the model: 28 for the spatial model and 1736 ($= 28 \times 62$) for the spatio-temporal model.

We now discuss an optimal choice for the decay parameters ϕ_s and ϕ_t by using grid search. To choose the optimal values we set aside all of 62 days data from the eight sites: 8, 11, 12, 14, 18, 21, 24 and 28 as shown in Figure 6.3. Thus, there are 496 (8×62) data points for model validation. We again remind the reader that all validations are done at the original scale although modeling is performed at the square-root scale. Data from the remaining 20 sites are used for model estimation and validation prediction. For ϕ_s we entertain the six values: (0.001, 0.005, 0.025, 0.125, 0.625) and for ϕ_t we entertain four possibilities: (0.05, 0.25, 1.25, 6.25). The combination $\phi_s = 0.005$ and

	Model M1				Model M2			
	mean	sd	2.5%	97.5%	mean	sd	2.5%	97.5%
β_0	2.22	0.24	1.75	2.69	0.19	1.74	−3.22	3.60
β_1	0.17	0.01	0.16	0.19	0.26	0.04	0.19	0.33
β_2	0.09	0.01	0.06	0.11	0.01	0.04	−0.06	0.08
β_3	−0.17	0.03	−0.23	−0.12	−0.01	0.11	−0.23	0.20
σ^2	0.55	0.02	0.52	0.59	12.19	0.41	11.40	13.03

TABLE 7.1: Parameter estimates from the two models M1 and M2 for the spatio-temporal air pollution data set. Here we have taken $\phi_s = 0.005$ and $\phi_t = 0.05$.

$\phi_t = 0.05$ provides the least RMSE with a high level of coverage. For this choice the RMSE for the spatio-temporal model is also better than the linear regression model. Hence we select this combination to be the best one. A function phichoicep has been provided in the R code file for running the examples in this chapter. The validation results are reported in Table 7.5 along with three other models. The spatio-temporal regression model performs quite a lot better according to RMSE, MAE and CRPS. However, the achieved coverage is slightly higher for the full spatio-temporal model M2 and this is a possible criticism against this model. It is possible to narrow down the credible intervals by further fine tuning the decay parameters and the parameters of the prior distributions. We, however, do not pursue that here. Instead, we turn our attention to spatio-temporal modeling using software packages in the subsequent sections.

7.3 Independent GP model with nugget effect

The main idea in this section is to assume a temporally independent GP for the spatio-temporal process $w(\mathbf{s}_i, t)$ in the general model

$$Y(\mathbf{s}_i, t) = \mathbf{x}'(\mathbf{s}_i, t)\boldsymbol{\beta} + w(\mathbf{s}_i, t) + \epsilon(\mathbf{s}_i, t) \qquad (7.10)$$

for all $i = 1, \ldots, n$ and $t = 1, \ldots, T$. In the above, for simplicity we have assumed a non-spatially varying regression effect of the covariates $\mathbf{x}(\mathbf{s}_i, t)$. The pure error term $\epsilon(\mathbf{s}_i, t)$ is assumed to follow the independent zero mean normal distribution with variance σ_ϵ^2. The stochastic process $w(\mathbf{s}, t)$ is assumed to follow a zero mean temporally independent GP with the Matèrn correlation function (2.2) with unknown parameters $\boldsymbol{v}$. Thus,

$$\text{Cov}(w(\mathbf{s}_i, t), w(\mathbf{s}_j, t)) = \sigma_w^2 \rho\left(d_{ij} | \boldsymbol{v}\right) \qquad (7.11)$$

where d_{ij} denotes the distance between the sites $\mathbf{s}_i$ and $\mathbf{s}_j$ and $\rho(\cdot | \cdot)$ has been defined in (2.2).

Using vectorised notation we write this model hierarchically as

$$\mathbf{Y}_t|\mathbf{w}_t \sim N(\boldsymbol{\mu}_t + \mathbf{w}_t, \sigma_\epsilon^2 I) \tag{7.12}$$

$$\mathbf{w}_t \sim N(\mathbf{0}, \sigma_w^2 S_w) \tag{7.13}$$

independently for $t = 1, \ldots, T$ where $\boldsymbol{\mu}_t = X_t\boldsymbol{\beta}$, I is the identity matrix of order n, S_w is the $n \times n$ correlation matrix whose elements are formed using the Matèrn correlation function given above. Note that the matrix S_w does not contain the variance parameter σ_w^2 but it is a complex non-linear function of the parameters $\boldsymbol{v}$ describing the correlation function of the GP.

A full Bayesian model specification requires specification of the prior distributions for all the unknown parameters $\boldsymbol{\theta} = (\boldsymbol{\beta}, \sigma_\epsilon^2, \sigma_w^2, \boldsymbol{v})$. We can continue to assume a conjugate multivariate normal prior distribution for $\boldsymbol{\beta} \sim N(\boldsymbol{\beta}_0, V)$ without the scaling factor σ_ϵ^2. For non-informative prior specification, V will be taken as a diagonal matrix with large values for the diagonal entries. We can also scale the prior variance matrix V by multiplying σ_ϵ^2 as well. We continue to work with precisions $\lambda_w^2 = 1/\sigma_w^2$ and $\lambda_\epsilon^2 = 1/\sigma_\epsilon^2$ and assign independent $G(a_w, b_w)$ and $G(a_\epsilon, b_\epsilon)$ prior distributions. Other type of prior distributions such as half-Cauchy for σ_ϵ and σ_w are also possible and will be illustrated with numerical examples. Let $\pi(\boldsymbol{\theta})$ denote the combined joint prior distribution for all the parameters. The log-posterior density from the above model specifications and prior distributions is now written as:

$$\log \pi(\boldsymbol{\theta}, \mathbf{w}|\mathbf{y}) \propto -\frac{nT}{2}\log(\sigma_\epsilon^2) - \frac{1}{2\sigma_\epsilon^2}\sum_{t=1}^{T}(\mathbf{y}_t - \boldsymbol{\mu}_t - \mathbf{w}_t)'(\mathbf{y}_t - \boldsymbol{\mu}_t - \mathbf{w}_t)$$

$$-\frac{T}{2}\log|\sigma_w^2 S_w| - \frac{1}{2\sigma_w^2}\sum_{t=1}^{T}\mathbf{w}_t'S_w^{-1}\mathbf{w}_t + \log \pi(\boldsymbol{\theta}). \tag{7.14}$$

This joint posterior distribution can be computed in several ways as described in the following subsections. There we also discuss how to predict at an unobserved location $\mathbf{s}_0$ at a time point t.

7.3.1 Full model implementation using spTimer

A simple linear transformation, called hierarchical centering by Gelfand et al. (1995a), can be applied to (7.14) to yield a more efficient MCMC implementation. The transformation is $\mathbf{o}_t = \boldsymbol{\mu}_t + \mathbf{w}_t$ and in that case the log-posterior density is written as:

$$\log \pi(\boldsymbol{\theta}, \mathbf{o}|\mathbf{y}) \propto -\frac{nT}{2}\log(\sigma_\epsilon^2) - \frac{1}{2\sigma_\epsilon^2}\sum_{t=1}^{T}(\mathbf{y}_t - \mathbf{o}_t)'(\mathbf{y}_t - \mathbf{o}_t)$$

$$-\frac{T}{2}\log|\sigma_w^2 S_w| - \frac{1}{2\sigma_w^2}\sum_{t=1}^{T}(\mathbf{o}_t - \boldsymbol{\mu}_t)'S_w^{-1}(\mathbf{o}_t - \boldsymbol{\mu}_t) + \log \pi(\boldsymbol{\theta}).$$

Bass and Sahu (2017) provide further discussion regarding the parameterisation issues in spatial data modeling with GPs. This last centering parameterisation has been applied in the Gibbs sampling implementation of this model in R package `spTimer` by Bakar and Sahu (2015). The package provides samples from the full posterior distribution $\pi(\boldsymbol{\theta}, \mathbf{o}|\mathbf{y})$ where $\mathbf{o}$ denotes the nT random effects $\mathbf{o}_t$ for $t = 1, \dots, T$. Thus, the conditional model and the sampler treats both $\boldsymbol{\theta}$ and $\mathbf{o}$ as unknowns and draws samples from $\pi(\boldsymbol{\theta}, \mathbf{o}|\mathbf{y})$.

Prediction of $Y(\mathbf{s}_0, t)$ at the location-time combination $\mathbf{s}_0$ and t requires calculation of the posterior predictive distribution (4.8) given the observations $\mathbf{y}$. For the centered model, we have

$$
\begin{aligned}
Y(\mathbf{s}_0, t) &= O(\mathbf{s}_0, t) + \epsilon(\mathbf{s}_0, t) \\
O(\mathbf{s}_0, t) &= \mathbf{x}'(\mathbf{s}_0, t)\boldsymbol{\beta} + w(\mathbf{s}_0, t).
\end{aligned}
$$

To find the required predictive distribution we work with the hierarchical specifications:

$$
\begin{aligned}
Y(\mathbf{s}_0, t)|o(\mathbf{s}_0, t), \boldsymbol{\theta}, \mathbf{o}, \mathbf{y} &\sim N(o(\mathbf{s}_0, t), \sigma_\epsilon^2) \\
O(\mathbf{s}_0, t)|\boldsymbol{\theta}, \mathbf{o}, \mathbf{y} &\sim N(\mu_{ot}, \sigma_{ot}^2)
\end{aligned}
$$

where μ_{ot} and σ_{ot}^2 are obtained below. Note that the distribution of $Y(\mathbf{s}_0, t)$ above is conditionally independent of $\mathbf{o}$ and $\mathbf{y}$ given $\boldsymbol{\theta}$. The distribution of $O(\mathbf{s}_0, t)$ given $\boldsymbol{\theta}$ and $\mathbf{o}$ is conditionally independent of $\mathbf{y}$. Moreover, this last distribution depends only on $\mathbf{o}_t$ since the GP at each time point t is assumed to be independent of other time points. Thus, we need to obtain the distribution of $O(\mathbf{s}_0, t)|\boldsymbol{\theta}, \mathbf{o}_t$. This distribution is found similarly as in Section 7.2.2. By exploiting the GP we first claim that given $\boldsymbol{\theta}$ the joint distribution of $O(\mathbf{s}_0, t)$ and $\mathbf{O}_t$ is the $n+1$ dimensional multivariate normal distribution:

$$
\begin{pmatrix} O(\mathbf{s}_0, t) \\ \mathbf{O}_t \end{pmatrix} \sim N \left[\begin{pmatrix} \mathbf{x}'(\mathbf{s}_0, t)\boldsymbol{\beta} \\ X_t\boldsymbol{\beta} \end{pmatrix}, \ \sigma_w^2 \begin{pmatrix} 1 & S_{w,12} \\ S_{w,21} & S_w \end{pmatrix} \right]
$$

where $S_{w,12}$ is $1 \times n$ with the ith entry given by $\rho(d_{i0}|\boldsymbol{v})$ defined in (2.2) where d_{i0} is the distance between the sites $\mathbf{s}_i$ and $\mathbf{s}_0$, for $i = 1, \dots, n$ and $S_{w,21} = S'_{w,12}$, i.e. the transpose. Using the distribution theory for the multivariate normal distribution in Section A.1, we conclude that:

$$
O(\mathbf{s}_0, t)|\boldsymbol{\theta}, \mathbf{o}_t \sim N\left(\mu_{ot}, \ \sigma_{ot}^2\right) \tag{7.15}
$$

where

$$
\begin{aligned}
\mu_{ot} &= \mathbf{x}'(\mathbf{s}_0, t)\boldsymbol{\beta} + S_{w,12}S_w^{-1}(\mathbf{o}_t - X_t\boldsymbol{\beta}), \\
\sigma_{ot}^2 &= \sigma_w^2\left(1 - S_{w,12}S_w^{-1}S_{w,21}\right).
\end{aligned}
$$

The above considerations allow us to write the density of the posterior predictive distribution (4.8) of $Y(\mathbf{s}_0, t)|\mathbf{y}$ as:

$$
f(y(\mathbf{s}_0, t)|\mathbf{y}) = \int_{-\infty}^{\infty} f\left(y(\mathbf{s}_0, t)|o(\mathbf{s}_0, t), \boldsymbol{\theta}\right) \pi\left(o(\mathbf{s}_0, t)|\mathbf{o}_t, \boldsymbol{\theta}\right) \pi(\boldsymbol{\theta}, \mathbf{o}_t|\mathbf{y}) do(\mathbf{s}_0, t) \, d\mathbf{o}_t d\boldsymbol{\theta}. \tag{7.16}
$$

This distribution can be estimated by using composition sampling as follows. For each $j = 1, \dots, J$:

Step 1: Draw $(\boldsymbol{\theta}^{(j)}, \mathbf{o}^{(j)})$ from the posterior distribution $\pi(\theta, \mathbf{o}|\mathbf{y})$.

Step 2: Draw $o^{(j)}(\mathbf{s}_0, t)$ from $N\left(\mu_{ot}^{(j)}, \sigma_{ot}^{2(j)}\right)$ in (7.15) where the mean and variance parameters are obtained by plugging in the current values $\boldsymbol{\theta}^{(j)}$ and $\mathbf{o}_t^{(j)}$.

Step 3: Finally draw $y^{(j)}(\mathbf{s}_0, t)$ from $N\left(o^{(j)}(\mathbf{s}_0, t), \sigma_w^{2(j)}\right)$.

Once J samples, $y^{(j)}(\mathbf{s}_0, t)$, for $j = 1, \ldots, J$, have been drawn we proceed to make any desired predictive inference by appropriately summarizing the drawn samples. Note that before forming summaries we need to apply the inverse transformation to make inference on the original scale of observations if the data were modeled on a transformed scale in the first place. For example, if the log transformation had been used in modeling then we simple take summaries of $\exp\left(y^{(j)}(\mathbf{s}_0, t)\right)$. This reverting back ensures that the uncertainties are appropriately estimated for the predictive inference on the original scale. No further action is needed for estimation of uncertainty.

The full model of this section has been implemented as the GP model in the spTimer package, see Bakar and Sahu (2015). This package is exclusively devoted to fitting and prediction for point referenced spatio-temporal data. The main model fitting function is spT.Gibbs which takes a formula, a data frame and a matrix of coordinate locations for the observation sites along with many options regarding the covariance function and distance calculation, the scale for model fitting, the prior distributions for various parameters and MCMC computing related parameters such as the number of iterations and tuning standard deviations for the proposal distributions for sampling parameters which have non-standard full conditional distributions, e.g. the spatial decay and smoothness parameters ϕ and ν. The package provides a demo which can be executed by issuing the command: demo(nyExample, package="spTimer"). No further introduction is provided here for brevity.

The Bsptime function in the bmstdr package can be called with a formula and data argument for model fitting and validation using the spTimer package. The R help file on this function explains the required arguments. For estimating the spatial decay parameter ϕ there are three options for prior.phi: "FIXED", "Unif" and "Gamm". The first option chooses ϕ to be fixed at the value 3/maximum distance. The option "Unif" assumes a discrete uniform prior distribution whose range is provided by the argument prior.phi.param. Default values are provided that specify an effective range between 1% and 100% of the maximum distance between the data locations. The number of support points if not provided by the additional argument phi.npoints is set at 10. For the gamma distribution choice "Gamm" a $G(2, 1)$ prior distribution is assumed by default and this is also the default if the prior.phi is not specified.

As before, the Bsptime function assumes that the data are sorted first by space and then by time within space as mentioned before in Section 7.1. The function Bsptime can then perform model fitting and validation for specific rows in the data set provided by the `validrows` argument as before. The code for fitting this model is given by:

```
M3 <- Bsptime(package="spTimer", formula=f2, data=nysptime,
    coordtype="utm", coords=4:5, scale.transform = "SQRT", N=5000)
```

where we continue to use the model formula `f2` as before. Note that the only difference between this command and the previous two commands lies in changing the package name to `"spTimer"`. The GP model is the default for the `spTimer` package.

The command `names(M3)` lists all the components of the model fit. The `spTimer` model fitted object is provided as the component `M3$fit`. Hence any investigation, e.g. prediction, that can be performed using the `spTimer` package can also be performed using the `bmstdr` model fit.

The command `plot(M3)` provides several plots for examining MCMC convergence and the quality of model fit. The first set of plots shows traces of the parameters along with estimates of the marginal posterior density. We do not include the plot here for brevity. The plot command also obtains the residuals against fitted values plot, which is also omitted for brevity.

The summary command `summary(M3, digits=3)` obtains the parameter estimates reported in Table 7.2. These estimates are comparable to those obtained from the two models M1 and M2 reported in Table 7.1. The GP model implies a strong spatial correlation with an effective range estimate of 214.3 (3/0.014) kilometres with a 95% credible interval of $(176.5, 272.7)$ obtained directly from the credible interval for ϕ. Note that the maximum distance between the data sites is about 591 kilometers. Hence these parameter estimates seem plausible.

	mean	sd	2.5%	97.5%
β_0	2.140	0.452	1.259	3.031
β_1	0.171	0.013	0.145	0.196
β_2	0.114	0.021	0.073	0.155
β_3	−0.126	0.052	−0.225	−0.019
σ_w^2	0.586	0.071	0.484	0.758
σ_ϵ^2	0.018	0.004	0.012	0.026
ϕ	0.014	0.002	0.011	0.017

TABLE 7.2: Parameter estimates from the independent GP model fitted by `spTimer`. The G(2, 1) distribution is adopted as the prior distribution for ϕ, σ_w^2, and σ_ϵ^2 and the model has been fitted in the square-root scale as before.

Continuing our discussion of the `spTimer` model fitting we now illustrate predictions by holding out some of the data rows. In this regard we follow an example in Chapter 11 of the book by Banerjee et al. (2015). We hold out 50% (i.e. 31) of the 62 time points at random for the three validation sites shown in Figure 7.1. The `Bsptime` function enhances the `spTimer` package in performing these out of sample predictions by simply including the additional argument `validrows` in the model fitting command M3 noted above. The argument `validrows` should contain the row numbers of 93 validation observations which may be picked using simple R commands. Once requested, `Bsptime` will calculate the four validation criteria (see Section 6.8) and also draw an observed versus prediction plot. One such plot is illustrated in Figure 7.5 below in a different context.

The output of the validation model fit can also be used to perform other types of investigation. For example in Figure 7.2 we plot the observed, fitted and the 95% credible intervals for the fitted values at the three sites. Notice that the uncertainties in the fitted values where the observations were held-out (filled circles in the plot) are higher than the ones which have been used in model fitting (open circles). The code to reproduce this figure is not included here for brevity but is available online from github[1].

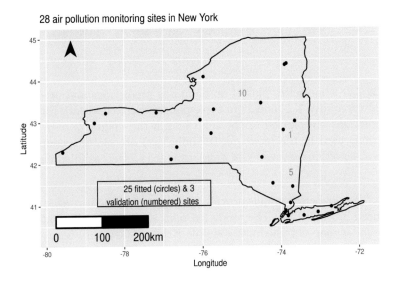

FIGURE 7.1: Three validation sites and twenty five model fitting sites in New York.

> ♣ **R Code Notes 7.1. Figure 7.2** This plot uses the function `fig11.13.plot` from the `bmstdr` library. The function `geom_ribbon` draws the ribbon in each plot.

Our final illustration of `spTimer` model fitting is the production of a predicted map of average pollution levels over the two months for the `nysptime`

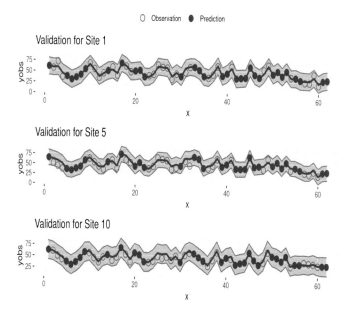

FIGURE 7.2: Time series plot of observations predictions and prediction intervals at three sites 1, 5, and 10 in three panels (top to bottom). Fitted values and predictions are plotted as the solid line in the middle of each plot. The observations used in the modeling (training set) are plotted as open circles and the observations set aside for validation are plotted as filled circles.

data set. The prediction map is provided in Figure 7.3, and its uncertainty map is produced in Figure 7.4. The predicted map is much more spatially varying than a similar map obtained using simple Kriging shown in Figure 3.4. Thus, where possible, it is worthwhile to consider spatio-temporal modeling as has been done here. The code to obtain these maps are not included for brevity but are made available online from github[2].

7.3.2 Marginal model implementation using Stan

The spatio-temporal random effects $\mathbf{w}_t$ can be integrated out from the hierarchical specifications (7.12) and (7.13). Thus, we have the marginal model for the data $\mathbf{Y}_t$:

$$\mathbf{Y}_t \sim N\left(\boldsymbol{\mu}_t, \sigma_\epsilon^2 I + \sigma_w^2 S_w\right) \tag{7.17}$$

for $t = 1, \ldots, T$ independently. Let $H = \sigma_\epsilon^2 I + \sigma_w^2 S_w$ denote the covariance matrix of the marginal distribution of $\mathbf{Y}_t$. The log-posterior density function

---

[2]https://github.com/sujit-sahu/bookbmstdr.git

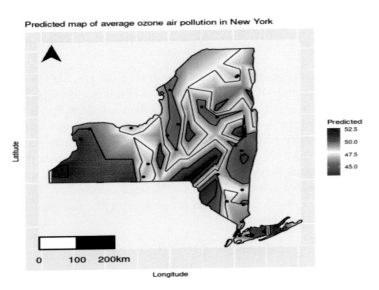

FIGURE 7.3: Predicted map of average air pollution over the months of July and August in 2006.

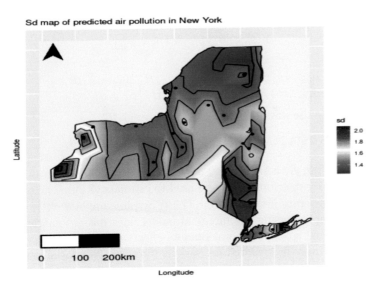

FIGURE 7.4: Sd of predicted map of average air pollution over the months of July and August in 2006.

of $\boldsymbol{\theta}$ is now written as:

$$\log \pi(\boldsymbol{\theta}|\mathbf{y}) \propto -\frac{T}{2}\log(|H|) - \frac{1}{2}\sum_{t=1}^{T}(\mathbf{y}_t - \boldsymbol{\mu}_t)'H^{-1}(\mathbf{y}_t - \boldsymbol{\mu}_t) + \log\pi(\boldsymbol{\theta}). \quad (7.18)$$

The above log-posterior density function is free of the random effects $\mathbf{w}_t$ as intended and as a result MCMC implementation of this model avoids having to sample them. Because of this reduced sampling demand and $\pi(\boldsymbol{\theta}|\mathbf{y})$ is a density on a lower dimensional space, the marginal model may lead to faster MCMC convergence. However, conjugacy is lost for sampling the variance components σ_ϵ^2 and σ_w^2 since the $\pi(\boldsymbol{\theta}|\mathbf{y})$ in (7.18) is a complex non-linear function of the variance components. Hence sampling of the variance components will require tuning if implemented using a Metropolis-Hastings sampler.

Prediction of $Y(\mathbf{s}_0, t)$ given $\mathbf{y}$ can proceed using the marginal model as follows. Note that, marginally

$$Y(\mathbf{s}_0, t) \sim N(\mathbf{x}'(\mathbf{s}_0, t)\boldsymbol{\beta}, \sigma_\epsilon^2 + \sigma_w^2).$$

But to calculate the posterior predictive distribution $f(y(\mathbf{s}_0, t)|\mathbf{y})$ we require the conditional distribution of $Y(\mathbf{s}_0, t)|\boldsymbol{\theta}, \mathbf{y}_t$. This can be obtained by noting that:

$$\begin{pmatrix} Y(\mathbf{s}_0, t) \\ \mathbf{Y}_t \end{pmatrix} \sim N\left[\begin{pmatrix} \mathbf{x}'(\mathbf{s}_0, t)\boldsymbol{\beta} \\ X_t\boldsymbol{\beta} \end{pmatrix}, \begin{pmatrix} \sigma_\epsilon^2 + \sigma_w^2 & \sigma_w^2 S_{w,12} \\ \sigma_w^2 S_{w,21} & H \end{pmatrix}\right]$$

where $S_{w,12}$ and $S_{w,21}$ have been defined before. Using the distribution theory for the multivariate normal distribution in Section A.1, we conclude that:

$$Y(\mathbf{s}_0, t)|\boldsymbol{\theta}, \mathbf{y}_t \sim N\left(\mu_{yt}, \sigma_{yt}^2\right) \quad (7.19)$$

where

$$\begin{aligned} \mu_{yt} &= \mathbf{x}'(\mathbf{s}_0, t)\boldsymbol{\beta} + \sigma_w^2 S_{w,12}H^{-1}(\mathbf{y}_t - X_t\boldsymbol{\beta}), \\ \sigma_{yt}^2 &= \sigma_\epsilon^2 + \sigma_w^2 - \sigma_w^4 S_{w,12}H^{-1}S_{w,21}. \end{aligned}$$

Now we re-write the posterior predictive density (7.16)

$$f(y(\mathbf{s}_0, t)|\mathbf{y}) = \int_{-\infty}^{\infty} f\left(y(\mathbf{s}_0, t)|\boldsymbol{\theta}, \mathbf{y}_t\right)\pi(\boldsymbol{\theta}|\mathbf{y})\, d\boldsymbol{\theta}, \quad (7.20)$$

where $f\left(y(\mathbf{s}_0, t)|\boldsymbol{\theta}, \mathbf{y}_t\right)$ is the normal density in (7.19). As before, samples $\boldsymbol{\theta}^{(j)}$ from the posterior distribution can be used to estimate (7.20). The extra step required is to generate $y^{(j)}(\mathbf{s}_0, t)$ from (7.19) by plugging in $\boldsymbol{\theta}^{(j)}$.

MCMC sampling using the marginal model does not automatically generate the centered spatio-temporal random effects $\mathbf{o}_t$ or the un-centered $\mathbf{w}_t$. However, these may be required for various purposes for spatial exploration and out of sampling predictions using the (7.16) if desired instead of the marginal (7.20). After sampling $\boldsymbol{\theta}^{(j)}$ from the marginal model we can generate $\mathbf{o}_t^{(j)}$'s using the full conditional distribution:

$$\mathbf{o}_t|\boldsymbol{\theta}, \mathbf{y} \sim N\left\{\Sigma_{[\mathbf{o}_t|\boldsymbol{\theta}, \mathbf{y}]}\left(\frac{1}{\sigma_\epsilon^2}\mathbf{y}_t + \frac{S_w^{-1}}{\sigma_w^2}X_t\boldsymbol{\beta}\right), \ \Sigma_{[\mathbf{o}_t|\boldsymbol{\theta}, \mathbf{y}]}\right\}$$

where

$$\Sigma_{[o_t | \boldsymbol{\theta}, \mathbf{y}]} = \left(\frac{I}{\sigma_\epsilon^2} + \frac{S_w^{-1}}{\sigma_w^2} \right)^{-1}.$$

The Stan software package can be used to fit the marginal model and perform predictions using either of the two predictive distributions (7.20) or (7.16). Model fitting and validation for this marginal model is available using the Bsptime function when the option package="stan" is passed on. The model fitting command is given below.

```
M4 <- Bsptime(package="stan",formula=f2, data=nysptime, coordtype="
    utm", coords=4:5, scale.transform = "SQRT", N=1500, burn.in=500,
    mchoice=T, verbose = F)
```

Options to choose different prior distributions are discussed in the documentation of the function Bsptime. The prior distribution for the spatial decay parameter ϕ, prior.phi in Bsptime, can be one of "Unif", "Gamm" or "Cauchy". The last choice "Cauchy" specifies a standard half-Cauchy prior distribution. The first choice "Unif" specifies a continuous uniform prior distribution whose limits are set by the argument prior.phi.param. Default values are provided that specify an effective range between 1% and 100% of the maximum distance between the data locations. The choice "Gamm" specifies a $G(2, 1)$ prior distribution by default and this is the default in case the prior distribution is omitted as in the above code for fitting M4.

Four additional arguments with default values control the sampling behavior of the Stan model fitting procedure. These are:

(i)   no.chains=1: how many parallel chains to run.

(ii) ad.delta = 0.8: adaptive delta parameter. Set it to higher value to aid MCMC convergence.

(iii) t.depth = 15: maximum tree depth. This should be increased too in case of problematic MCMC run.

(iv) s.size = 0.01: step size during the leap frog iterations.

These parameters should be changed in case there are problems with MCMC convergence.

Model validation is performed automatically if the additional argument validrows is provided. The resulting output is similar to that for the spTimer package. Moreover, the S3 methods **print**, **plot** and **summary** are available for this model fitting too. Table 7.3 provides the parameter estimates for the fitted model M4. The estimates are similar to the earlier ones for the models M1-M3.

We are now in a position to compare the four models M1-M4. Table 7.4 provides the model choice criteria to facilitate this. Note that as in Section 6.7 the model choice criteria are calculated using the conditional distribution of each data point at each spatial location given all other spatial observations at

	mean	sd	2.5%	97.5%
β_0	2.838	0.680	1.512	4.204
β_1	0.137	0.019	0.099	0.174
β_2	0.101	0.024	0.053	0.149
β_3	−0.019	0.070	−0.155	0.113
σ_ϵ^2	0.095	0.007	0.081	0.110
σ_w^2	0.538	0.058	0.449	0.670
ϕ	0.003	0.000	0.003	0.004

TABLE 7.3: Parameter estimates from the independent GP model fitted by `stan`. The $G(2, 1)$ distribution is adopted as the prior distribution for σ_w^2 and σ_ϵ^2. A uniform prior distribution has been specified for the decay parameter ϕ. The model has been fitted in the square-root scale as before.

that time point. The assumption of temporal independence allows us to use time as independent replications.

According to Table 7.4, model M4, fitted using `Stan`, is chosen by all the criteria. Indeed, the marginal model with fewer number of parameters than the full `spTimer` model provides stable model fitting. This is also evident when we perform validation using the four models. Here we select the same eight sites as in Figure 6.3 in Section 6.8.3. We select all 62 time points for validation and hence, the validation results, provided in Table 7.5, are based on $496 \, (= 8 \times 62)$ observations. Model M4 performs slightly better than the other models. The 496 prediction intervals show a coverage value of 92.83%, which is compared to 99.59% for the `spTimer` model M3. A plot of the predictions against the observed values is provided in Figure 7.5. The red colored open circles indicate the points for which the 95% prediction intervals do not contain the $y = x$ line plotted in blue color in the figure.

The above investigation shows that M4 is the best model so far. We now examine the quality of the model fit by obtaining a time series plot of the residuals for each of the 28 sites in Figure 7.6. This plot can be obtained from the fitted model object `M4` by issuing the command `residuals(M4)`. As in usual linear model fitting, we expect this plot to be a random scatter about the horizontal line at 0, and it should not show any patterns if the model provides a good fit. In this plot, however, the residuals do seem to show the overwhelming temporal pattern of the data seen in Figure 3.6 in the exploratory data analysis Chapter 3. Hence, we must continue our search for structured time series models that can reduce the serial correlations seen in this plot.

	M1	M2	M3	M4
p_{DIC}	5.06	5.07	78.65	30.70
$p_{DIC\ alt}$	5.38	5.38	841.96	31.37
DIC	3912.25	3214.02	3132.10	2695.77
DIC alt	3912.88	3214.64	4658.72	2697.11
$p_{waic\ 1}$	4.95	14.11	48.53	9.32
$p_{waic\ 2}$	4.97	14.18	132.90	10.31
WAIC 1	3912.14	2449.15	2603.86	2088.81
WAIC 2	3912.18	2449.30	2772.60	2090.79
gof	963.68	285.75	216.75	328.83
penalty	967.10	240.63	873.84	361.70
PMCC	1930.78	526.38	1090.59	690.54

TABLE 7.4: Model choice criteria for the four models M1 to M4. M1 is the linear model, M2 is the spatio-temporal regression model with default fixed values of ϕ_s and ϕ_t, M3 is the independent GP model implemented in `spTimer`, and M4 is the independent GP model implemented using STAN.

	M1	M2	M3	M4
RMSE	9.37	6.50	6.40	6.41
MAE	7.55	5.01	4.94	4.84
CRPS	5.63	10.67	6.79	3.33
CVG	97.95	99.59	99.59	93.03

TABLE 7.5: Model validation criteria for the four models M1 to M4. The statistics are based on 496 observations from the 8 chosen sites.

7.4 Auto regressive (AR) models

7.4.1 Hierarchical AR Models using `spTimer`

The independent GP model of the previous section ignores any temporal correlation that may be present in the spatio-temporal data. A first order dependence, where the current value depends only on the value at the previous time point, is perhaps the simplest of the time series dependence structure that we can practically model. This type of models are called auto-regressive (AR) models in the traditional time series literature and can be generalized to include contributions from more distant past. In this section we discuss spatial first order auto-regressive models as developed by Sahu et al. (2007).

Continuing with the centered model of Subsection 7.3.1, we specify the auto regression on the centered random effects $\mathbf{O}_t$ and not on data $\mathbf{Y}_t$ directly. In

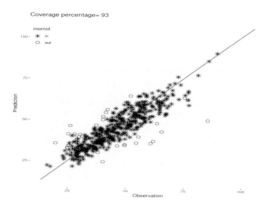

FIGURE 7.5: Observed versus predicted plot for validations performed using the independent GP model fitted using the `Stan` package. The prediction intervals for the data points plotted, as open circles do not include the 45 degree line.

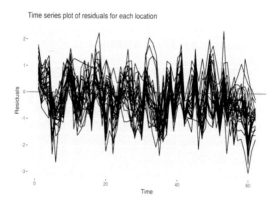

FIGURE 7.6: Time series plot of the residuals for each of the 28 air pollution monitoring sites obtained from fitted model M4.

particular, we assume

$$\mathbf{Y}_t = \mathbf{O}_t + \boldsymbol{\epsilon}_t, \tag{7.21}$$

$$\mathbf{O}_t = \rho \mathbf{O}_{t-1} + X_t \boldsymbol{\beta} + \mathbf{w}_t, \tag{7.22}$$

for $t = 1, \ldots, T$ where ρ denotes the unknown temporal correlation parameter assumed to be in the interval $(-1, 1)$. We continue to assume that $\boldsymbol{\epsilon}_t \sim N(\mathbf{0}, \sigma_\epsilon^2 I)$ as in (7.12) and $\mathbf{w}_t \sim N(\mathbf{0}, \sigma_w^2 S_w)$ as in (7.13). When $\rho = 0$, these models reduce to the GP models described above in Section 7.3. Hence the additive effect of the auto-regression can easily investigated and interpreted. There is another crucial modeling advantage in specifying the auto-regression for $\mathbf{O}_t$ instead of $\mathbf{Y}_t$. The advantage comes from the ability to handle missing observations using model (7.21). Any observation $Y(\mathbf{s}_i, t)$ can be allowed to be missing in the Bayesian model but the model $O(\mathbf{s}_i, t)$ will not be missing since there is a model (7.22) for it. Straightforward auto-regression $\mathbf{Y}_t = \rho \mathbf{Y}_{t-1} + \boldsymbol{\epsilon}_t$ runs into problems in the presence of missing observations in $\mathbf{Y}_t$.

The auto-regressive models require specification of the initial term $\mathbf{O}_0 = (O(\mathbf{s}_1, 0), \ldots, O(\mathbf{s}_n, 0))$. Here we specify an independent GP for $\mathbf{O}_0$ with mean $\mu_0 \mathbf{1}$, so that μ_0 is the overall mean at all locations, and the covariance matrix $\sigma_0^2 S_0$ where the correlation matrix S_0 is obtained using the Matérn correlation function in Equation (2.2) with the same set of correlation parameters $\boldsymbol{v}$ for $\mathbf{O}_t$ where $t > 1$. The collection of parameters $\boldsymbol{\theta}$ is now augmented with the new parameters ρ, μ_0 and σ_0^2. We can specify a flat prior for μ_0 and a gamma prior distribution for $\lambda_0^2 = \frac{1}{\sigma_0^2}$. We continue to write $\boldsymbol{\theta}$ to denote all the parameters $(\boldsymbol{\beta}, \rho, \sigma_\epsilon^2, \sigma_w^2, \boldsymbol{v}, \mu_0, \sigma_0^2)$ and $\pi(\boldsymbol{\theta})$ as the joint prior distribution. We also suppose that $\mathbf{O}$ contains all the random effects $\mathbf{O}_t$ for $t = 0, 1, \ldots, T$. The logarithm of the joint posterior distribution of the parameters is now given by:

$$\begin{aligned}
\log \pi(\boldsymbol{\theta}, \mathbf{o}|\mathbf{y}) \propto \; & -\frac{nT}{2} \log(\sigma_\epsilon^2) - \frac{1}{2\sigma_\epsilon^2} \sum_{t=1}^{T} (\mathbf{y}_t - \mathbf{o}_t)'(\mathbf{y}_t - \mathbf{o}_t) - \frac{T}{2} \log |\sigma_w^2 S_w| \\
& - \frac{1}{2\sigma_w^2} \sum_{t=1}^{T} (\mathbf{o}_t - \rho \mathbf{o}_{t-1} - \boldsymbol{\mu}_t)' S_w^{-1} (\mathbf{o}_t - \rho \mathbf{o}_{t-1} - \boldsymbol{\mu}_t) \\
& - \frac{1}{2} \log |\sigma_0^2 S_0| - \frac{1}{2} \frac{1}{\sigma_0^2} (\mathbf{o}_0 - \boldsymbol{\mu}_0)' S_0^{-1} (\mathbf{o}_0 - \boldsymbol{\mu}_0) + \log \pi(\boldsymbol{\theta}).
\end{aligned}$$

This model has been implemented in the spTimer package by Bakar and Sahu (2015) and we illustrate the model below with a practical example. For predicting $Y(\mathbf{s}_0, t)$ we modify the earlier method as follows.

Prediction of $Y(\mathbf{s}_0, t)$ at the location-time combination $\mathbf{s}_0$ and t requires calculation of the posterior predictive distribution (4.8) given the observations $\mathbf{y}$. For the centered model, we have

$$Y(\mathbf{s}_0, t) = O(\mathbf{s}_0, t) + \epsilon(\mathbf{s}_0, t) \tag{7.23}$$

$$O(\mathbf{s}_0, t) = \rho O(\mathbf{s}_0, t - 1) + \mathbf{x}'(\mathbf{s}_0, t)\boldsymbol{\beta} + w(\mathbf{s}_0, t). \tag{7.24}$$

Also

$$O(\mathbf{s}_0, 0) \sim N(\mu_0, \sigma_0^2).$$

From this it is clear that $O(\mathbf{s}_0, t)$ can only be sequentially determined using all the previous $O(\mathbf{s}_0, t)$ up to time t. Hence, we introduce the notation $\mathbf{O}(\mathbf{s}, [t])$ to denote the vector $(O(\mathbf{s}, 1), \ldots, O_l(\mathbf{s}, t))$ for $t \geq 1$.

The posterior predictive distribution of $Y(\mathbf{s}_0, t)$ is obtained by integrating over the unknown quantities in (7.23) with respect to the joint posterior distribution, i.e.,

$$
\begin{aligned}
\pi\left(y(\mathbf{s}_0, t) | \mathbf{y}\right) \;=\; & \int \pi\left(y(\mathbf{s}_0, t) | o(\mathbf{s}_0, [t]), \sigma_\epsilon^2\right) \pi\left(o(\mathbf{s}_0, [t]) | o(\mathbf{s}_0, 0), \boldsymbol{\theta}, \mathbf{o}\right) \\
& \pi\left(o(\mathbf{s}_0, 0) | \boldsymbol{\theta}, \mathbf{O}_0\right) \pi(\boldsymbol{\theta}, \mathbf{o} | \mathbf{y}) \, do(\mathbf{s}_0, [t]) \, dO(\mathbf{s}_0, 0) \, d\boldsymbol{\theta} \, do.
\end{aligned}
\tag{7.25}
$$

When using MCMC methods to draw samples from the posterior, the predictive distribution (7.25) is sampled by composition; draws from the posterior distributions, $\pi(\boldsymbol{\theta}, \mathbf{o} | \mathbf{y})$ enable draws from the above component densities, details are provided below.

In (7.25) we need to generate the random variables $O(\mathbf{s}_0, 0)$ and $O(\mathbf{s}_0, t)$ conditional on the posterior samples at the observed locations $\mathbf{s}_1, \ldots, \mathbf{s}_n$ and at the time points $1, \ldots, T$. We draw $O(\mathbf{s}_0, t)$ from its conditional distribution given all the parameters, data and $O(\mathbf{s}_0, [t-1])$. For $t = 1$ we need to sample $O(\mathbf{s}_0, 0)$. For this we have

$$
\begin{pmatrix} O(\mathbf{s}_0, 0) \\ \mathbf{O}_0 \end{pmatrix} \sim N\left[\begin{pmatrix} \mu_0 \\ \mu_0 \mathbf{1} \end{pmatrix}, \sigma_0^2 \begin{pmatrix} 1 & \Sigma_{w,12} \\ \Sigma_{w,21} & \Sigma_w \end{pmatrix} \right].
$$

Therefore,

$$
O(\mathbf{s}_0, 0) | \boldsymbol{\theta}, \mathbf{O}_0 \sim N\left(\mu_0 - \Sigma_{w,12} \Sigma_w^{-1} (\mathbf{o}_0 - \mu_0 \mathbf{1}), \sigma_0^2 \left(1 - \Sigma_{w,12} \Sigma_w^{-1} \Sigma_{w,21} \right) \right).
\tag{7.26}
$$

We now obtain for $t > 1$

$$
\begin{pmatrix} O(\mathbf{s}_0, t) \\ \mathbf{O}_t \end{pmatrix} \sim N\left[\begin{pmatrix} \rho O(\mathbf{s}_0, t-1) + \mathbf{x}'(\mathbf{s}_0, t)\boldsymbol{\beta} \\ \rho \mathbf{O}_{t-1} + X_t \boldsymbol{\beta} \end{pmatrix}, \sigma_w^2 \begin{pmatrix} 1 & \Sigma_{w,12} \\ \Sigma_{w,21} & \Sigma_w \end{pmatrix} \right].
$$

Hence,

$$
O(\mathbf{s}_0, t) | O(\mathbf{s}_0, 0), \mathbf{O}_t, \boldsymbol{\theta}, \mathbf{o} \sim N(\mu_{ot}, \sigma_{ot}^2)
\tag{7.27}
$$

where

$$
\sigma_{ot}^2 = \sigma_w^2 \left(1 - \Sigma_{w,12} \Sigma_w^{-1} \Sigma_{w,21} \right)
$$

and

$$
\mu_{ot} = \rho o(\mathbf{s}_0, t-1) + \mathbf{x}'(\mathbf{s}_0, t)\boldsymbol{\beta} + \Sigma_{w,12} \Sigma_w^{-1} \left(\mathbf{O}_t - \rho \mathbf{O}_{lt-1} - X_t \boldsymbol{\beta} \right).
$$

In summary, we implement the following algorithm to predict $Y(\mathbf{s}_0, t)$.

1. Draw a sample $\boldsymbol{\theta}^{(j)}$, and $\mathbf{o}^{(j)}$, for $j \geq 1$ from the posterior distribution $\pi(\boldsymbol{\theta}, \mathbf{o}|\mathbf{y})$.

2. Draw $o^{(j)}(\mathbf{s}_0, 0)$ using (7.26).

3. Draw $\mathbf{o}^{(j)}(\mathbf{s}, [t])$ sequentially from (7.27) starting with $t = 1$.

4. Finally draw $y^{(j)}(\mathbf{s}_0, t)$ from $N\left(o^{(j)}(\mathbf{s}_0, t), \sigma_\epsilon^{2(j)}\right)$.

Again we note that the inverse of the data transformation must be applied if we intend to predict on the original scale.

The bmstdr code for fitting the AR model from the spTimer package is same as that for fitting the GP model M3 but with the model option changed to model="AR". Hence we omit the full code. The prior distribution for the spatial decay parameter ϕ has been discussed in Section 7.3.1.

The parameter estimates from this AR model are provided in Table 7.6. The estimates are comparable to the ones presented in Table 7.2 for the independent GP model M3. However, here the effect of wind speed β_2 is no longer significant. There is significant amount of temporal correlation as seen by the estimate of ρ. Estimates of the variance components are similar.

	mean	sd	2.5%	97.5%
β_0	1.437	0.540	0.383	2.513
β_1	0.091	0.015	0.060	0.122
β_2	0.031	0.023	-0.014	0.078
β_3	-0.191	0.060	-0.304	-0.074
ρ	0.512	0.021	0.471	0.554
σ_ϵ^2	0.014	0.002	0.010	0.019
σ_w^2	0.570	0.032	0.511	0.638
ϕ	0.009	0.001	0.008	0.010

TABLE 7.6: Parameter estimates from the AR model, referenced by M5, fitted by spTimer.

7.4.2 AR modeling using INLA

An auto-regressive model using INLA can be easily implemented by adapting the details provided in Section 6.6.3. The hierarchical models are written as:

$$Y(\mathbf{s}_i, t) = N(O(\mathbf{s}_i, t), \sigma_\epsilon^2) \tag{7.28}$$
$$O(\mathbf{s}_i, t) = \mathbf{x}'(\mathbf{s}_i)\boldsymbol{\beta} + \xi(\mathbf{s}_i, t) \tag{7.29}$$
$$\xi(\mathbf{s}_i, t) = \rho\xi(\mathbf{s}_i, t - 1) + w(\mathbf{s}_i, t) \tag{7.30}$$

where $w(\mathbf{s}_i, t)$ is assumed to follow an independent zero-mean GP with the Matèrn covariance function (2.1).

Model fitting using INLA requires the following main steps. As in Section 6.6.3 we create a mesh using the command

```
mesh <- inla.mesh.2d(loc=coords, offset=offset, max.edge=max.edge)
```

where the offset and `max.edge` are parameters which can be set by the user, but default values are provided. These determine the size and density of the mesh. See the documentation for the INLA function `inla.mesh.2d`. We then set up the prior distributions by the commands:

```
spde <- inla.spde2.pcmatern(mesh = mesh, alpha = 1.5, prior.range =
    prior.range, prior.sigma = prior.sigma)
hyper  <- list(prec = list(prior = "loggamma",
param = c(prior.tau2[1], prior.tau2[2])))
```

where the user is able to choose the values for `prior.range`, `prior.sigma`, and `prior.tau2`. The implied prior is a penalized complexity prior for constructing Gaussian random fields using the Matèrn covariance function, see e.g. Fuglstad et al. (2018) and Simpson et al. (2017). The `prior.range` is a length 2 vector, with (range0, Prange) specifying that $P(\rho < \rho_0) = p_\rho$, where ρ is the spatial range of the random field. If Prange is NA, then range0 is used as a fixed range value. If this parameter is unspecified then range0 is taken as 0.90 times the maximum distance and Prange is set at 0.95. If instead a single value is specified then the range is set at the single value. The parameter `prior.sigma` is a length 2 vector = (sigma0, Psigma) specifying that $P(\sigma > \sigma_0) = p_\sigma$, where σ is the marginal standard deviation of the field. If Psigma is NA, then sigma0 is taken as the fixed value of this parameter. A similar discussion is provided in Section 6.6.3.

After creating the `mesh` by triangulation and `spde` we create a grouping index to identify time by issuing the command:

```
s_index  <- inla.spde.make.index(name="spatial.field",n.spde=spde$
    n.spde, n.group=tn)
```

where `tn` is the number of time points. The modified estimation stack is formed by issuing:

```
stack_est <- inla.stack(data=list(y=y), A=list(A_est, 1), effects=list(c(
    s_index,list(Intercept=1)), Xcov=X, tag="est")
```

The object `newformula` is created

```
newformula <- update(formula, y ~Intercept -1 + . + f(
    spatial.field, model = spde,
group=spatial.field.group, control.group=list(model="ar1")))
```

to send to the `inla` function. Finally, the INLA function is invoked by a call such as:

```
ifit <- inla(newformula, data=inla.stack.data(stack, spde=spde),
    family="gaussian", control.family = list(hyper = hyper),
    control.predictor=list(A=inla.stack.A(stack), compute=TRUE),
    control.compute = list(config = T, dic = T, waic = T)
```

To perform model validation it is sufficient to assign NA to all the y values of the validation data points as before. The `inla` fitted model returns the estimated values for each of the NA value in y along with their uncertainties. Like the `bmstdr` function `Bspatial` with `package="inla"` for spatial models the `Bsptime` function `package="inla"` fits the spatio-temporal AR model of this section. As before, it requires a data frame, a formula specifying the regression model, co-ordinates of the locations and the `scale.transform`. For validation the optional vector argument `validrows` giving the row numbers of the data frame to validated must be provided. The return values of the `Bsptime` with `package="inla"` is a list like before which can be explored with the S3 methods functions `print`, `plot` and `summary`.

With all the default values for the prior distributions and mesh sizes already provided fitting the INLA model for the running example requires only changing the package name to `package="inla"` and model option to `model ="AR"` to fit the INLA AR model. We reference this model by M6. Table 7.7 provides the parameter estimates for the model M6. Again, we obtain similar significant parameter estimates for the regression coefficients. According to this model the temporal correlation is stronger than the spatial correlation since the effective spatial range is estimated to be 9.06 kilometers $(3/0.331)$ only.

	mean	sd	2.5%	97.5%
β_1	0.241	0.003	0.235	0.247
β_2	0.052	0.012	0.028	0.077
β_3	-0.044	0.018	-0.077	-0.009
ρ	0.889	0.049	0.771	0.960
σ_ϵ^2	0.442	0.016	0.412	0.474
σ_w^2	0.764	0.518	0.197	2.086
ϕ	0.348	0.144	0.155	0.715

TABLE 7.7: Parameter estimates from the AR model, M6, fitted by INLA.

Table 7.8 compares the model choice criteria, as measured by PMCC, and model validation statistics for the two auto-regressive models M5, using `spTimer` , and the current model M6 using INLA . For the adopted choice of the hyper parameters the results show that M6 is better according to the PMCC but M5 is better at out of sample prediction. One worrisome statistics here is the coverage value for the 95% prediction intervals. The coverage value is low for the INLA based model M6. Perhaps different prior distributions and mesh sizes can improve the situation. We, however, do not explore this any further. In passing we note that the `spTimer` independent GP model M3 performs similarly to the AR model M5.

Package	gof	penalty	PMCC	RMSE	MAE	CRPS	CVG
spTimer	321.33	607.31	928.64	6.46	4.99	5.97	99.39
INLA	736.71	21.35	758.06	9.72	7.64	2.64	65.16

TABLE 7.8: Model choice and model validation statistics using two different AR models. The three model choice statistics, gof, penalty and PMCC have been calculated using data from all 28 sites and the four model validation statistics have been calculated for 8 validation sites shown in Figure 6.3.

7.5   Spatio-temporal dynamic models

7.5.1   A spatially varying dynamic model spTDyn

None of the previously discussed modeling implementations allow spatially varying regression coefficients. In this section we address this limitation using methodology developed by Bakar et al. (2015) and Bakar et al. (2016). The general regression term has been assumed as:

$$\mathbf{x}'(\mathbf{s}_i, t)\boldsymbol{\beta}(\mathbf{s}_i, t) = \sum_{j=1}^{p} x_j(\mathbf{s}_i, t)\beta_j(\mathbf{s}_i, t),$$

see for example (7.1). So far we have implemented the models by assuming

$$\beta_j(\mathbf{s}_i, t) = \beta_j, \; j = 1, \dots, p, \tag{7.31}$$

for all locations $\mathbf{s}_i, i = 1, \dots, n$ and time points $t = 1, \dots, T$. This constraint is now relaxed so that

$$\beta_j(\mathbf{s}_i, t) = \beta_{j0} + \beta_j(\mathbf{s}_i) \tag{7.32}$$

for some or all of the coefficients, $j = 1, \dots, p$ and for all $t = 1, \dots, T$. Clearly, the varying coefficient model admits the fixed coefficient model as a special case when $\beta_j(\mathbf{s}_i) = 0$ for all possible i and j. In (7.32), β_{j0} measures the overall effect of regressor j over the whole study region on the response and $\beta_j(\mathbf{s}_i)$ measures the incremental effect in location $\mathbf{s}_i$. Thus, $\beta_j(\mathbf{s}_i)$ defines infinitely many parameters since the study region being continuous in space contains uncountably infinitely many locations. From a Bayesian perspective the parameters $\beta_j(\mathbf{s})$ are handled by assuming an independent GP prior for each j for which spatially varying regression is assumed, $j = 1, \dots, p$. The GP prior specification for each $\beta_j(\mathbf{s})$ is given by $GP\left(\mathbf{0}, C\left(\cdot | \boldsymbol{\psi}_{\boldsymbol{\beta}_j}\right)\right)$. The parameters in $\boldsymbol{\psi}_{\boldsymbol{\beta}_j}$ will contain a variance term $\sigma_{\beta,j}^2$ and the correlation parameters $\phi_{\beta,j}$ and $\nu_{\beta,j}$ when the Matèrn family of correlation function is assumed, see Section 2.7. Candidate models for obvious simplifications assume $\sigma_{\beta,j}^2 = \sigma_\beta^2$ and $\phi_{\beta,j} = \phi_\beta$ and $\nu_{\beta,j} = \nu_\beta$ for all j.

The overall effects β_{j0} are assigned independent normal prior distributions with zero mean and large variances. Prior distributions for all the $\sigma_{\beta,j}^2$, $\phi_{\beta,j}$ and $\nu_{\beta,j}$ must be specified as before. The full Bayesian model specification is thus completed.

The idea of spatially varying coefficients of the previous subsection is now extended to the temporal case. The fixed coefficient model (7.31) is extended to:

$$\beta_j(\mathbf{s}_i, t) = \beta_j(t), \quad \text{and} \quad \beta_j(t) = \rho\beta_j(t-1) + \delta_j(t) \qquad (7.33)$$

for those j's for which temporally varying coefficients are to be assumed, $j = 1, \ldots, p$ where $0 \leq \rho_j \leq 1$ is an auto-regressive parameter and $\delta_j(t) \sim N(0, \sigma_\delta^2)$ independently for all j and t. The extreme value of $\rho = 1$ in (7.33) implies a random walk model. The initial condition involves the parameters $\beta_j(0)$ and we assume that $\beta_j(0) \sim N(0, \sigma_0^2)$ for each j independently. The Bayesian model is completed by assuming a Gamma prior distribution for all the inverse variance parameters $1/\sigma_\delta^2$ and $1/\sigma_0^2$.

The above two modeling innovations can be combined into a single model where some covariate effects are assumed spatially varying and some others are temporally dynamic. Methodology for spatial interpolation and temporal forecasting for this model has been detailed in the cited reference Bakar et al. (2016) and is not reproduced here. We will now illustrate model fitting for the running New York air pollution example using the `GibbsDyn` function of the R package `sptDyn`.

The `bmstdr` model fitting function `Bsptime` with the option package=" spTDyn" is able to fit and validate using the spatially varying dynamic model. The other required arguments are mostly similar to that for the option `package="spTimer"` with the following exceptions. Any covariate term, say `xmaxtemp`, which requires a spatially varying coefficient should be written twice in the right hand side of the formula: once as usual and then within an opening and closing parenthesis with the word sp upfront. For example, to include the incremental spatial effects of `xmaxtemp` the right hand side of the formula should include the terms `xmaxtemp + sp(xmaxtemp)`. Any covariate which should have a dynamic effect should be wrapped around the word `tp`. For temporally varying coefficients it is not necessary to include the term as in the spatially varying case. Some covariates can have both spatially varying and dynamic effects. In addition, a valid formula with mixture of covariate terms can be provided. For example,

```
library(spTDyn)
f3 <- y8hrmax ~ xmaxtemp+sp(xmaxtemp)+tp(xwdsp)+xrh
```

declares a model with spatially varying effects for maximum temperature, dynamic effects for wind speed and fixed effects for relative humidity. Thus, the `bmstdr` code line for this model fitting is given by:

```
M7 <- Bsptime(package="sptDyn", model="GP", formula=f3, data=
    nysptime, coordtype="utm", coords=4:5, scale.transform = "SQRT",
    mchoice=T)
```

The prior distribution for the spatial decay parameter ϕ has been discussed in Section 7.3.1.

Parameter estimates for this model M7 are provided in Table 7.9. The parameter β_1 in the second row corresponds to the first xmaxtemp in the formula. The third row for β_3 corresponds to the last term xrh in the formula. The fourth row ρ is the auto-regressive parameter ρ for wind speed in (7.33) and the row for σ_δ^2 provides the estimate for variance. The rows for ϕ, σ_w^2 and σ_ϵ^2 are interpreted as before, viz. the spatial variance and the nugget effect respectively. The estimate for σ_β^2 is the estimate of variance for the spatially varying effects of maximum temperature. Finally, the row for σ_0^2 provides the estimates of σ_0^2 for the initial condition $\beta_j(0)$.

	mean	sd	2.5%	97.5%
β_0 (Intercept)	1.887	0.597	0.727	3.042
β_1 (maxtemp)	0.192	0.015	0.163	0.221
β_3 (rh)	-0.133	0.064	-0.258	-0.006
ρ (wdsp)	0.263	0.231	-0.182	0.715
σ_ϵ^2	0.015	0.003	0.011	0.021
σ_w^2	0.272	0.028	0.230	0.340
σ_β^2	0.067	0.018	0.040	0.110
σ_δ^2	0.048	0.009	0.033	0.068
σ_0^2	0.853	1.051	0.190	3.271
ϕ	0.018	0.002	0.014	0.022

TABLE 7.9: Parameter estimates from the spatially varying dynamic model fitted by spTDyn.

The fitted model M7 can be explored further graphically by obtaining various plots of the random effects as illustrated in the papers Bakar et al. (2015) and Bakar et al. (2016). For example, Figure 7.7 provides a boxplot of the spatial effects at the 28 modeling sites and Figure 7.8 depicts the dynamic effects of wind speed. The code for drawing these plots are provided online from github[3].

7.5.2   A dynamic spatio-temporal model using spBayes

In this section we introduce a dynamic spatio-temporal model following the development in Section 11.5 of Banerjee et al. (2015). The top-level model continues to be the first one (7.1) introduced at the beginning of the chapter. But the crucial difference is that the regression coefficients are allowed to have a dynamic prior distribution. In addition, the spatio-temporal random effect is assigned a GP with different correlation structure at each time point. Here are the modeling details.

---

[3]https://github.com/sujit-sahu/bookbmstdr.git

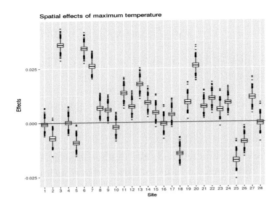

FIGURE 7.7: Boxplot of MCMC simulated incremental spatial effects of maximum temperature at the 28 data modeling sites.

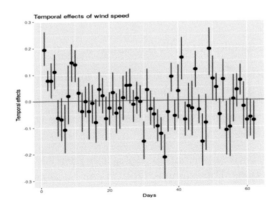

FIGURE 7.8: Point (filled circles) and 95% interval estimates (end points of the line segments) of dynamic effects of wind speed over the 62 days.

The models are written hierarchically as:

$$Y(\mathbf{s}_i, t) = \mathbf{x}'(\mathbf{s}_i, t)\boldsymbol{\beta}_t + O(\mathbf{s}_i, t) + \epsilon(\mathbf{s}_i, t), \quad \epsilon(\mathbf{s}_i, t) \sim N(0, \sigma^2_{\epsilon, t}) \tag{7.34}$$

$$\boldsymbol{\beta}_t = \boldsymbol{\beta}_{t-1} + \boldsymbol{\eta}_t, \quad \boldsymbol{\eta}_t \sim N(\mathbf{0}, \Sigma_\eta) \tag{7.35}$$

$$O(\mathbf{s}_i, t) = O(\mathbf{s}_i, t-1) + w(\mathbf{s}_i, t), \quad w(\mathbf{s}_i, t) \sim GP(\mathbf{0}, C_t(\cdot|\boldsymbol{\psi}_t)) \tag{7.36}$$

independently for all $i = 1, \ldots, n$ and $t = 1, \ldots, T$. The model components are explained as follows.

1. The top level model (7.34) is of the same general form as (7.1) but now the pure error model has been provided with a time varying variance $\sigma^2_{\epsilon, t}$.

2. The regression coefficients $\boldsymbol{\beta}_t$ are made dynamic in (7.35) and they are given multivariate normal prior distribution $N(\mathbf{0}, \Sigma_\eta)$. The hyper-parameter Σ_η is given an inverse Wishart prior distribution.

3. The spatio-temporal random effects are given the random-walk dynamic model (7.36). The innovation, $w(\mathbf{s}_i, t)$, at each time point has been assigned a time varying GP with a time varying covariance function $C_t(\cdot|\boldsymbol{\psi}_t)$. This covariance function will accommodate a time varying variance component $\sigma^2_{w, t}$ and correlation parameters $\boldsymbol{v}_t$.

The models are completed by assuming distributions to initialize the dynamic parameters and prior distributions for the model parameters as follows.

- The initial regression parameter $\boldsymbol{\beta}_0$ is assumed to follow $N(\mathbf{m}_0, \Sigma_0)$ where $\mathbf{m}_0$ and Σ_0 are hyper-parameters suitably chosen to represent vague prior distribution.

- The initial random effect $O(\mathbf{s}_i, 0)$ is assumed to be $\mathbf{0}$.

- Usual gamma prior distributions are assumed for the inverse of the variance components $\sigma^2_{\epsilon, t}$ and $\sigma^2_{w, t}$ as in the previous sections.

- Also similar prior distributions are assumed for the correlation structure parameters $\boldsymbol{v}_t$.

The posterior predictive distribution $f(y(\mathbf{s}_0, t)|\mathbf{y})$ can be computed similarly as in previous sections. For example, $O(\mathbf{s}_0, t)$ can be sequentially determined by using Kriging as in the AR model, see (7.27). We omit the details here. This model has been implemented as the `spDynLM` function in the software package `spBayes`. The `bmstdr` model fitting function `Bsptime` with the option `package="spBayes"` is able to fit and validate using this model. The model fitting command using the same prior distributions noted in Section 11.5 of the book by Banerjee et al. (2015) is given by:

```
M8 <- Bsptime(package="spBayes", formula=f2, data=nysptime,
    prior.sigma2=c(2, 25), prior.tau2 =c(2, 25), prior.sigma.eta =c(2,
    0.001), coordtype="utm", coords=4:5, scale.transform = "SQRT",
    mchoice=T)
```

In this model all parameters are either spatially varying or dynamic. Hence it is not possible to obtain a table of parameter estimates. However, code provided in the documentation of the spBayes package can be used to explore the parameter estimates graphically. Figure 7.9 illustrates the dynamic regression coefficients while Figure 7.10 plots the θ parameters σ_w^2, σ_ϵ^2 and the effective ranges. Here the parameter plots are not very informative but in other examples those may provide interesting conclusions.

In Table 7.10 we compare the two dynamic models using the spTDyn and spBayes packages. The spBayes package does not seem to perform well. This may be due to difficulties in assigning suitable prior distributions for the dynamic parameters. However, the prediction intervals do seem to provide the right level of coverage as claimed in Section 11 of the book by Banerjee et al. (2015).

Package	gof	penalty	PMCC	RMSE	MAE	CRPS	CVG
spTDyn	166.93	434.90	601.83	5.07	3.87	2.88	96.77
spBayes	3583.68	7288.88	10872.56	22.38	18.65	11.67	98.92

TABLE 7.10: Model choice and validation statistics for two dynamic models implemented using spTDyn and spBayes for the New York air pollution example. The model choice statistics are based on data from all 28 sites and validation statistics are for the three validation sites 1, 5, and 10 shown in Figure 7.1.

7.6 Spatio-temporal models based on Gaussian predictive processes (GPP)

All the previously discussed models with the fully specified GP are problematic to fit when the number of locations n is moderately large. Matrix inversions, storage and other operations involving the large dimensional matrices become prohibitive and the GP based models become impractical to use, see Banerjee et al. (2008). To overcome these computational problems for spatio-temporal data Sahu and Bakar (2012b) propose a lower dimensional GP which has been shown to produce very accurate results requiring only a small fraction of computing resources. The main idea here is to define the random effects $O(\mathbf{s}_i, t)$ at a smaller number, m, of locations, called the knots, and then use Kriging

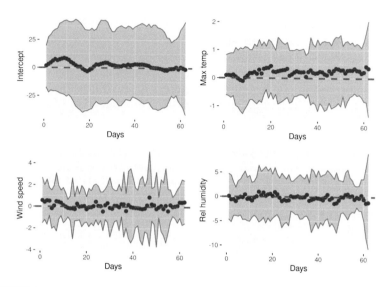

FIGURE 7.9: Plots of β-parameter estimates for the spatio-temporal dynamic model implemented in `spBayes` package. The command `geom_hline` adds the dashed horizontal line in each plot.

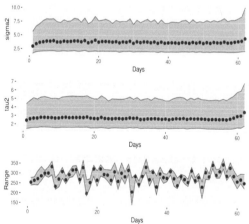

FIGURE 7.10: Plots of θ parameter estimates for the spatio-temporal dynamic model implemented in `spBayes` package.

to predict those random effects at the data and prediction locations. Here, an AR model, see previous section, is only assumed for the random effects at the knot locations and not for all the random effects at the observation locations. At the top level we assume the model:

$$\mathbf{Y}_t = X_t \boldsymbol{\beta} + A\mathbf{O}_t + \boldsymbol{\epsilon}_t, \tag{7.37}$$

for $t = 1, \ldots, T$, where $A = CS_{w*}^{-1}$ and C denotes the n by m cross-correlation matrix between the random effects at the n observation locations $\mathbf{s}_1, \ldots, \mathbf{s}_n$ and m knot locations, $\mathbf{s}_1^*, \ldots, \mathbf{s}_m^*$, and S_{w*} is the m by m correlation matrix of the m random effects $\mathbf{o}_t$. That is, C has the element

$$C_{ij} = C(d_{ij*}|\boldsymbol{v})$$

at the ith row and jth column for $i = 1, \ldots, n$ and $j = 1, \ldots, m$ where d_{ij*} is the distance between the sites $\mathbf{s}_i$ and $\mathbf{s}_j^*$ and $C(\cdot|\cdot)$ is the Matèrn correlation given in (2.2). Also, the elements of S_{w*} are obtained from the Matèrn correlation function evaluated for the locations $\mathbf{s}_i^*$ and $\mathbf{s}_j^*$ for $i = 1, \ldots, m$ and $j = 1, \ldots, m$.

We specify an AR model for $\mathbf{O}_t$ at the knot locations:

$$\mathbf{O}_t = \rho\, \mathbf{O}_{t-1} + \mathbf{w}_t, \tag{7.38}$$

for $t = 1, \ldots, T$, where $\mathbf{w}_t \sim N(\mathbf{0}, \sigma_w^2 S_{w*})$ independently. Note that, here S_{w*} is an $m \times m$ matrix, which is of much lower dimensional than the same for two previous models GP and AR since we assume that $m \ll n$.

The auto-regressive models are completed by the assumption for the initial conditions, $\mathbf{o}_0 \sim N(\mathbf{0}, \sigma_0^2 S_0)$, where the correlation matrix S_0 is obtained by using the Matérn correlation function in Equation (2.2). Let $\mathbf{o}$ denote the random effects $\mathbf{o}_t$ for $t = 0, 1, \ldots, T$. Let $\boldsymbol{\theta}$ denote all the parameters $\boldsymbol{\beta}$, ρ, σ_ϵ^2, σ_w^2, $\boldsymbol{v}$ and σ_0^2. The logarithm of the joint posterior distribution of the parameters and the missing data is given by:

$$\log \pi(\boldsymbol{\theta}, \mathbf{o}|\mathbf{y}) \;\propto\; -\frac{nT}{2}\log(\sigma_\epsilon^2) - \frac{1}{2\sigma_\epsilon^2}\sum_{t=1}^{T}(\mathbf{y}_t - X_t\boldsymbol{\beta} - A\mathbf{o}_t)'(\mathbf{y}_t - X_t\boldsymbol{\beta} - A\mathbf{o}_t)$$

$$-\frac{T}{2}\log(|\sigma_w^2 S_{w*}|) - \frac{1}{2\sigma_w^2}\sum_{t=1}^{T}(\mathbf{o}_t - \rho\mathbf{o}_{t-1})'S_{w*}^{-1}(\mathbf{o}_t - \rho\mathbf{o}_{t-1})$$

$$-\frac{1}{2}\log(|\sigma_0^2 S_0|) - \frac{1}{2}\frac{1}{\sigma_0^2}\mathbf{o}_0' S_0^{-1}\mathbf{o}_0 + \log \pi(\boldsymbol{\theta}), \tag{7.39}$$

where we adopt the same prior distribution $\pi(\boldsymbol{\theta})$ as before.

Prediction of $Y(\mathbf{s}_0, t)$ at the location-time combination $\mathbf{s}_0$ and t requires calculation of the posterior predictive distribution (4.8) given the observations $\mathbf{y}$. From the top level model in this section, we have

$$Y(\mathbf{s}_0, t) = \mathbf{x}'(\mathbf{s}_0, t)\boldsymbol{\beta} + \mathbf{a}_0' S_{w*}^{-1}\mathbf{O}_t + \epsilon(\mathbf{s}_0, t) \tag{7.40}$$

where $\mathbf{a}_0'$ is a $1 \times m$ vector with the jth entry given by $C(d_{0j}|\boldsymbol{v})$ in (2.2) where d_{0j} is the distance between the sites $\mathbf{s}_0$ and $\mathbf{s}_j$. Note that $Y(\mathbf{s}_0, t)$ and $\mathbf{Y}_t = (Y(\mathbf{s}_1, t), \ldots, Y(\mathbf{s}_n, t))$ are conditionally independent given $\mathbf{O}_t$ and $\boldsymbol{\theta}$. This is because the distribution of $Y(\mathbf{s}_0, t)$ in (7.40) is completely determined given the values of the parameters in $\boldsymbol{\theta}$ and $\mathbf{O}_t$. Hence the posterior predictive distribution of $Y(\mathbf{s}_0, t)$ is obtained using the simpler version (4.9). Now the posterior predictive distribution is obtained by integrating over the unknown quantities in (7.40) with respect to the joint posterior distribution, i.e.,

$$\pi\left(y(\mathbf{s}_0, t)|\mathbf{y}\right) = \int_{-\infty}^{\infty} \pi\left(y(\mathbf{s}_0, t)|\mathbf{o}_t, \boldsymbol{\theta}\right) \pi(\boldsymbol{\theta}, \mathbf{o}|\mathbf{y}) \, d\boldsymbol{\theta} \, d\mathbf{o}. \tag{7.41}$$

When using MCMC methods to draw samples from the posterior, the predictive distribution (7.41) is sampled by composition; draws from the posterior distributions, $\pi(\boldsymbol{\theta}, \mathbf{o}|\mathbf{y})$ enable draws from (7.40). Unlike in the previous two cases, there is no need to perform any further Kriging like conditional distributional calculations since conditioning has been built into the likelihood models (7.37) and (7.40) already. Thus, sampling for predictions is much cheaper computationally while using the GPP model. Again we note that the inverse of the data transformation must be applied if we intend to predict on the original scale.

The GPP models of this section can be fit and validated using the Bsptime model fitting engine with the options package="spTimer" together with model ="GPP". The other options like formula, data, coordtype, coords must be provided as before. Additionally, the required knot locations can be specified in either of two ways: (i) by specifying the g_size argument or (ii) the knots.coords argument. The g_size argument, if passed on as a scalar chooses a square grid of knot locations within the rectangular box of the coordinates of the model fitting locations. Otherwise, if it is vector of length 2 then a rectangular grid is chosen. The knots.coords argument specifies the knot locations chosen by the user already. An error occurs if either none or both of g_size and knots.coords are specified. The chosen knot locations are returned in the model fitted object.

The GPP model is the last model of this chapter and this is fitted by the bmstdr command:

```
M9 <- Bsptime(package="spTimer", model="GPP", g_size=5, formula=f2
    , data=nysptime, coordtype="utm", coords=4:5,
  scale.transform = "SQRT")
```

The argument g_size can be changed to experiment with the size of the grid. Table 7.11 provides the model choice criteria and model validation statistics for three grid sizes. Grid sizes 4 and 5 are better than grid size 3 according to these statistics. The model with grid size 5 gives a better model fit according to the PMCC. Hence, we proceed with grid size 5. Figure 7.11 provides a plot of the knot locations.

Table 7.12 presents the parameter estimates for this GPP model. The estimates are comparable to the previously tabled estimates for the other models. We do not explore this model any further for this small data set with only 28 spatial locations. Section 8.5 in the next chapter provides a substantial practical example where only the GPP model is viable because of the huge number of spatial locations.

Grid	gof	penalty	PMCC	RMSE	MAE	CRPS	CVG
3 × 3	200.13	953.99	1154.12	6.57	5.05	7.80	99.80
4 × 4	145.39	836.05	981.44	6.34	4.82	7.02	99.39
5 × 5	146.31	815.13	961.44	6.40	4.87	7.30	99.39

TABLE 7.11: Model choice and validation statistics for the GPP models with different knot sizes for the New York air pollution example. The model choice statistics are based on data from all 28 sites and validation statistics are for the same 8 validation sites in this running example.

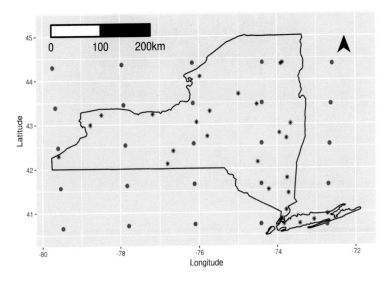

FIGURE 7.11: Data and 25 knot locations (in red) in New York.

We now examine the residuals of the GPP model M9 to see if these still contain overwhelming temporal dependencies seen earlier in Figure 7.6. Figure 7.12 plots the residuals and we can see that this plot does not show the temporal patterns present in Figure 7.6. There are only few influential data points but the magnitudes of the absolute values of the corresponding residuals are much lower than what has been seen earlier in Figure 7.6. Hence the GPP model does seem to do a good job of model fitting.

	mean	sd	2.5%	97.5%
β_0	3.122	0.711	1.695	4.353
β_1	0.126	0.021	0.078	0.162
β_2	0.111	0.030	0.057	0.173
β_3	-0.101	0.079	-0.253	0.047
ρ	0.188	0.041	0.107	0.270
σ_ϵ^2	0.142	0.009	0.128	0.161
σ_w^2	0.821	0.138	0.647	1.200
ϕ	0.007	0.001	0.004	0.009

TABLE 7.12: Parameter estimates for the GPP model, M9, with a 5×5 grid for the knot locations.

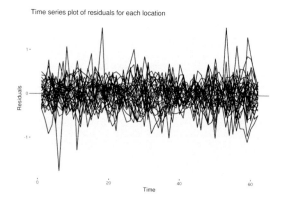

FIGURE 7.12: Time series plot of the residuals for each of the 28 air pollution monitoring sites obtained by fitting the GPP model M9.

7.7 Performance assessment of all the models

As the final act of this chapter, we compare all the previous models by the four chosen validation statistics and also the PMCC. The basic regression model, implemented in the square-root scale, is given by the same formula

```
f2 <- y8hrmax ~xmaxtemp+xwdsp+xrh
```

except for the spatio-temporal dynamic model, M7 where there are two additional terms one for spatially varying effect of maximum temperature and the other an auto-regressive term for wind speed. The regression model for M7 is given by:

```
f3 <- y8hrmax ~xmaxtemp+sp(xmaxtemp)+tp(xwdsp)+xrh
```

We set aside data from eight sites shown in Figure 6.3. This gives us 496

$(= 8 \times 62)$ data points the validation set and the model is fitted with remaining 1240 space time observations from the 20 modeling sites. Table 7.13 provides the results. The models M1-M9 have been noted before and are described below:

M1 Independent error regression model, see Section 7.2.5.

M2 Spatio-temporal model with a separable covariance structure, see Section 7.2.

M3 Temporally independent GP model fitted using the `spTimer` package, see Section 7.3.1

M4 Marginal model implemented using `Stan`, see Section 7.3.2.

M5 First order auto-regressive model fitted using `spTimer`, see Section 7.4.1.

M6 First order auto-regressive model fitted using `INLA`, see Section 7.4.2.

M7 Spatio-temporal dynamic models fitted using `spTDyn`, see Section 7.5.

M9 Spatio-temporal models based on GPP, fitted using `spTimer`, see Section 7.6.

Note that we have omitted M8, the model based on `spBayes` because we were not able to produce comparable results. We do not report the WAIC and the DIC since those are not available in the `bmstdr` package for *all* the models in Table 7.13. The PMCC, see Section 4.16.3, will choose the separable model M2 which does not require any iterative model fitting. But note that a grid search method has been used to obtain the optimum spatial and temporal decay parameters assumed in model fitting. The `Stan` fitted model M4 is the next one based according to the PMCC. But recall that the residual plot in Figure 7.6 is not satisfactory for M4. The GPP model M9 seems to be the best for validation as it provides the least values of RMSE, MAE and CRPS. However, it does provide a slight over fitting since the coverage is higher than the expected 95%.

We end this comparison with some words of caution. The comparison should not be generalized to make statements like package A performs better than package B. For example, the marginal GP model, M4, implemented using `Stan` performed slightly worse than M9. But there may be another model, e.g. auto-regressive, implemented using `Stan`, that may perform better than the `spTimer` models. The worth of this illustration lies in the comparison itself. Using the `bmstdr` package it is straightforward to compare different models implemented in different packages without having to learn and program the individual packages.

	M1	M2	M3	M4	M5	M6	M7	M9
RMSE	9.35	6.49	6.40	6.42	9.73	6.46	6.59	6.36
MAE	7.54	5.00	4.94	4.85	7.64	4.99	5.11	4.85
CRPS	5.67	10.56	6.79	3.57	2.63	5.97	5.12	7.47
CVG	98.36	99.59	99.59	92.83	65.37	99.39	99.39	99.39
G	728.91	218.49	181.71	173.45	527.67	185.76	71.30	146.69
P	731.61	195.37	935.42	266.23	17.19	718.47	467.46	815.85
G+P	1460.52	413.86	1117.13	439.67	544.86	904.23	538.76	962.54

TABLE 7.13: Validation statistics and PMCC for different models implemented using different software packages. M1 is the independent error regression model.

7.8　Conclusion

This chapter has put together different spatio-temporal models for spatially point referenced temporal data which have been assumed to follow the Gaussian distribution. Theoretical descriptions of the models have been included to keep the rigor in presentation of the models. However, the theory can be skipped if one is only interested in fitting and validation using these models. The `bmstdr` code lines for implementing these models have been provided for the running example of the New York air pollution data set.

The discussion in this chapter has mainly focused on model fitting comparison and validation. No discussion has been provided on studying prior sensitivity and MCMC convergence for the sake of brevity. The reader is encouraged to perform such studies themselves in order to learn more regarding the nature of the models and their fitting procedures. As always, the reader is also encouraged to run the models and replicate the results first before embarking on the task of modeling their own data. Further examples of real life spatio-temporal data modeling, validation and forecasting are provided in the following two chapters 8 and 9.

This chapter does not discuss non-linear models at all. There are many successful non-linear models, for example, the Kriged-Kalman filter model proposed by Sahu and Mardia (2005a). Non-linear modeling requires much different ways of handling and methods for generalizations when compared to linear models versions of which can be fitted by the lm command.

Another important limitation of the presentation in this chapter is the total omission of modeling of discrete, i.e. count, data using generalized linear models. This is mainly because of paucity of reliable MCMC based model fitting software packages for such data. Moreover, the exact model fitting methods for the base linear model (M1) and the separable model are no longer feasible since the associated posterior distributions are not available in closed form. However, the non-MCMC based `INLA` package can fit such models using techniques outlined in Section 7.4.2. Bespoke code needs to be written to fit the models using `Stan`, although there are helpful websites, e.g.https://mc-stan.org/, which the interested reader is referred to.

7.9 Exercises

1. Verify all the theoretical results reported in Section 7.2.

2. The `bmstdr` fitting of the models M1 and M2 estimates the missing observations by the grand mean of the data. Discuss how this limitation can be removed by adopting a Bayesian solution. One such solution is to estimate those alongside the parameters and Gibbs sampling can be used to perform this task. Using the conditional distributions deduced in Section 7.2 write your own Gibbs sampling routines for performing model fitting and validation.

3. The G(2, 1) has been used as the prior distribution in Table 7.2. Study sensitivity of these parameter estimates when the prior distribution is changed. The documentation of the function, `?Bsptime` details what other prior distributions are allowed for ϕ.

4. Reproduce Figures 7.1 and 7.2 for your choice of three different validation sites.

5. Blangiardo and Cameletti (2015) use the `piemonte` data set to illustrate the `INLA` software package. Perform modeling and validation of this data set available from github[4]. Undertake several validation investigations including leave-one-out as there are data from only 24 air pollution monitoring sites. Construct a table of validation statistics like Table 7.8.

6. Fit the GPP model to the `piemonte` data set. Select knot points falling within the borders of the Piemonte region only. The borders are available from Blangiardo and Cameletti (2015) and also from github.

7. Reproduce Table 7.13 for the `nysptime` data set and then obtain such a table for the `piemonte` data set.

---

[4]https://github.com/sujit-sahu/bookbmstdr.git

8

Practical examples of point referenced data modeling

8.1 Introduction

This chapter showcases point referenced spatio-temporal modeling using five practical examples. The examples highlight the practical use of such modeling and extend the methodologies where necessary. The examples build on the basic concepts introduced in the earlier chapters, especially Chapter 7, on spatio-temporal modeling. Some of these examples have already been published in the literature.

The data sets used in the examples have been introduced previously in Chapter 1. The examples serve various purposes, e.g., spatial, temporal or spatio-temporal aggregation, assessment of trends in various settings and evaluating compliance with respect to air pollution regulations. The reader is able to choose the topic they are interested in and experience the methodologies offered by the examples. The data sets and the code to reproduce the results are provided online on github[1].

8.2 Estimating annual average air pollution in England and Wales

This example is based on the data set introduced in Section 1.3.2 where we have daily data from $n = 144$ air pollution monitoring sites in England and Wales for $T = 1826$ days during the 5-year period from 2007 to 2011. As mentioned there we only illustrate modeling of NO_2 data for the 365 days in the year 2011. See Mukhopadhyay and Sahu (2018) for exploratory data analysis and modeling and validation for the full data set.

Following Mukhopadhyay and Sahu (2018) we model on the square-root scale and we let $Y(\mathbf{s}_i, t)$ denote the square-root data to be modeled. The site type classifier, which takes three possible values: Rural, Urban or RKS, see

---

[1]https://github.com/sujit-sahu/bookbmstdr.git

DOI: 10.1201/9780429318443-8

Section 1.3.2, is used as a covariate. As an additional covariate, the models use the estimated daily concentrations from the Air Quality Unified Model (AQUM, Savage et al., 2013), available on the corners of a 12 kilometer square grid covering England and Wales. AQUM is a 3-dimensional weather and chemistry transport model used by the Met Office to deliver air quality forecast for the Department for the Environment, Food and Rural Affairs (DEFRA) and for scientific research.

The general spatio-temporal model for these data is the one written down as Equation (7.10) but without the independence assumption for the $w(\mathbf{s}_i, t)$'s. The mean function is provided by the regression model:

$$\mu(\mathbf{s}_i, t) = \gamma_0 + \gamma_1 x(\mathbf{s}_i, t) + \sum_{\ell=2}^{r} \delta_\ell(\mathbf{s}_i) \left(\gamma_{0\ell} + \gamma_{1\ell} x(\mathbf{s}_i, t) \right),$$

a site type specific regression on the modeled square-root AQUM concentrations $x(\mathbf{s}_i, t)$. We take $r = 3$, corresponding to the three site types (Rural, Urban, RKS), and the rural site type corresponds to $\ell = 1$ and is the base line level. Thus, (γ_0, γ_1) are respectively the slope and intercept terms for the Rural sites, while $(\gamma_{0j}, \gamma_{1\ell})$ are the incremental adjustments for site type ℓ, $\ell = 2, 3$. Finally, $\delta_\ell(\mathbf{s}_i)$ is an indicator function, equaling one if site $\mathbf{s}_i$ is of the ℓth site type and zero otherwise.

We fit models with the default prior distributions for all the parameters. The underlying GP is assumed to have the exponential covariance function and the decay parameter is given the G(2, 1) prior distribution. Both the spatial and pure error variances, σ_w^2 and σ_ϵ^2, are given the G(2, 1) prior distribution. The regression coefficients and the autoregressive parameter rho are given the flat prior distribution.

Following, Mukhopadhyay and Sahu (2018) we only consider the GPP model discussed in Section 7.6. In order to fit the GPP model we need to define a set of knot locations. We investigate with three different sets of knot locations corresponding to three regular grids of sizes 100, 225 covering the map of England and Wales. The grid locations falling inside the boundaries of the map are proposed as knots. This is done using over function in the package sp. This process gives us three sets of knots having 25, 58 and 106 points respectively from the corresponding grids of 100, 225 and 400 locations. Figure 8.1 shows the grid with 225 locations and the corresponding 58 knot locations which fall inside the map. To select the number of knots we perform validation by setting aside 1000 randomly selected observations out of the 52,560 (144×365) space-time data points. The model with the 58 knot locations performs better than the other two with 25 and 106 knot locations according to the RMSE criterion. Henceforth, we model with the 58 knot locations.

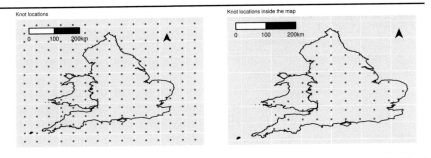

FIGURE 8.1: A grid with 225 locations (left) and the 58 knots falling inside the map (right) of England and Wales.

♣ **R Code Notes 8.1. Figure 8.1** For plotting the outline map see the code for drawing Figure 1.2. To select the knot locations inside the land boundary we use code like the following.

```
delta <- 100000
a <- c(min(p2011data$easting)-1.5*delta, max(p2011data$
    easting)+delta)
b <- c(min(p2011data$northing)-delta, max(p2011data$northing)
    +1.5*delta)
knots.coords <- spT.grid.coords(Lon=a, Lat=b, by=c(15, 15))
knots <- data.frame(easting=knots.coords[,1], northing=
    knots.coords[,2])
pnts <- knots
colnames(pnts) <- c("long", "lat")
coordinates(pnts)<- ~long + lat
pnts <- SpatialPoints(coords = pnts)
proj4string(pnts) <- proj4string(ewmap)
identicalCRS(ewmap, pnts)
a1 <- sp::over(pnts, ewmap, fn = NULL) #, returnList = TRUE)
knots.inside <- knots[!is.na(a1$geo_code), ]
dim(knots.inside)
knots.inside.58 <- knots.inside
```

These commands assume that the pollution data frame p2011data and map polygon ewmap are present in the current work space.

The model is fitted by using the commands:

```
f2 <- obs_no2 ~type + sqrt(aqum_no2) + type:sqrt(aqum_no2)
M9 <- Bsptime(package="spTimer", model="GPP", formula=f2, data=
    p2011data, coordtype="utm", coords=4:5,
```

```
scale.transform = "SQRT", knots.coords=knots.inside.58, n.report=10, N
   =1000, burn.in = 500)
```

These commands assume the presence of the data frame `p2011data` and the knot locations `knots.inside.58`.

Parameter estimates of the fitted model are presented in Table 8.1. The site type RKS (Road and Kerb Side) is taken as the base level. Both the Rural and Urban sites are seen to have significantly lower intercepts than the RKS site types. The interaction effect is significant as well. The estimate of the autoregressive parameter shows moderate levels of temporal correlation while the estimate of the spatial decay parameter ϕ shows presence of spatial correlation upto a range of about 51 kilometers $(3/\hat{\phi})$. The variance components show higher level of spatial variability than the residual pure error.

	mean	sd	2.5%	97.5%
Intercept	4.715	0.062	4.590	4.836
Rural	−2.172	0.071	−2.317	−2.033
Urban	−1.258	0.054	−1.364	−1.149
AQUM	0.513	0.012	0.490	0.538
Rural:AQUM	−0.054	0.014	−0.081	−0.027
Urban:AQUM	0.022	0.009	0.003	0.039
ρ	0.453	0.012	0.429	0.476
σ_ϵ^2	1.557	0.019	1.528	1.602
σ_w^2	10.670	1.357	8.540	13.531
ϕ	0.058	0.004	0.051	0.067

TABLE 8.1: Parameter estimates of the fitted model for the daily NO_2 data in 2011.

We now illustrate spatio-temporal prediction using the fitted model. The predictions are to be made for a large number of locations so that we are able draw a heat map of annual NO_2 levels. In order to do this we predict at the 2124 locations, see Figure 8.2. These locations have been obtained by sampling from a larger 1-kilometer covering England and Wales used by Mukhopadhyay and Sahu (2018). Thus, we generate MCMC samples for the 775,260 $(= 2124 \times 365)$ space-time random variables. This large number limits the number of MCMC samples to 500 that can be worked with in R without causing memory problems. From these predictions we find the 2124 annual averages for each of the 500 MCMC samples. These annual MCMC samples are then averaged to obtain the annual predictions and their uncertainties.

The prediction command is given by:

```
gpred <- predict(M9$fit, tol.dist =0.0, newdata=predgrid, newcoords=
   ~ easting+northing)
```

where the data frame `predgrid` contains the covariate values and the location information for the 2124 locations and 365 days. The annual predictions at

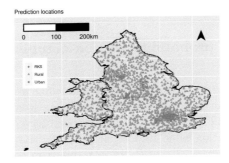

FIGURE 8.2: Map showing 2124 predictive locations.

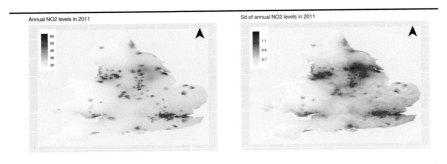

FIGURE 8.3: Maps showing annual predictions and their standard deviations for NO$_2$ in 2011.

each MCMC iteration are obtained by averaging over the suitable quantities. These MCMC iterates are summarized to obtain the annual estimates and their standard deviations. These annual values are then processed using the `interp` function of the `akima` library (Akima, 2013) to obtain the map surfaces. The interpolations outside the boundaries have been eliminated using the `over` function as in Figure 8.1.

Figure 8.3 plots the map of annual average NO$_2$ levels along with their uncertainties. There is considerable spatial variation in the map as expected. The NO$_2$ levels are higher in London and other urban areas than the rural areas. The predictive uncertainties are higher where the NO$_2$ levels are higher. This is also well known – higher pollutant levels are generally associated with higher uncertainty levels. It is possible to summarize the prediction samples to any desired administrative geography levels as has been illustrated by Mukhopadhyay and Sahu (2018). We, however, do not consider that here.

8.3 Assessing probability of non-compliance in air pollution

We now return to the air pollution data modeling example introduced in Section 1.3.3. The analysis presented here is taken from our earlier work in Sahu and Bakar (2012b). Let $Y_l(\mathbf{s}_i, t)$ denote the observed value of the square-root ozone concentration value at location $\mathbf{s}_i$, $i = 1, \ldots, n$ on day t within year l for $l = 1, \ldots, r$ and $t = 1, \ldots, T$. As introduced in Section 1.3.3, we have data from $n = 691$ sites for $T = 153$ days during May to September for each of $r = 10$ years from 1997 to 2006. For this moderately large data modeling problem we use the GPP based model discussed in Section 7.6.

We assume the model (7.37) for $Y_l(\mathbf{s}_i, t)$:

$$\mathbf{Y}_{lt} = X_{lt}\boldsymbol{\beta} + A\mathbf{O}_{lt} + \boldsymbol{\epsilon}_{lt}, l = 1, \ldots r, \ t = 1, \ldots, T, \qquad (8.1)$$

with $A = CS_w^{-1}$ where C denotes the n by m cross-correlation matrix between the random effects at the n observation locations $\mathbf{s}_1, \ldots, \mathbf{s}_n$ and m knot locations, $\mathbf{s}_1^*, \ldots, \mathbf{s}_m^*$. The lower-dimensional random effects, $\mathbf{O}_{lt}$ are assumed to follow the auto-regressive model

$$\mathbf{O}_{lt} = \rho\,\mathbf{O}_{lt-1} + \mathbf{w}_{lt}, \qquad (8.2)$$

for $t = 1, \ldots, T$ and for each $r = 1, \ldots, r$. The $\mathbf{w}_{lt}$ for each t and l are assumed to be m-dimensional realization of the independent GP with mean 0 and Matérn correlation function in Equation (2.2).

To initialize the auto-regressive models, we assume that $\mathbf{O}_{l0} \sim N(\mathbf{0}, \sigma_l^2 S_0)$ independently for each $l = 1, \ldots, r$, where the elements of the covariance matrix S_0 are obtained using the correlation function, $\psi(d; \phi_0)$, i.e. the same correlation function as previously but with a different variance component for each year and also possibly with a different decay parameter ϕ_0 in the correlation function. Further modeling details are provided in Section 7.6.

To illustrate we use the daily maximum temperature, which is known to correlate very well with ozone levels. We model on the square-root scale and hence use the same transformation for the covariate daily maximum temperature. We center and scale this covariate to have better computational stability which encourages faster MCMC convergence. The article by Sahu and Bakar (2012b) discusses modeling results based on using two further covariates wind speed and relative humidity but those are not used in the illustration here since those data are not available any more.

We use the default flat prior distributions for the regression coefficients and independent proper G(2, 1) distributions for the variance components. We have experimented with three choices for handling the spatial decay parameter ϕ in the assumed exponential covariance function. The first of these fixes the value of ϕ at the program default value of three over the maximum distance between the 691 site locations. The other two are the gamma distribution and the uniform distribution with the default values of the hyper-parameters. Preliminary investigation using validation statistics for 1000 randomly selected space-time observations does not show any major differences in model performance. Henceforth, we proceed with the gamma prior distribution for the decay parameter ϕ. Below is an example model fitting command for fitting the model with a $G(2, 1)$ distribution for the decay parameter ϕ.

```
library(spTimer)
 f2 <- o8hrmax ~ xsqcmax
 time.data <- spT.time(t.series =153, segments=10 )
M2 <- Bsptime(formula=f1, data=euso3, package="spTimer", model="GPP
     ", coordtype = "lonlat", coords=2:3, time.data = time.data,
     prior.phi="Gamm", scale.transform = "SQRT", knots.coords =
     knots.coords.151)
```

The `time.data` statement declares that at each location there is data for 153 time points (days) in each of the 10 segments (years). The model fitting code assumes that the data set `euso3` contains all the necessary information, e.g. the columns `o8hrmax` and `xsqcmax` for the response and the scaled covariate on the square-root scale. The coordinates of the $m = 151$ knot locations constitute the matrix `knots.coords.151`. These locations are shown in Figure 8.4.

The 151 knot locations used in the above code have been chosen by first creating a 21×21 rectangular grid using the command:

```
knots.coords <- spTimer::spT.grid.coords(Lon=c(-98.2,-66), Lat=c(49
    .5,24),by=c(21,21))
```

Out of these 441 locations only 151 locations fall within the boundary of our study region and these 151 locations are chosen as the knot locations in this example. Sahu and Bakar (2012b) suggest a recipe to choose the number of knots based on minimizing the validation RMSE. There are many other methods to choose the knots, see e.g. Banerjee et al. (2008). While using the `bmstdr` package the user can provide the `g_size` argument, e.g. `g_size=10`, instead of the `knots.coords` argument. In such a case a uniform grid of 100 knots will be used within a rectangular box containing the study region. Some of those knot points may fall outside the boundary of the study region.

Parameter estimates of the fitted model with 151 knot points are provided in Table 8.2. The covariate maximum temperature remain significant in the spatio-temporal model with a positive effect. The auto-regressive parameter ρ is also significant and the pure error variance σ_ϵ^2 is estimated to be smaller than the spatial variance σ_w^2. The spatial decay parameter is estimated to be 0.0034 which corresponds to an effective spatial range (Sahu (2012)) of 882 kilometers that is about half of the maximum distance between any two locations inside the study region. The estimates of σ_l^2 and μ_l can be extracted from the `spTimer` model fitted object `M2$fit`. We omit those for brevity.

We extract the fitted values of the model to examine the behavior of the annual *4th* highest values of ozone concentrations and their 3-year rolling

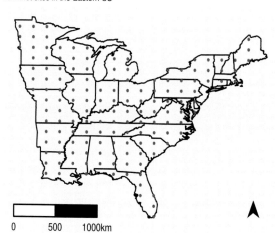

FIGURE 8.4: A map of the eastern US showing 151 knot locations.

	mean	sd	2.5%	97.5%
β_0	6.7372	0.0247	6.7085	6.8112
β_1	0.3114	0.0035	0.3041	0.3183
ρ	0.3554	0.0029	0.3497	0.3609
σ_ϵ^2	0.2637	0.0004	0.2630	0.2645
σ_w^2	0.7472	0.0043	0.7385	0.7555
ϕ	0.0034	0.0000	0.0033	0.0034

TABLE 8.2: Parameter estimates for the GPP model fitted to the eastern US air pollution data set.

averages. These values can be constructed from the MCMC samples $Y_l^{(j)}(\mathbf{s}_i, t)$ for $j = 1, \ldots, J$ where J is the number of MCMC iterations. To obtain the values on the original scale these samples must be squared and then averaged. Thus, for each year $l = 1, \ldots, r$ and at any location $\mathbf{s}_0$ (which can be one of the data locations), we obtain the annual $4th$ highest maximum ozone value denoted by:

$$f_l^{(j)}(\mathbf{s}_0) = \text{4th highest value } \left[Y_l^{(j)}(\mathbf{s}_0, t)\right]^2, \tag{8.3}$$

over $t = 1, \ldots, T$ at the jth MCMC iteration. We square the simulated $Y_l^{(j)}(\mathbf{s}_0, t)$ values to revert to the original scale of the ozone concentration values. The 3-year rolling averages of these are given by

$$g_l^{(j)}(\mathbf{s}_0) = \frac{f_{l-2}^{(j)}(\mathbf{s}_0) + f_{l-1}^{(j)}(\mathbf{s}_0) + f_l^{(j)}(\mathbf{s}_0)}{3}, \, l = 3, \ldots, r. \tag{8.4}$$

The `spTimer` package, however, does not save the samples $Y_l^{(j)}(\mathbf{s}_i, t)$ required to calculate (8.3) because of memory problems in storing a huge data set that may arise for all data locations (i), time points given by t and l, and iteration j. Instead, it provides the mean and standard deviations for all values of i, l and t. The additional `spTimer` option `fitted.values="ORIGINAL"` does the square transformation required in (8.3) when the square-root is scale used in modeling. If this option has not been used then we may square the fitted values obtained by the default transformed scale of data modeling.

We use these fitted values summaries in our illustration to estimate the annual summaries. For example, to obtain the fitted value of the annual $4th$ highest ozone level we simply find the $4th$ highest value of $\hat{Y}_l^2(\mathbf{s}_i, t)$ for each given location i and year l, where $\hat{Y}^2$ denotes the fitted value returned by the `spTimer` package on the original scale. Figure 8.5 plots the fitted values of the annual $4th$ highest maximum and their rolling averages corresponding to the observed values plotted in Figure 1.4. As expected, the fitted values plots are much more smooth than the observed values plot.

The trends seen in Figure 8.5 are raw overall trends which may have been caused by trends in the meteorological variable. That is, a rising trend in ozone levels is perhaps due to a rising trend in temperature. To investigate the trends in ozone levels after adjusting for meteorological variables we follow Sahu et al. (2007). These adjusted trends are estimated by considering the random effects, $AO_l^{(j)}(\mathbf{s}_0, t)$ since these are the residuals after fitting the regression model $\mathbf{x}_l(\mathbf{s}_0, t)'\boldsymbol{\beta}^{(j)}$. These realizations are first transformed to the original scale by obtaining $(AO_l^{(j)}(\mathbf{s}_0, t))^2$. These transformed values are then averaged over a year to obtain the covariate adjusted value, denoted by $h_l(\mathbf{s}_0)$, for that year, i.e.

$$h_l^{(j)}(\mathbf{s}_0) = \text{4th highest value of } (AO_l^{(j)}(\mathbf{s}_0, t))^2. \tag{8.5}$$

A further measure of trend called the relative percentage trend (RPCT) for any year l relative to the year l' is defined by

$$100 \times (h_l^{(j)}(\mathbf{s}_0) - h_{l'}^{(j)}(\mathbf{s}_0))/h_{l'}^{(j)}(\mathbf{s}_0) \qquad (8.6)$$

is also simulated at each MCMC iteration j. The MCMC iterates $f_l^{(j)}(\mathbf{s}_0)$, $g_l^{(j)}(\mathbf{s}_0)$, $h_l^{(j)}(\mathbf{s}_0)$ and also any relative trend values are summarized at the end of the MCMC run to obtain the corresponding estimates along with their uncertainties.

The MCMC iterates required to evaluate (8.5) and (8.6) are not saved by the `spTimer` package. Hence, we use a similar summary based approach to obtain these trends as we have done to obtain Figure 8.5. Here we estimate the squared-residuals $(\hat{\mathbf{Y}} - X\hat{\boldsymbol{\beta}})^2$ for all the sites, years and days where $\hat{\mathbf{Y}}$ are the fitted values on the square-root scale used to model the data. Now the meteorologically adjusted values $\hat{h}_l(\mathbf{s}_0)$ are obtained directly by calculating the $4th$ highest value of $(\hat{Y}_l(\mathbf{s}_0, t) - \mathbf{x}_l(\mathbf{x}_0, t)'\hat{\boldsymbol{\beta}})^2$. These $\hat{h}_l(\mathbf{s}_0)$ values are then used to obtain 3-year rolling averages. Figure 8.6 plots these summaries. This plot does not show the predominant downward trend seen in Figure 8.5. Hence, the downward trend in absolute ozone values can perhaps be due to meteorology, although there may be other explanations.

♣ R Code Notes 8.2. Figure 8.7 The outline map is the same one used in Figure 1.3. The `ggplot` function `geom_circle` has been used to draw a circle at each of the 691 sites. The radius of the circle at each location is proportional to the relative percentage trend defined in (8.6). The constant of proportionality for the positive trends (plotted in green) has been chosen to be twice the same for the ones showing positive trend in red. This is because the negative trends were much smaller in magnitude than the positive trends and the choice of different constants of proportionality allowed us to clearly see the locations with the negative trends. The red and green colors are chosen by the `ggplot` functions `scale_fill_manual` and `scale_color_manual`.

We now return to the main objective of this example, which is to assess compliance with respect to the primary ozone standard which states that the 3-year rolling average of the annual $4th$ highest daily maximum 8-hour average ozone concentration levels should not exceed 85 ppb as noted in Section 1.3.3. The compliance is assessed by evaluating the exceedance probability at each location at each year, see e.g. Sahu et al. (2007). In order to estimate the probability for each location and year combination we need to have the samples, $Y_l^{(j)}(\mathbf{s}_0, t)$, from the posterior predictive distribution so that we can have the $g_l^{(j)}(\mathbf{s}_0)$ as defined in (8.4) and then can estimate the probability of non-compliance as

$$p_l(\mathbf{s}_0) = \frac{1}{J}\sum_{j=1}^{J} I\left(g_l^{(j)}(\mathbf{s}_0) > 85\right) \qquad (8.7)$$

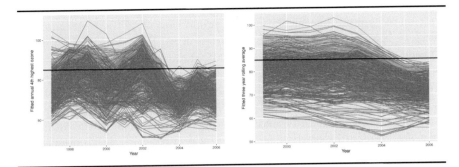

FIGURE 8.5: Time series plots of the model fitted ozone concentration summaries from 691 sites. Left panel: annual 4*th* highest maximum and right panel: 3-year rolling average of the annual 4*th* highest maximum.

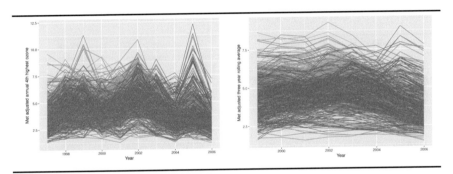

FIGURE 8.6: Time series plots of the meteorologically adjusted ozone values. Left panel: annual 4*th* highest maximum and right panel: 3-year rolling average of the annual 4*th* highest maximum.

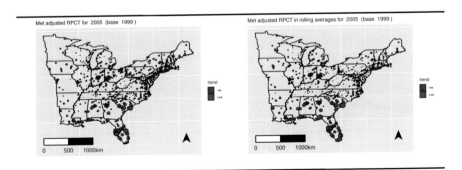

FIGURE 8.7: Meteorlogically adjusted relative percentage trend. Left panel: annual 4*th* highest maximum and right panel: 3-year rolling average of the annual 4*th* highest maximum.

where $I(\cdot)$ is the indicator function. Thus, the non-compliance probability is estimated simply by calculating the proportion of $g_l^{(j)}(s_0)$ exceeding 85 over $j = 1, \ldots, J$ for each year l and location s_0. But as noted before, the spTimer package does not save the MCMC iterates $Y_l^{(j)}(s_0, t)$ but it returns the mean and standard deviation of the posterior predictive distribution of $Y_l(s_0, t)$. Hence, intuitively we approximate the posterior predictive distribution by generating random samples $Y_l^{(j)}(s_0, t)$, $j = 1, \ldots, J$ from the normal distribution with mean and standard deviation being equal to the estimated mean and standard deviation of the posterior predictive distribution. This necessary computation requirement is quite burdensome and hence we code this procedure in C++ and use the Rcpp package to generate the samples and estimate $p_l(s_0)$.

The probabilities of non-compliance at the data modeling sites are shown in Figure 8.8 for the years 2005 and 2006. The plots show that many areas are out of compliance and the spatial patterns in two successive years are very similar.

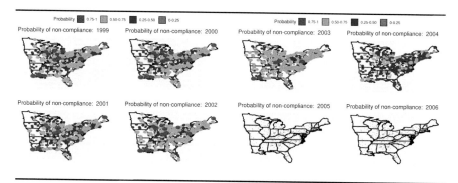

FIGURE 8.8: Probability of non-compliance in years 2005 and 2006.

♣ **R Code Notes 8.3. Figure 8.8** These maps are plotted using very similar code as in Figure 8.7. The plotted circles have radii proportional to the probability of non-compliance in (8.7). The constants of proportionality have been chosen to be different in the three color classes to make the sites visible similarly as in Figure 8.7.

8.4 Analyzing precipitation data from the Hubbard Experimental Forest

This example, based on Sahu et al. (2020), concerns modeling of precipitation volume observed at a network of monitoring sites in the Hubbard Brook Experimental Forest in New Hampshire, USA. There are several practical modeling objectives regarding studying of long-term trend in precipitation and estimating those trend for spatially aggregated ecological areas of interests, called watersheds. Here we present a modeling extension to model the mixture of discrete and continuously observed precipitation values. Output from the modeling effort is the pre-requisite for studying the response of forest ecology to long term trend in precipitation volumes.

Measuring total precipitation volume in aggregated space and time is important for many environmental and ecological reasons such as air and water quality, the spatio-temporal trends in risk of flood and drought, forestry management and town planning decisions. Here we model observed precipitation data from the Hubbard Brook Ecosystem Study (HBES) in New Hampshire, USA. Established in 1955 this study continuously observes many environmental outcome variables such as temperature, rainfall volume, nutrient volumes in water streams. HBES is based on the 8,000-acre Hubbard Brook Experimental Forest (see e.g. https://hubbardbrook.org/) and is a valuable source of scientific information for policy makers, members of the public and students and scientists in many educational institutes and programs.

8.4.1 Exploratory data analysis

The data set used in this study consists of weekly precipitation measurements in millimeter (mm) recorded by 25 precipitation gauges, denoted by RG1 to RG25, located in the Hubbard Brook Experimental Forest in New Hampshire, USA, see Figure 1.5. Data from gauge RG18 is not included in this analysis because it was decommissioned in 1975. We also note that RG22 is located near the head quarters of the forest administration, which is far from the other gauges. This makes it an ideal gauge to test the spatio-temporal modeling methods employed here.

This analysis considers weekly precipitation data collected from January 1, 1997 to December 31, 2015. Over these 19 years, there are $T = 991$ weeks, giving a total of $24 \times 991 = 23,784$ weekly observations from the 24 precipitation gauges. Table 8.3 provides the average weekly and annual precipitation volumes along with the standard deviations. The watershed number column provides the watershed number in which the particular rain gauge either belongs to or the closest to. None of the rain gauges have been allocated to Watersheds 2 and 5 and gauge RG22, located at the head quarters of the forest administration, is not typical of the other rain gauges. In fact, RG22 records the lowest amount of precipitation. The highest volume is recorded in the gauge RG24, see Table 8.3.

Id	Gauge	Watershed	mean (weekly)	sd (weekly)	mean (annual)	sd (annual)
1	RG1	1	27.80	25.72	1450.02	176.36
2	RG2	1	28.43	26.30	1483.07	186.28
3	RG3	1	26.91	24.67	1403.48	165.08
4	RG4	3	27.48	25.12	1433.26	166.40
5	RG5	3	27.68	24.98	1443.84	164.10
6	RG6	4	28.39	25.76	1480.65	177.31
7	RG7	4	28.19	25.71	1470.43	171.42
8	RG8	4	28.18	25.85	1469.97	173.39
9	RG9	6	28.58	26.11	1490.80	187.80
10	RG10	6	29.23	26.57	1524.58	184.56
11	RG11	6	28.25	26.14	1473.36	174.79
12	RG12	7	29.14	26.48	1519.93	192.31
13	RG13	7	29.33	26.83	1529.91	193.39
14	RG14	7	29.72	27.38	1550.19	208.74
15	RG15	7	29.37	26.96	1531.77	197.50
16	RG16	7	29.59	26.69	1543.55	192.02
17	RG17	8	29.10	26.52	1517.77	193.40
19	RG19	8	29.29	27.45	1527.55	203.24
20	RG20	8	29.28	27.17	1527.14	194.59
21	RG21	9	29.81	27.13	1554.59	195.25
22	RG22	–	24.63	23.07	1284.78	161.22
23	RG23	9	29.61	27.62	1544.50	189.77
24	RG24	9	30.83	28.27	1608.17	207.25
25	RG25	9	28.87	26.33	1505.89	186.65

TABLE 8.3: Average weekly and annual precipitation volumes and the standard deviations at the 24 rain gauges over the years 1997 to 2015. The gauge RG22 does not belong to any of the 9 watersheds.

The precipitation values range from 0 to 197.30 millimeter (mm) and the majority of the measurements falls below 120mm (on the original scale), as shown in the left panel of Figure 8.9. The right panel of this figure shows that the observation distribution is more symmetric in the square-root scale. However, there is a spike at the zero value which comes from zero precipitation in the dry weeks. There are 1247 observations with zero precipitation measurements, corresponding to roughly 5.24% of the full data set. Thus, there is a positive probability of observing zero precipitation while the values are measured in a continuous scale. This causes a problem in modeling since a mixture distribution, accommodating the discrete values at zero and continuous values greater than zero, must be adopted to model these data. Modeling extension discussed in this section takes care of this issue. There is no missing observation in the data.

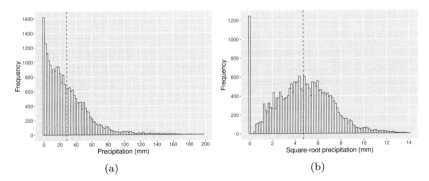

(a) (b)

FIGURE 8.9: Histogram of the precipitation data on original scale (a) and square-root scale (b) with the mean indicated by the red dotted line. The command `geom_vline` adds the vertical lines.

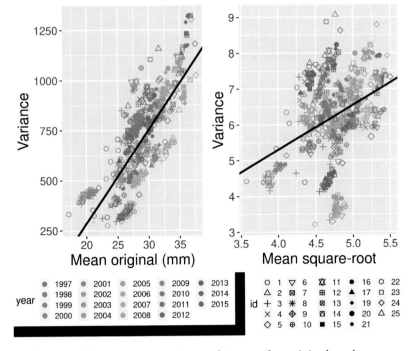

FIGURE 8.10: Mean versus variance plots on the original and square-root scales.

To identify the best modeling scale we explore the mean-variance relationships of the mean weekly data (grouped by rain gauge and year) on the original scale (a), and on the square-root scale (b) in Figure 8.10. We do not use the log scale due to the presence of zero precipitation volumes during the dry weeks. The figure shows that there is a much stronger mean-variance relationship on the original scale than on the square-root scale. This is also evident from the best fit regression line superimposed on the two plots. Therefore, we adopt the square-root scale in our modeling, although there seems to be a positive mean-variance relationship on this scale as well. These two plots also show clustering effect of annual means, which, in turn, hints that there may be variation due to the different years.

Besides facing (see Figure 1.5), values of three covariates: elevation, slope and aspect are available for modeling precipitation values. Figure 8.11 provides a map of elevation values. Similar plots for slope and aspect are omitted for brevity.

The average square-root precipitation increases slightly as elevation increases, see Figure 8.12. However, RG3 and RG24 have lower and higher averages, respectively, compared to gauges at similar elevation.

Figure 8.13 (a) plots the site-wise averages of weekly data against latitude, and it shows two distinct clusters of north and south facing sites where the north facing ones receive higher levels of precipitation. Sites at the higher latitude receive less precipitation but note that all of those also happen to be south facing. A similar story is seen in the plot of site-wise averages against longitude in panel (b) of this figure. The south facing gauges at higher longitude values receive less precipitation on average than the north facing ones in the south. However, note that unlike the latitude values of the two types of north and south facing gauges, the longitude values overlap considerably. Hence, the north and south facing gauges are not much clustered in panel (b) as they were in panel (a) of this figure.

To explore temporal variation we plot the weekly data grouped by months within each of the years 1997-2015 in Figure 8.14. Precipitation generally peaks during the summer months with a higher level of variation than in the winter months. In particular, February seems to be the driest and June seems to be the wettest in most years. However, there seems to be no common monthly patterns over the years, although seasonality must be investigated in the modeling stage. Precipitation values are much more stable during the spring months of March and April. More extreme precipitation levels are seen during the summer months of June-August and also during February and October.

There is much annual variation in the precipitation levels, see Figure 8.15. In this figure the plot of the 3-year rolling averages (where the rolling average for a particular year is defined as the average of the previous two years and that year) show site-wise trend in annual precipitation, where cycles of higher and lower precipitation volumes are observed. Model based inference regarding these trend patterns will be pursued below in the modeling section.

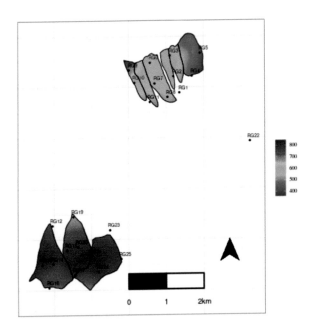

FIGURE 8.11: A map showing elevation and the precipitation gauges.

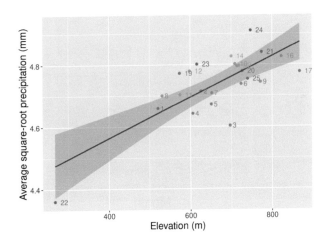

FIGURE 8.12: A plot of the square-root weekly rainfall averages at the 23 gauges against elevation. A linear regression line with a 95% confidence interval is superimposed.

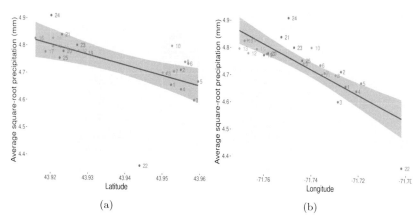

FIGURE 8.13: Plot of the square-root weekly rainfall averages at the 23 gauges against latitude (a) and longitude (b). A linear regression line with a 95% confidence interval is superimposed in both the plots.

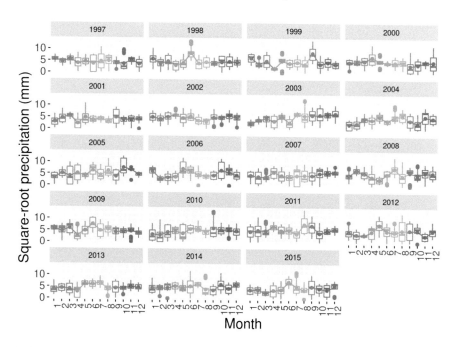

FIGURE 8.14: Boxplot of square-root precipitation against months over the years. The line running through the middle of the boxes connects the mean values.

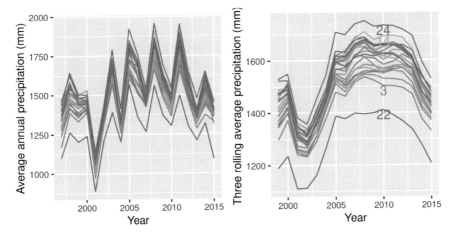

FIGURE 8.15: Mean annual total precipitation observed in the 24 gauges (left panel) and their 3-year rolling averages (right panel).

8.4.2 Modeling and validation

Non-zero precipitation volumes are often rounded to the nearest 10th of a millimeter. These discrete values occur with non-zero probabilities, when the actual volume is a continuous measurement falling between two discrete endpoints. This is a very common problem in Bayesian modeling of rainfall data with zero rainfall, see e.g. Sansó and Guenni (1999, 2000). A common approach is to model the zeros by the values of a latent continuous variable below a threshold value (censoring) as has been done by Jona Lasinio et al. (2007). The general model proposed there is described as follows.

Let $Z(\mathbf{s}, t)$ denote the observed precipitation volume at location $\mathbf{s}$ at time t and let $g(z(\mathbf{s}, t))$ be a scaling transformation e.g. square root or log that stabilizes the variance. In the current example, we will use the square-root transformation as has been justified previously in the exploratory data analysis Section 8.4.1. Thus, we take,

$$g(z(\mathbf{s}, t)) = \sqrt{z(\mathbf{s}, t)}.$$

To define the censoring mechanism, we introduce a latent random variable $Y(\mathbf{s}, t)$ whose observed value is $g(z(\mathbf{s}, t))$ when the observed $z(\mathbf{s}, t) > 0$ and that observation has not been rounded. But when $z(\mathbf{s}, t) = 0$, or it has been rounded, we assume that the corresponding $Y(\mathbf{s}, t)$ value is unobserved (i.e. latent) but we force it to lie in an appropriate interval so that the positive probability of $Y(\mathbf{s}, t)$ lying in that interval is the probability of $Z(\mathbf{s}, t)$ taking that discrete value (0 or the rounded value). This extends the censoring mechanism and allows the observed random variable to take multiple discrete values.

The above discussion is mathematically presented by the modeling formulation:

$$g(Z(\mathbf{s},t)) = \begin{cases} \lambda_1 & \text{if } Y(\mathbf{s},t) < c_1, \\ \lambda_2 & \text{if } c_1 \le Y(\mathbf{s},t) < c_2, \\ \vdots & \vdots \\ \lambda_k & \text{if } c_{k-1} \le Y(\mathbf{s},t) < c_k, \\ Y(\mathbf{s},t) & \text{otherwise.} \end{cases}$$

Here $k > 0$ is the number of discrete values that $Z(\mathbf{s},t)$ can take – for the current precipitation volume example we will assume $k = 1$. The constants $\lambda_1 < \lambda_2 < \cdots < \lambda_k$ are the transformed values of the k possible discrete values (in ascending order) that $Z(\mathbf{s},t)$ can take. The constants $c_1, \ldots, c_k$, such that $c_i > \lambda_i$ for all $i = 1, \ldots, k$, are the threshold values which can be set as the mid-point of $(\lambda_i, \lambda_{i+1})$ for $i = 1, \ldots, k-1$ and the last value c_k is chosen as the mid point of the interval with end points λ_k and the transformed value of the observed minimum positive precipitation value, see e.g. Jona Lasinio et al. (2007).

In the current example, we have $k = 1$ and we take $c_1 = 0$ so that the $P(Z(\mathbf{s},t) = 0) = P(Y(\mathbf{s},t) < 0)$. Thus, the positive probability of having zero precipitation is the same as the transformed latent variable $Y(\mathbf{s},t)$ taking a negative value. Henceforth, we model the transformed and censored random variable $Y(\mathbf{s},t)$. The only software packages that can handle censoring and truncation for spatio-temporal data are the `spTimer` and `spTDyn` through the use of various truncated model options, see Bakar (2020). We only illustrate with the `truncatedGP` model in the `spTimer` package. The exploratory data analysis performed above dictates us to use the square root transformation as the choice for the $g(\cdot)$ function and the truncation is set at $c_1 = 0$ as mentioned above.

Different spatio-temporal linear models from Chapter 7 can be assumed for $Y(\mathbf{s},t)$. Such models can be chosen and validated with the methodology described previously in this book. A full discussion on covariate selection and model choice is beyond the scope of this book due to limitations in space. Here we illustrate the following model that performed well in validation. Guided by the previously conducted exploratory data analysis and preliminary model exploration we have chosen to keep elevation and scaled UTMX values (equivalent to longitude) in the model. All other site characteristic variables such as UTMY (latitude), slope, aspect, and facing did not contribute significantly to model fitting and validation after including elevation and UTMX.

To capture seasonality we use the Fourier representations using pairs of sine and cosine functions as discussed in Chapter 8 of West and Harrison (1997). See also Sansó and Guenni (1999) and Jona Lasinio et al. (2007) who adopted those functions for practical spatio-temporal Bayesian modeling. Let m be the known periodicity of the data and define $K = m/2$ if m is even and $(m-1)/2$ otherwise. The seasonal term at time t and at any location $\mathbf{s}$ is assumed to be:

$$S_t(\mathbf{s}) = \sum_{r=1}^{K} c_r(\mathbf{s}) \cos\left(\frac{2\pi tr}{m}\right) + d_r(\mathbf{s}) \sin\left(\frac{2\pi tr}{m}\right). \tag{8.8}$$

where the unknown coefficients $c_r(\mathbf{s})$ and $d_r(\mathbf{s})$ may depend on the location $\mathbf{s}$ as well. When m is even we may set $d_K(\mathbf{s}) = 0$ so that $m - 1$ free seasonal parameters are kept in the model; the remaining parameter is obtained through the requirement that the seasonal effects cancel each other. Or we may adopt a sum to zero constraint. If there are no justifications for having spatially varying seasonal effects we work with the simpler model $c_r(\mathbf{s}) = c_r$ and $d_r(\mathbf{s}) = d_r$, as we do here in our example. In this example we choose $m = 365$ to adopt annual seasonal variation.

All the K pairs of terms in (8.8) may not significantly contribute to modeling and validation. Bayesian model selection criteria can be adopted to choose a minimum number of pairs to reduce model complexity. In our example we use $K = 1$ which corresponds to just using the first pair of sine and cosine functions. We scale these variables for ease in model fitting and to help achieve faster MCMC convergence. Thus, the mean structure of the model has five parameters: an overall intercept and coefficients for scaled values of four covariates: UTMX, elevation, $\sin\left(\frac{2\pi t}{m}\right)$ and $\cos\left(\frac{2\pi t}{m}\right)$ where $m = 365$.

We fit the `truncatedGP` model in `spTimer` using the exponential covariance function. The choice of the prior and the estimation method for the spatial decay parameter, ϕ, have an effect on the prediction results. In a classical inference setting it is not possible to consistently estimate both ϕ and σ_w^2 in a typical model for spatial data with a covariance function belonging to the Matérn family, see Zhang (2004). Moreover, Stein (1999) shows that spatial interpolation is sensitive to the product $\sigma_w^2\phi$ but not to either one individually.

The default gamma $G(2, 1)$ prior distribution for the decay parameter ϕ leads to very slow convergence and hence we assume ϕ to be fixed and choose the optimal value of ϕ by validating 991 precipitation measurements in the gauge RG22, which is far away from the other 23 gauges. This is an empirical Bayes (EB) approach which minimizes the RMSE and then estimates the variance parameter σ_w^2 conditionally on the fixed estimated value of ϕ.

Figure 8.16 plots the values of RMSE, MAE and CRPS for different values of ϕ in discrete grid starting from 0.01 to 1.5. The implied range for the effective range parameter $3/\phi$ is between 2 to 300 kilometers. The three criteria do not agree on a single value of ϕ as they try to optimize on different predictive characteristics. However, it is clear that all three criteria point to an optimal value of ϕ somewhere between 0.5 and 1. In fact, spatio-temporal interpolation is not very sensitive to a particular ϕ value in this range and in the rest of the illustration of this example we choose $\phi = 1$ which minimizes the RMSE. We note that the value of coverage of the 95% prediction intervals is around 99% does not change much for different values of ϕ and hence coverage has been omitted from the discussion in this paragraph.

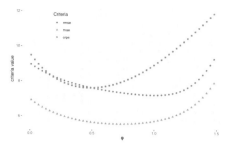

FIGURE 8.16: Plot of values validation criteria for different values of ϕ for validating 991 weekly precipitation volumes.

We now discuss a typical `bmstdr` model fitting command with the option to validate. The precipitation data set is called `ds3` which contains all the data for the 24 sites and at 991 time points. To identify the validation rows for RG22 we use the following commands:

```
vs <- getvalidrows(sn=24, tn=991, valids=c(21), allt=T)
```

Note that RG22 is marked as site 21 in the data set. The model fitting commands are given by the following commands.

```
library(bmstdr)
f2 <- rainfall ~xutmx + xelevation + xsin1 + xcos1
library(spTimer)
time.data <- spT.time(t.series = as.numeric(table(ds3$ts, ds3$id)
    [,1]),segments = length(unique(ds3$ts)))


M1 <- Bsptime(package="spTimer", model="truncatedGP", data=ds3,
    formula=f2, scale.transform ="NONE", time.data=time.data,
    prior.phi = "FIXED", prior.phi.param =1,
truncation.para = list(at = 0,lambda =2), validrows=vs, coordtype="utm",
    coords=3:4, n.report= 2, N=6000, burn.in =1000, mchoice = T)
```

We now explain the above commands and arguments. The model formula is obtained as `f2` and the `time.data` command obtains the definition of the temporal structure, days within years, in the data set. More explanation regarding this is provided in the help file for `spT.time` function. The arguments for the model fitting command are explained below.

- Arguments `package="spTimer"` and `model="truncatedGP"` choose the package `spTimer` and the truncated model suitable for fitting precipitation data.

- The pair of arguments `prior.phi = "FIXED"` and `prior.phi.param =1` fixes the spatial decay parameter ϕ at the value 1.

- Here the argument `scale.transform ="NONE"` instructs the program not to apply any transformation to the precipitation values. However, the square-root transformation is forced through the option `truncation.para = list(at = 0,lambda =2)`. The value `at=0` sets the truncation parameter c_1 to 0, defined previously in this section.

- The other arguments, e.g. `data`, are the usual arguments in model fitting discussed previously in Chapter 7.

Parameter estimates for the chosen model, given by the formula `f2` above and with $\phi = 1$, are presented in Table 8.4. The scaled variables UTMX and the seasonal terms (the sine and cosine pair) remain significant in the spatio-temporal model. Elevation is not seen to be significant in the model but we still keep it in the predictive model since it adds to the spatial variation in the predictions. The estimates of the variance components σ_ϵ^2 and σ_w^2 show a high level of precision because of the large numbers of the data points used to fit the model.

	mean	sd	2.5%	97.5%
Intercept	10.6824	0.0101	10.6625	10.7019
UTMX	−0.0466	0.0089	−0.0645	−0.0294
Elevation	0.0049	0.0060	−0.0066	0.0167
Sin1	0.3090	0.0353	0.2391	0.3784
Cos1	0.2093	0.0354	0.1398	0.2785
σ_ϵ^2	0.0051	0.0001	0.0050	0.0052
σ_w^2	0.5130	0.0047	0.5037	0.5225

TABLE 8.4: Parameter estimates for the truncated GP model assuming $\phi = 1$.

8.4.3 Predictive inference from model fitting

Bayesian computation methods enable a vast range of predictive inference capabilities once a good model has been fitted and MCMC (or Monte Carlo) samples for parameters and missing data are available. In this section we illustrate with a few inference problems.

8.4.3.1 Selecting gauges for possible downsizing

One of the goals of our analysis was to select rain gauges for possible decommissioning as a cost-saving measure. Monitoring agencies are interested in reducing the number of gauges without losing much predictive accuracy. This reduction would require selection of gauges with precipitation volumes that can be predicted by the remaining gauges with the lowest possible error rates. We estimated the leave-one-out model validation criteria by removing each of the 24 gauges, in turn. Table 8.5 shows the values of the first three model validation criteria (see Section 6.8) by leaving out one gauge at a time. RG15 has the lowest RMSE and MAE values and would, therefore, be the first gauge to decommission based on these criteria. The next gauge to remove would be RG7 which has the second lowest RMSE value. This removal process may be continued to eliminate additional gauges from the network.

Note that RG22 has the highest RMSE, MAE, and CRPS values. This rain gauge is located away from the other rain gauges clustered in and around the monitored watersheds and therefore is the most difficult gauge for spatio-temporal interpolation. Because this rain gauge is at a lower elevation and receives less precipitation, the model suggests that it should be retained. This rain gauge was not included in the earlier analysis by Green et al. (2018), which focused on the gauges used to characterize precipitation in the monitored watersheds.

The approach used by Green et al. (2018) was to omit each gauge individually and calculate the effect on the long-term average (1998 to 2012) precipitation estimate for each watershed using inverse-distance weighting for the spatial interpolation. There is some agreement between the two approaches; for example, RG3 is shown to be important both in Table 8.5 and in Table 2 in Green et al. (2018), because this gauge consistently collects low amounts of precipitation relative to the others. However, the approaches are not directly comparable.

Another reason that our results do not correspond with the decisions to decommission gauges reported by (Green et al., 2018) is that other practical operational factors affect these decisions, such as ease or safety of access and the history of other measurements at the same location. If indeed such practical considerations require definite retention (or removal) of a subset of gauges then the statistical method shown here should be modified accordingly. For example, the method will simply calculate the leave-one-out cross-validation statistics only for the subset of candidate gauges which are available to be removed from the network. In addition, it is possible to define and maximize a utility function that combines the leave-one-out cross-validation statistics and some measure of desirability for all the gauges. This approach to network optimization can be easily implemented and should be of value to monitoring design.

8.4.3.2 Spatial patterns in 3-year rolling average annual precipitation

Recall the 3-year rolling averages plotted in Figure 8.15. This figure only shows the rolling averages for the 24 gauges. The fitted spatio-temporal model can be used to obtain predictions of these rolling averages at any location away from the 24 precipitation gauges. This, in turn, enables us to investigate spatial patterns in the de-trended annual averages as described subsequently.

At a new location s_0 we first obtain the posterior predictive draws (Bakar and Sahu, 2015) $Y^{(j)}(s_0, t)$ given all the observed data for $j = 1, \ldots, J$ and for all the weeks in $t = 1, \ldots, 991$. These basic weekly predictions are averaged

Id	Gauge	RMSE	MAE	CRPS
1	RG1	2.83	2.15	6.60
2	RG2	3.27	2.07	6.36
3	RG3	3.72	2.67	6.65
4	RG4	2.85	2.14	6.74
5	RG5	3.16	2.41	7.18
6	RG6	3.05	2.19	6.36
7	RG7	2.31	1.73	6.44
8	RG8	2.60	1.97	6.36
9	RG9	3.04	2.27	6.81
10	RG10	3.05	2.10	6.62
11	RG11	2.78	2.06	6.63
12	RG12	5.33	2.49	6.76
13	RG13	2.77	2.02	6.65
14	RG14	2.81	2.07	6.46
15	RG15	2.24	1.70	6.48
16	RG16	3.06	2.34	7.01
17	RG17	2.78	2.20	6.70
19	RG19	3.87	2.56	6.40
20	RG20	2.82	1.96	6.57
21	RG21	2.39	1.80	6.43
22	RG22	7.19	5.69	8.90
23	RG23	3.88	2.58	6.91
24	RG24	3.75	2.68	6.89
25	RG25	3.43	2.58	7.18

TABLE 8.5: Leave one out cross-validation statistics for the 24 rain gauges in Hubbard Brook experimental forest during the years 1997 to 2015.

to obtain annual predictions for each given year. For example,

$$\bar{Y}^{(j)}_{\text{year}}(\mathbf{s}_0) = \frac{1}{52} \sum_{t=1}^{52} Y^{(j)}(\mathbf{s}_0, t), \ j = 1, \ldots, J,$$

where the summation is over all the weeks in the given year. The 3-year rolling average for a particular year, say 2005 is then calculated as:

$$\bar{\bar{Y}}^{(j)}_{2005}(\mathbf{s}_0) = \frac{1}{3} \sum_{\text{year}=2003}^{2005} Y^{(j)}_{\text{year}}(\mathbf{s}_0), \ j = 1, \ldots, J.$$

Thus, the rolling average for a particular year is the average for that year and the two previous years. These calculated annual values, $\bar{\bar{Y}}^{(j)}_{\text{year}}(\mathbf{s}_0)$, for $j = 1, \ldots, J$ are averaged to estimate the prediction and the associated uncertainty for the 3-year rolling average at any location s0.

To illustrate, Figure 8.17 shows a spatially interpolated map of the 3-year rolling average annual precipitation volumes for 2010. There is a clear east-west gradient in this map which was expected based on model fitting results in Table 8.4. The Bayesian computation methods also allow estimation of uncertainties for this map. The uncertainty map is provided in the same figure and shows higher uncertainty levels for the locations that are farther away from the rain gauges.

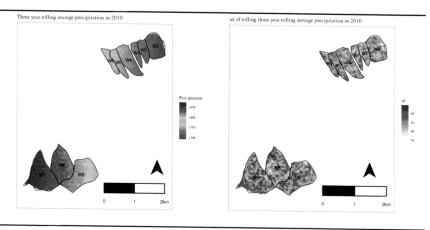

FIGURE 8.17: A predicted map of 3-year annual rolling average precipitation and their standard deviations in 2010.

Change over time in annual precipitation can be evaluated at a location $\mathbf{s}_0$ by using the quantities calculated $\bar{\bar{Y}}_{\text{year}}^{(j)}(\mathbf{s}_0)$ for any two years. For example, to estimate the annual percentage change in precipitation between 2005 and 2015 at a given location $\mathbf{s}_0$ we first evaluate

$$e^{(j)}(\mathbf{s}_0)_{2005,2015} = 100 \times \frac{\bar{\bar{Y}}_{2015}^{(j)}(\mathbf{s}_0) - \bar{\bar{Y}}_{2005}^{(j)}(\mathbf{s}_0)}{\bar{\bar{Y}}_{2005}^{(j)}(\mathbf{s}_0)}, \quad j = 1, \ldots, J.$$

These last MCMC iterates can be averaged to produce maps of the annual percentage change in precipitation volumes between any two years.

8.4.3.3 Catchment specific trends in annual precipitation

Catchment-specific trends in annual values can be estimated by spatially aggregating $e^{(j)}(\mathbf{s}_0)_{2005,2015}$ for any two specific years. For the jth catchment ($j = 1, \ldots, 9$) this can be calculated as:

$$h(W_j)_{2005,2015}^{(j)} = \frac{1}{n(W_j)} \sum_{i=1}^{n(W_j)} e^{(j)}(\mathbf{s}_{0i})_{2005,2015}, \quad j = 1, \ldots, J.$$

where the summation is over all $n(W_j)$ locations $\mathbf{s}_{0i}$, for example, within each watershed W_j, $j = 1, \ldots, 9$. Then $h(W_j)_{2005,2015}^{(j)}, j = 1, \ldots, J$ are averaged to estimate the trend and its uncertainty for the jth watershed, $j = 1, \ldots, 9$.

We complete our illustration by showing watershed specific trends and their uncertainties in Figure 8.18. This figure indicates a strong negative trend which seems plausible according to the 3-year rolling averages in Figure 8.17. Uncertainties in these watershed specific trends are plotted in the bottom panel of this figure.

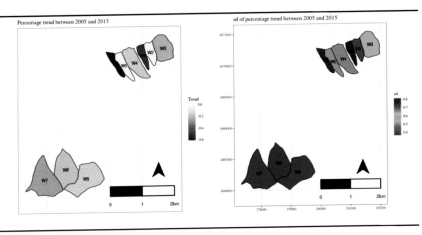

FIGURE 8.18: Trend for each watershed between 2005 and 2015 and standard deviation.

♣ **R Code Notes 8.4. General computing notes** In this example, we performed a relatively small number of MCMC iterations to demonstrate the methodology with a workable amount of data in R without causing memory problems. The large scale of the data set used in this analysis posed a challenge because of the massive computing requirements necessary to re-produce the results. The main issue was in generating the posterior predictive realizations for each of the 991 weekly measurements at each of the 1500 grid locations. The 200 MCMC replications implemented in this analysis resulted in a total of 297 million numbers. The exact code used has been provided online on github. Specific code lines to reproduce the graphs using model output are also provided online on github.

8.4.3.4   A note on model fitting

Our preliminary investigation shows that the truncated model in the spTimer package is the most viable tool currently available for solving this problem. The truncated models in the spTDyn package failed to converge, and modeling with INLA (Blangiardo and Cameletti, 2015) is not possible because of the large size of the data set. The R modeling software package `rstan` (Stan Development Team, 2020) could potentially be used; however, further coding would be required to produce acceptable results.

8.5   Assessing annual trends in ocean chlorophyll levels

Return to the ocean chlorophyll (chl) example introduced in Section 1.3.5. The analysis here is based on 196 monthly data collected from `http://www.esa-oceancolour-cci.org/`, which is version 2.0 of the European Space Agency's OC-OCI project. Hammond et al. (2017) provide further details regarding the source and quality of the data. Our data set comprises monthly data for 23,637 locations in 23 (out of 54) Longhurst (Longhurst, 1995) regions of the world's oceans. Each of the 54 Longhurst regions represents a unique set of environmental conditions such as climate, and as a result presents a unique ecosystem of growing environment for phtoplankton and other living organisms. Longhurst regions are named after Alan R. Longhurst, the author of the book Longhurst (1998). The distribution of 23,637 locations in among 23 regions is provided in the first two columns of Table 8.6. A map of the regions is provided in Figure 8.19 which also provides the main result of this modeling example.

Our data set includes level of chl as measured by the color of the ocean and sea surface temperature (SST). SST has been used as a continuous covariate in the model. Monthly seasonal levels are also included in the spatio-temporal model. The objective of the study is to evaluate region-specific long term trends after accounting for seasonality and SST. For this purpose, we include the variable time counting the number of months beginning September 1997 in the model. Thus, the regression part of the model is specified by the formula

```
f1 <- chl ~ time + SST + month
```

where month is declared as a factor. This model is fitted separately independently for each of the 23 Longhurst regions in our study. As a result we are able to estimate region specific estimates of trend, effect of SST and seasonality. This model thus explicitly accounts for regional seasonal variation and effect of SST when estimating long-term trend in chl levels.

The regression model of the previous paragraph is to be encompassed in a suitable spatio-temporal model discussed previously in Chapter 7 to account for expected spatio-temporal correlation in the data. This modeling is seen to be a challenging task since the number of space time observations to be modeled varies from 22,736 (for region 15) to 741,860 (for region 22). The spatio-temporal model based on GPP, presented in Section 7.6 is the only viable method for modeling these large data sets. Henceforth we adopt this approach as has been done by Hammond et al. (2017). Following these authors we also fix the decay parameter ϕ and in this implementation we fix the value at 0.006 which corresponds to an effective range of 500 kilometers. Our inference is conditional on this fixed parameter, although we remark that it is possible to experiment further regarding this choice as has been done in Hammond et al. (2017). For each of the 23 regions we fit the model:

```
Model <- Bsptime(package="spTimer", data=dataj, formula =f1, model=
    "GPP", coordtype="lonlat", coords=2:3, N=N, burn.in=burn.in,
knots.coords=knotgrid, n.report=10, prior.phi ="FIXED", prior.phi.param
    =0.006)
```

where `dataj` is the data frame and `knotgrid` are the collections of the knot locations for the *j*th region. In our illustration, we have taken N=2000 and `burn.in`=1000 and these were deemed to be adequate for achieving convergence. It takes a long time to fit these models for all the 23 regions. Hence, model fitting results are saved after fitting the model for each region and then all the results are gathered to produce the tables and graphs reported below.

Tables 8.6 and 8.7 provide the parameter estimates for the 23 regions. The parameter estimates show high degree of accuracy, as measured by the uncertainties in the 95% limits or standard deviations, due to the very large number of observations used to fit these models. SST has similar significant negative effects for all but region 7. The auto-regressive parameter, estimated between 0.50 and 0.93, shows high temporal dependence after accounting for all other effects including monthly seasonal variations. In Table 8.7 we see that the intercept estimates are broadly similar and spatial variance σ_w^2 estimates are higher than the nugget effect estimates σ_ϵ^2. Hammond et al. (2017) provide further results and their explanations including a comparison of results from a multiple linear regression model without the spatio-temporal random effects as modeled by the GPP.

We now discuss the estimates of the region specific trends presented in Table 8.6. These trend values show a very broadly similar picture of significant negative trends in chl levels, although there are exceptions most notably in region 13 in the North Atlantic. Percentage annual trends, which are values of β(trend) multiplied by 1200 to convert the monthly values to annual percentages, are plotted in Figure 8.19. This figure gives an overall impression of the scale of annual decline in ocean chlorophyll levels. However, there are noticeable regional disparities in both positive and negative trends present across the globe. Hammond et al. (2017) discuss and compare these trends with similar estimates published in the literature in much more detail. They conclude the importance of accounting for spatio-temporal correlation in studying such trends, which is the main subject matter of this book.

Region	n	β(trend)	95% limits		β(sst)	95% limits		ρ	95% limits	
1	173	−0.11	−0.14	−0.07	−0.05	−0.05	−0.04	0.68	0.63	0.73
2	1586	−0.05	−0.07	−0.03	−0.08	−0.08	−0.08	0.86	0.85	0.87
3	1526	−0.07	−0.08	−0.06	−0.02	−0.02	−0.02	0.87	0.86	0.87
4	118	−0.12	−0.14	−0.10	−0.06	−0.06	−0.06	0.78	0.77	0.79
5	1263	−0.15	−0.17	−0.13	−0.06	−0.06	−0.06	0.82	0.81	0.83
6	412	0.01	−0.10	0.11	−0.03	−0.03	−0.02	0.78	0.76	0.80
7	406	0.02	−0.04	0.07	0.04	0.04	0.04	0.88	0.87	0.90
8	307	−0.20	−0.32	−0.08	−0.22	−0.22	−0.21	0.75	0.73	0.78
9	441	−0.05	−0.10	0.00	−0.23	−0.23	−0.22	0.81	0.79	0.83
10	408	0.18	0.12	0.26	−0.08	−0.08	−0.07	0.50	0.46	0.53
11	671	−0.00	−0.04	0.03	−0.05	−0.05	−0.05	0.87	0.86	0.89
12	1311	−0.03	−0.04	−0.02	0.01	0.01	0.01	0.85	0.84	0.86
13	703	0.38	0.26	0.48	−0.24	−0.25	−0.24	0.88	0.87	0.90
14	556	−0.04	−0.05	−0.03	−0.10	−0.11	−0.10	0.70	0.68	0.72
15	116	0.01	−0.03	0.06	−0.17	−0.18	−0.17	0.54	0.48	0.60
16	439	−0.05	−0.10	0.01	−0.07	−0.07	−0.07	0.51	0.48	0.54
17	1794	−0.01	−0.03	0.02	−0.01	−0.01	−0.01	0.82	0.82	0.83
18	3222	0.01	−0.00	0.01	−0.02	−0.02	−0.02	0.93	0.93	0.93
19	294	0.11	0.03	0.18	−0.21	−0.22	−0.20	0.69	0.66	0.72
20	767	0.04	0.01	0.07	−0.07	−0.07	−0.07	0.66	0.64	0.68
21	1259	−0.07	−0.07	−0.06	−0.06	−0.06	−0.06	0.67	0.65	0.68
22	3785	0.09	0.08	0.10	−0.02	−0.02	−0.02	0.90	0.87	0.92
23	1018	−0.02	−0.02	−0.01	−0.01	−0.01	−0.01	0.91	0.90	0.92

TABLE 8.6: Number of sites in each area (first column) and parameter estimates for β(trend), β(sst), and ρ from the GPP model.

♣ **R Code Notes 8.5. Figure 8.19** Plotting of this map starts from the blank map plotted in Figure 1.6, see also the code notes 1.4. The estimates of the annual percentage trends values are manually assigned to the 23 Longhurst regions for which models were fitted. This map can be plotted by using the saved model fitting results. Detailed code lines are made available online on github.

8.6 Modeling temperature data from roaming ocean Argo floats

Return to the Argo float data introduced in Section 1.3.6. For this data, Sahu and Challenor (2008) adopt a kernel convolution approach (Higdon, 1998)

	intercept	sd	σ_ϵ^2	95% interval		σ_w^2	95% interval	
1	0.2903	0.0017	0.0095	0.0093	0.0096	0.0164	0.0148	0.0181
2	0.1664	0.0008	0.0020	0.0020	0.0020	0.0053	0.0052	0.0055
3	0.1153	0.0004	0.0005	0.0005	0.0005	0.0020	0.0019	0.0020
4	0.1846	0.0010	0.0022	0.0022	0.0023	0.0060	0.0059	0.0062
5	0.1805	0.0009	0.0035	0.0035	0.0035	0.0076	0.0074	0.0078
6	0.2813	0.0051	0.0363	0.0360	0.0366	0.0634	0.0601	0.0669
7	0.2817	0.0023	0.0161	0.0159	0.0162	0.0201	0.0190	0.0214
8	0.2900	0.0055	0.0643	0.0636	0.0651	0.0776	0.0729	0.0826
9	0.2594	0.0022	0.0128	0.0127	0.0130	0.0219	0.0209	0.0230
10	0.3666	0.0037	0.0352	0.0349	0.0355	0.0896	0.0847	0.0948
11	0.1584	0.0019	0.0122	0.0121	0.0123	0.0200	0.0192	0.0209
12	0.0876	0.0004	0.0003	0.0003	0.0003	0.0014	0.0014	0.0015
13	0.3117	0.0045	0.0669	0.0664	0.0674	0.0732	0.0702	0.0764
14	0.1041	0.0005	0.0008	0.0008	0.0008	0.0044	0.0043	0.0046
15	0.3015	0.0026	0.0111	0.0109	0.0113	0.0368	0.0334	0.0405
16	0.4320	0.0035	0.0225	0.0223	0.0228	0.1078	0.1025	0.1130
17	0.1987	0.0009	0.0046	0.0046	0.0047	0.0178	0.0174	0.0182
18	0.0910	0.0004	0.0007	0.0007	0.0007	0.0015	0.0014	0.0015
19	0.4002	0.0038	0.0749	0.0741	0.0759	0.3637	0.3405	0.3875
20	0.2352	0.0017	0.0055	0.0055	0.0055	0.0170	0.0165	0.0177
21	0.0774	0.0003	0.0003	0.0003	0.0003	0.0021	0.0020	0.0021
22	0.2147	0.0003	0.0150	0.0149	0.0150	0.6205	0.5002	0.7377
23	0.0837	0.0004	0.0005	0.0005	0.0006	0.0014	0.0014	0.0015

TABLE 8.7: Parameter estimates for β_0, σ_ϵ^2, and σ_w^2 from the GPP model.

for bivariate modeling of temperature and salinity. In this section we do not attempt bivariate modeling but instead focus on modeling of deep ocean temperature data alone. Following Sahu and Challenor (2008) we use the longitude and latitude of the recording locations as possible covariates. Indeed, Figure 8.20 confirms correlation between each of these and temperature. This plot also shows possible non-linear and interaction effects. To capture such effects we include the square of latitude, and the interaction term longitude × latitude, and also the square of this product interaction term. We have also included the seasonal harmonic terms as in the precipitation volume modeling example. All the covariates are scaled to achieve faster MCMC based computation. However, the seasonal harmonic terms are not significant in any of our model fits below. Hence, we work with the regression model formed using the remaining covariates based on longitude and latitude of the Argo float locations.

The main difficulty in modeling Argo float data lies in the fact that any particular location in the ocean is never re-sampled by the Argo floats. Figure 1.7 illustrates the moving Argo float locations in each of the 12 months in 2003. Hence all the spatio-temporal models for data from fixed monitors (such as for air pollution, precipitation volumes etc.) are not suitable for modeling

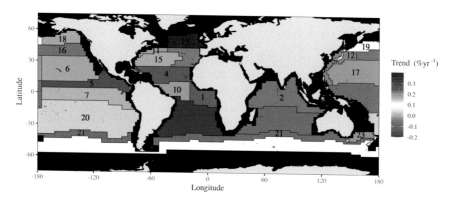

FIGURE 8.19: Annual percentage trend in ocean chlorophyll levels.

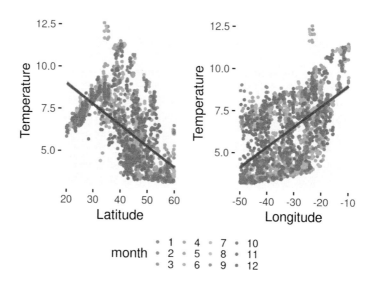

FIGURE 8.20: Scatter plot of temperature at the deep ocean against latitude and longitude.

Argo float data. Analysis methods based on some sort of spatial aggregation, e.g. averaging over some grid-areas, are also unlikely to work for the modeling problem since there may only be a few (much less than 10) data points observed in any particular day, which is the time unit for modeling such data.

The full spatio-temporal GP model (7.10) is difficult to fit because of the large number of unique locations where the Argo floats record the measurements through time. However, at one given time point there are only few locations, n_t say, where the measurements are recorded. This is because each Argo float is programmed independently and their measurement recording is not synchronized. Let $\mathbf{Y}_t = (Y(\mathbf{s}_1, t), \ldots, Y(\mathbf{s}_{n_t}, t))$ denote the response variable of interest at time t whereby $\mathbf{Y}_t$ is of n_t dimensional. Note that some of these n_t may be zero corresponding to the time points where none of the Argo floats recorded any measurement. In the following discussion let $\mathbf{X}(\mathbf{s}_j, t)$ denote the covariate vector recorded at location $\mathbf{s}_j$ at time t where $j = 1, \ldots, n_t$.

The GPP model described in Section 7.6 is a suitable methodology here since it assumes a GP in a much smaller grid of knot locations and then it uses Kriging method to interpolate at other locations, including the data locations, at all time points. In this section we assume an independent in time reduced dimensional GP as follows. The model is given by:

$$\mathbf{Y}_t = \mathbf{X}_t \boldsymbol{\beta} + A_t \mathbf{w}_t + \boldsymbol{\epsilon}_t, t = 1, \ldots, T \tag{8.9}$$
$$\boldsymbol{\epsilon}_t \sim N(0, \Sigma_\epsilon) \tag{8.10}$$

where:

- $\mathbf{X}_t$ is $n_t \times p$. The jth row of $\mathbf{X}_t$ is the row vector $\mathbf{X}(\mathbf{s}_j, t)$.

- $\boldsymbol{\beta} = (\beta_1, \ldots, \beta_p)$ is $p \times 1$ vector of un-known regression coefficients.

- $\mathbf{w}_t = (w(\mathbf{s}_1^*, t), \ldots, w(\mathbf{s}_m^*, t))$ is $m \times 1$ spatio-temporal random effects anchored at the m-knot locations $\mathbf{s}_1^*, \ldots, \mathbf{s}_m^*$. We assume $\mathbf{w}_t \sim GP(0, C(\Psi))$ independently for $t = 1, \ldots, T$ instead of the auto-regressive model for $\mathbf{w}_t$ assumed in the spTimer package.

- to define A_t we first define C_t to be $n_t \times m$ having the jth row and kth column entry $\exp(-\phi|\mathbf{s}_j - \mathbf{s}_k^*|)$ for $j = 1, \ldots, n_t$ and $k = 1, \ldots, m$. Thus, this matrix captures the cross-correlation between the observation locations at time t and the knot locations. By using the Kriging equations we set $A_t = C_t S_w^{-1}$, where S_w is $m \times m$ and has elements induced by the GP specification $GP(0, C(\Psi))$.

- we assume $\Sigma_\epsilon = \sigma_\epsilon^2 I$ where I is the identity matrix of order n_t.

The marginal model for data is written as (after integrating w_t out):

$$\mathbf{Y}_t \sim N(\mathbf{X}_t \boldsymbol{\beta}, \ A_t S_w A_t' + \Sigma_\epsilon), \tag{8.11}$$

independently for each $t = 1, \ldots, T$ for which data have been observed.

The package `bmstdr` includes a new function `Bmoving_sptime` to fit and validate using this model. The model fitting function requires arguments similar to the ones for the `Bsptime` function described in Chapter 7. The full list of arguments are described in the help file for this function. The example data set `argo_floats_atlantic_2003` is also included in the `bmstdr` package, see ? `argo_floats_atlantic_2003` for further details. From this built-in data set we prepare a data set `deep` containing the data for the deep ocean along with the scaled covariates. The model fitting commands are given below:

```
f2 <- temp ~xlon + xlat + xlat2+ xinter + x2inter
M2 <- Bmoving_sptime(formula=f2, data = deep, coordtype="lonlat",
    coords = 1:2, N=1100, burn.in=100, mchoice = T)
```

The model fitted object `M2` can be examined using the S3 methods commands such as summary, fitted and so on. Table 8.8 reports the parameter estimates. All the covariates are seen to be significant and the spatial decay parameter estimate point to a reasonable effective spatial range $(3/\hat{\phi})$ of about 1154 kilometers.

Parameter	mean	sd	2.5%	97.5%
Intercept	5.5190	0.0367	5.4472	5.5881
Lon	−1.7607	0.1267	−2.0050	−1.5039
Lat	6.6283	0.2431	6.1765	7.0984
Lat^2	−5.6962	0.1947	−6.0664	−5.3147
Lon × Lat	4.0838	0.1742	3.7440	4.4343
$(\text{Lon} \times \text{Lat})^2$	0.7548	0.0281	0.7029	0.8086
σ_ϵ^2	0.6379	0.0358	0.5712	0.7088
σ_w^2	1.0852	0.2105	0.7246	1.5325
ϕ	0.0026	0.0003	0.0021	0.0031

TABLE 8.8: Parameter estimates for the independent GPP model for the Argo floats temperature data for the deep ocean.

8.6.1   Predicting an annual average temperature map

It is of interest to produce an annual prediction surface along with an uncertainty surface for temperature. In order to do this we follow Sahu and Challenor (2008) by averaging the daily model (8.11) evaluated at the posterior samples. More details are provided now.

We only perform predictions at the data locations and then we use a deterministic smoother to produce a predictive map as has been done before in other examples. The independence assumption in model (8.11) implies that the model for annual average, $\bar{Y}(s_j)$ at the jth data location observed at any particular time t is given by

$$N \left(\frac{\sum_{t=1}^{T} \mathbf{X}_t(\mathbf{s}_j)}{T} \boldsymbol{\beta}, \quad \frac{\sum_{t=1}^{T} \sigma_{jt}^2}{T} \right) \tag{8.12}$$

where σ_{jt}^2 is the jth diagonal entry of the covariance matrix $A_t S_w A'_t + \Sigma_\epsilon$ in (8.11). We average this annual average model over the MCMC samples to obtain annual prediction and its uncertainty at $\mathbf{s}_j$. That is we evaluate the mean and variance in (8.12) at each MCMC iteration, j, and then generate a random sample from that distribution which we denote by $\bar{Y}(\mathbf{s}_j)^{(j)}$. We then average the $\bar{Y}(\mathbf{s}_j)^{(j)}, j = 1, \ldots, J$ to obtain the annual prediction $\hat{\bar{Y}}(\mathbf{s}_j)$ and the MCMC samples also allow us to obtain the associated uncertainty.

The annual prediction of mean temperature and the standard deviation of the predictions are plotted in Figure 8.21. As in Sahu and Challenor (2008) the predictions show two distinct ocean currents: the cooler polar currents and the warmer equatorial currents. We also note small variability of the annual predictions at the deep ocean as would be expected.

♣ **R Code Notes 8.6. Figure 8.21** To produce the maps, we start coding after fitting the model M2 as detailed above. The MCMC samples for the parameters $\boldsymbol{\beta}$, σ_ϵ^2 and Ψ are retrieved by using the command `listofdraws <-rstan::extract(M2$fit)`. Once we have the samples, we evaluate (8.12) using a set of R code lines made available online on github. The fitted values are then interpolated using the `interp` function in the `akima` library. The interpolated values are plotted using the `ggplot` function. The ggplot function draws various layers using the functions such as `geom_raster`, `geom_contour`, `geom_polygon` and `geom_point` and `scale_fill_gradientn`. The colour scheme used is the one given by the output of the command `topo.colors(5)`.

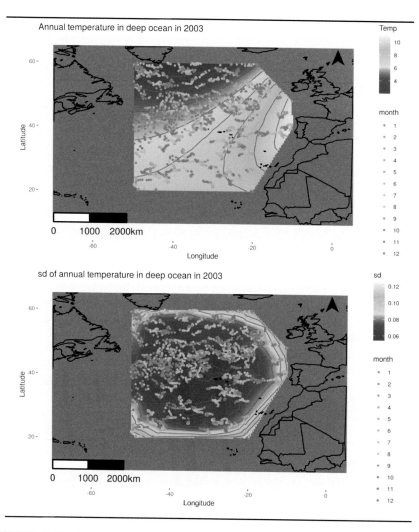

FIGURE 8.21: Annual prediction map (left panel) and sd of the predictions (right panel) of temperature at the deep ocean in 2003.

8.7 Conclusion

This chapter has showcased five examples on spatio-temporal Bayesian modeling for point referenced Gaussian data. The first example is on modeling daily air pollution data in England and Wales. Using the model based interpolations on a fine grid the example shows how to obtain a temporally aggregated annual air pollution surface along with its uncertainty. It also demonstrates the use of the over function from the sp library (Pebesma et al., 2012) to identify points within a given map polygon. The over function has been used to select knot locations within the boundaries of mainland England and Wales.

The second example analyzes a bigger data set assessing trend in ground level ozone air pollution levels in the eastern United States. Various concepts of meteorlogically adjusted trends in 3-year rolling averages of annual 4th highest daily 8-hour maximum ozone levels are introduced and their uncertainty levels are assessed. The methodology then evaluates probability of site-wise non-compliance with respect to air pollution regulations.

The third example is on modeling point-level precipitation data by using truncated models to accommodate zero precipitation in a model for continuous data. The trends in 3-year rolling averages are also assessed in this example. Several associated ecological problems, e.g. network size reduction and assessment of trends in catchment wise accumulated precipitation, are also addressed. This example also illustrates empirical Bayesian estimation of the decay parameter ϕ in the exponential covariance function.

The fourth example assesses trends in ocean chlorophyll levels in 23 areas of the world oceans known as Longhurst regions. The main contribution here is the assessment of annual trends after accounting for spatio-temporal correlation and other relevant covariates e.g. sea-surface temperature. A chloropeth map of the annual trends in the 23 regions shows a long-term declining trend that is a cause of considerable concerns for the primary food production levels in the oceans.

The last example, based on oceanography, illustrates estimation of an annual ocean temperature map based on actual readings from moving argo-floats in the north Atlantic Ocean. Unlike the fixed monitoring of air pollution or precipitation, here the main methodological challenge is due to the presence of the moving sensors. A model for daily data is aggregated to the annual level which enables us to simulate from the posterior distribution of the annual aggregates. This, in turn, allows us to estimate the annual map and its uncertainty levels.

Rather than being definitive, all the examples presented in this chapter are illustrative of the model fitting methodologies. In each case it is possible to further fine tune the methods, monitor MCMC convergence and thereby obtain better MCMC based inference. However, in each case it is also challenging to handle such huge data sets. The illustrations here merely demonstrate the feasibility of the methods. These should not be taken as definitive answers to the problems described.

8.8   Exercises

1. The data set used in Section 8.2 for air pollution data from England and Wales contains values of other pollutants such as PM_{10} and $PM_{2.5}$. Using these as response variables perform estimation and model validation using the spTimer and INLA packages. Obtain estimates of annual averages and their uncertainties at the modeling sites.

2. Reproduce all the results included in the US air pollution example in Section 8.3. Experiment with different knot sizes to select an optimal one. Use cross-validation to obtain optimal values of the spatial decay parameter ϕ as has been illustrated through Figure 8.16. Advanced researchers are invited to write their own C++ code to process MCMC output, possibly to calculate different statistics, as discussed in Section 8.3.

3. Reproduce all the results for the precipitation modeling example in Section 8.4. Advanced users are again invited to write their own C ++ code to process large volumes of MCMC output more efficiently. Even reproducing the results reported, or obtaining somewhat comparable results will be a great achievement.

4. Reproduce all the results reported in Section 8.5. Also obtain all the results reported in the original paper Hammond et al. (2017).

5. Reproduce all the results reported in Section 8.6. The argo_floats_atlantic_2003 data set contains values if temperature and salinity at the surface and mid-layer as well. Obtain annual prediction maps for these as well.

9

Bayesian forecasting for point referenced data

9.1   Introduction

The topic of forecasting has been avoided so far in this book. However, Bayesian forecasting methods are very much in demand in many application areas in environmental monitoring and surveillance. Consequently, model based forecasting has attracted much attention in the literature, see e.g., West and Harrison (1997); Mardia et al. (1998); Bauer et al. (2001); Damon and Guillas (2002); Feister and Balzer (1991); Huerta et al. (2004); Kumar and Ridder (2010); McMillan et al. (2005); Sahu and Mardia (2005a,b); Sahu et al. (2009, 2011); Sahu and Bakar (2012a); Sousa et al. (2009); Stroud et al. (2001) and Zidek et al. (2012). Some of these papers also consider space-time modeling for the purposes of forecasting. However, many of the methods proposed in these articles are not able to handle the computational burden associated with large space-time data sets that we model in this chapter for forecasting purposes.

Conceptually, forecasting presents no extra challenge in model based Bayesian inference settings. The key theoretical tool to perform forecasting is still the posterior predictive distribution (4.8) of a future observation given the observations up-to the current point in time. This basic predictive distribution is generalized for the adopted spatio-temporal model to perform forecasting for spatio-temporal data. Forecasting on the basis of the full probability distribution of the random variable to be forecasted automatically delivers calculated uncertainties for the forecasts made.

If we are interested in forecasting $k(>0)$ time steps ahead of the current time then the forecasts are called k-step ahead forecasts. In the usual time-series literature k-step ahead forecasts relate to one particular time series which may have been observed at one particular point or areal spatial location. However, for spatio-temporal data the k-step ahead forecasts can also be obtained at previously unobserved sites within the study region. The spatio-temporal dependency structures built into the Bayesian model can be exploited to perform forecasting at both observed and unobserved locations in the study region of interests.

The main challenge in Bayesian forecasting in time and interpolation in space comes from the complex nature of spatio-temporal dependencies that need to be modeled for better understanding and description of the underlying

DOI: 10.1201/9780429318443-9

processes. Also the forecaster must obey the underlying conditional distribution theory in the forecast distribution (4.8). As noted before in Section 7.2.2, Kriging is automatically performed in the Bayesian setting. Also, as before we continue to perform Kriging independently when we perform forecasting at more than one locations where we do not have any past observations. Also, the Bayesian forecasting methods based on the posterior predictive distribution ensure performing time series filtering as has been discussed in the text book by West and Harrison (1997).

Limitations in out of sample forecasting in space arise sometimes due to the unavailability of the covariate values at the unobserved locations. For example, when forecasting ozone concentration values using a model involving maximum temperature we require the availability of maximum temperature values at the prediction sites as well. Obviously, there is another immediate problem that crops up is that we also need to forecast the covariate values if those are time varying. Forecasting the covariate values will introduce another layer of uncertainty that should be accumulated in the uncertainty for the forecasted response.

Forecasting using hierarchical Bayesian models is further limited by the lack of suitable software packages. There are a few available packages for forecasting using variants of the dynamic linear models (West and Harrison, 1997), see e.g., Petris et al. (2010). However, these packages do not allow incorporation of rich spatial covariance structure for the modeled data. On the other hand, spBayes, Finley et al. (2015), can model and forecast for short-length time series data. The spTimer package can model and forecast for large volumes of spatio-temporal data as well illustrate in this chapter.

Bayesian forecasting may be performed by using a simple model validation calculation trick while model fitting as has been used previously in Chapters 7 and 8. The trick involves setting up the fitting data set to include the future time points for forecasting but by making the response variable unavailable, i.e. NA in R. Most Bayesian model fitting software packages will automatically estimate the NA's in the data set with the respective uncertainties. Indeed, examples in this chapter will demonstrate this. Additionally, some packages, e.g. spTimer, provide explicit forecasting methods which can be used in practical problems as has been demonstrated in Section 9.3.

The forecasting problem exacerbates further when the geographical study region, such as the one in the Eastern United States considered in this chapter, is vast and the training data set for forecasting, and modeling, is rich in both space and time. For point referenced spatial data from a large number of locations, exact likelihood based inference becomes unstable and infeasible since that involves computing quadratic forms and determinants associated with a high dimensional variance-covariance matrix (Stein, 2008). Besides the problem of storage (Cressie and Johannesson, 2008), matrix inversion, at each iteration of the model fitting algorithm, such as the EM algorithm, is of $O(n^3)$ computational complexity, which is prohibitive, where n is a large number of modeled spatial locations. This problem also arises in the evaluation of the joint or conditional distributions in Gaussian process based models under a hierarchical Bayesian setup, see e.g., Banerjee et al. (2015).

To address this problem, several easy to use and scalable forecasting methods are discussed in this chapter. The first of these is an exact Bayesian method based on the separable spatio-temporal model discussed previously in Section 7.2. In this case it is possible to evaluate the exact forecasting distribution as has been done in Section 7.2.2. Such a model provides a fast Bayesian forecasting method that has been exploited by Sahu et al. (2011) to forecast ozone concentration levels in the eastern United States. We discuss some of their results in Section 9.2 below.

Forecasting for more general models starts with the independent in time GP based regression model that is simple to implement. Indeed, this is discussed below in Section 9.3.1. Forecasting using the spatio-temporal autoregressive model is discussed in Section 9.3.2. This method has been shown to be better in out of sample validation than some versions of dynamic linear models (Sahu and Bakar, 2012a) and also a wide class of models (Cameletti et al., 2011). The third and final forecasting method is the one based on the GPP models of Section 7.6 originally developed by Sahu and Bakar (2012b). This forecasting method, described in Section 9.3.3, is able to model and forecast for large volumes of data as we demonstrate in this chapter.

Forecasting methods must be compared rigorously using sound forecast calibration methods. Towards this end we develop Markov chain Monte Carlo (MCMC) implementation methods for several forecast calibration measures and diagnostic plots that have been proposed to compare the skills of the Bayesian forecast distributions, see e.g., Gneiting et al. (2007). The measures include the four model validation methods, RMSE, MAE, CRPS and coverage previously described in Chapter 6. In addition, in this chapter we introduce the hit and false alarm rate and three diagnostic plots: sharpness diagram, probability integral transform and a marginal calibration plot in Section 9.4. These measures and plots enable us to compare the implied Bayesian forecast distributions fully – not just their specific characteristics, e.g., the mean forecast, as would be done by simple measures such as the RMSE and MAE.

The substantial example in this chapter is based on the work of Sahu et al. (2015) who have detailed some of the methodologies with an illustrative example on ground-level ozone in the Eastern US. Ground-level ozone is a pollutant that is a significant health risk, especially for children with asthma and vulnerable adults with respiratory problems. It also damages crops, trees and other vegetation. It is a main ingredient of urban smog. Because of these harmful effects, air pollution regulatory authorities are required by law to monitor ozone levels and they also need to forecast in advance, so that at risk population can take necessary precaution in reducing their exposure. In the United States of America, a part of which is our study region in this chapter, the forecasts are issued, often, up to 24-hours in advance by various mass-media, e.g. newspapers and also the website `airnow.gov`. However, ozone concentration levels, and also other air pollutants, are regularly monitored by only a finite number of sites. Data from these sparse network of monitoring sites need to be processed for developing accurate forecasts. In this chapter, we compare the forecasts of ground-level ozone, based on three models using a three-week test data set on daily maximum ozone concentration levels observed over a large region in the Eastern US. Forecast validations, using several moving windows, find a model developed using the GPP to be the best, and it is the only viable method for large data sets when computing speed is also taken into account. The methods are implemented using the previously introduced `spTimer` package.

9.2 Exact forecasting method for GP

We start with the spatio-temporal regression model (7.2)

$$Y(\mathbf{s}_i, t) = \beta_1 x_1(\mathbf{s}_i, t) + \cdots + \beta_p x_p(\mathbf{s}_i, t) + \epsilon(\mathbf{s}_i, t)$$

with the separable covariance structure (7.3) for the spatio-temporal Gaussian Process error distribution. Assuming the same conjugate prior distributions, we recall that the posterior predictive distribution is the t-distribution obtained using (7.7). The forecast distribution is the t-distribution with $nT + 2a$ degrees of freedom having mean

$$E(Y(\mathbf{s}_0, t_0)|\mathbf{y}) = \mathbf{x}'(\mathbf{s}_0, t_0)\boldsymbol{\beta}^* + \Sigma_{12} H^{-1}(\boldsymbol{v})(\mathbf{y} - X\boldsymbol{\beta}^*),$$

as given in (7.8). The scale parameter of this distribution is:

$$\frac{2b^*}{nT + 2a} \left(\delta^2(\mathbf{s}_0, t_0) + \mathbf{g}'(\mathbf{s}_0, t_0)(M^*)^{-1}\mathbf{g}(\mathbf{s}_0, t_0) \right),$$

where

$$\delta^2(\mathbf{s}_0, t_0) = 1 - \Sigma_{12} H^{-1}(\boldsymbol{v}) \Sigma_{21},$$

as noted in (7.6). All other parameters and expressions are noted in Section 7.2.

For forecasting using these posterior predictive distributions, we first note that if $\mathbf{s}_0$ is a data location, i.e. one of $\mathbf{s}_1, \ldots, \mathbf{s}_n$ then $\delta^2(\mathbf{s}_0, t_0) = 1$ for any $t_0 > T$ since each row of Σ_{12} will be like a row of $H(\boldsymbol{v})$ and we know that HH^{-1} is the identity matrix.

For forecasting at a new location $\mathbf{s}_0$ for a time point $t_0 > T$, we recall the simplifications of the expression $\Sigma_{12} H^{-1}$ previously noted in Section 7.2.3. Using those simplified expressions we can easily evaluate the parameters of the above t-distribution.

The exact distribution for forecasting does not deliver the exact forecasts when a data transformation such as square root or log has been used. In such a case the forecast distribution is on the transformed scale. Often, the forecasts on the transformed scale are transformed back to the original scale but that is not satisfactory at all, especially in the Bayesian case since, e.g. $E(Y^2(\mathbf{s}_0, t_0)|\mathbf{y})$ is not equal to $E(Y(\mathbf{s}_0, t_0)|\mathbf{y})^2$ in general. Estimating the uncertainties on the original scale will also pose further problems.

Sampling based approaches provide an easy recipe to solve the scale transformation problem. Using this method a large number of samples are to be generated from the forecast t-distribution. These samples are then transformed back to the original scale of the data and then Monte Carlo averages are formed to estimate the forecasts and their uncertainties as discussed in the Bayesian computation Chapter 5. This is the method we recommend if the log transformation has been used in modeling the observations.

There is an exact analytical solution to the problem, however, if the square-root transformation has been used. This exact method comes from the identity

$$E(Y^2(\mathbf{s}_0, t_0)|\mathbf{y}) = \{E(Y(\mathbf{s}_0, t_0)|\mathbf{y})\}^2 + \mathrm{Var}\{Y(\mathbf{s}_0, t_0)|\mathbf{y})\}.$$

Thus, to estimate the forecast at location $\mathbf{s}_0$ and time t_0 in the original scale we simply evaluate:

$$\{\mathbf{x}_0'\boldsymbol{\beta}^* + \Sigma_{12}H^{-1}(\mathbf{y} - X\boldsymbol{\beta}^*)\}^2 + 2b^* \frac{\delta^2(\mathbf{s}_0, t_0) + \mathbf{g}'(\mathbf{s}_0, t_0)(M^*)^{-1}\mathbf{g}(\mathbf{s}_0, t_0)}{nT + 2a - 2}.$$

The variance of the prediction, $\mathrm{Var}(Y^2(\mathbf{s}_0, t_0)|\mathbf{y})$, is calculated using the expression for the variance of the square of the t–distribution noted in (A.27) in Appendix A. This method, however, will not yield the exact forecast intervals. To obtain the forecast intervals we can adopt one of the two following approaches. The first method approximates the forecast t-distribution using a normal distribution which can be justified because of the high value of the degrees of freedom $nT + 2a$. Using this normal approximation an approximate 95% prediction interval is given by

$$E(Y^2(\mathbf{s}_0, t_0)|\mathbf{y}) \pm 1.96 \times \sqrt{\mathrm{Var}(Y^2(\mathbf{s}_0, t_0)|\mathbf{y})}.$$

The second method is to use the Monte Carlo method suggested above for any general transformation.

9.2.1 Example: Hourly ozone levels in the Eastern US

Air quality changes very fast in space and time as airborne particles and harmful gases are transported by the prevailing weather conditions and human activity, such as motoring, in the immediate neighborhood and beyond. For example, dust particles originating from the Sahara desert have been known to pollute the air in the UK and Europe in 2014 and 2015. Thus, episodes in air pollution can occur in a study region for activities and natural phenomena taking place in areas even 1000s of miles apart. How then can air pollution levels be forecast accurately so that at risk people, i.e. children and those suffering from respiratory illnesses can be alerted to exposure risk?

Air quality models have been developed based on chemical transport models and those for atmospheric air dispersion systems. In the United State of America (USA), national air quality forecasts and near real-time predictive spatial maps are provided to the general public through the EPA-AIRNow web site: https://www.airnow.gov/ Current and next day particulate matter and ozone (O_3) air quality forecasts for over 200 U.S. cities are now provided on a daily basis. These forecast maps, however, are based primarily on the output of a computer simulation model known as the Eta-CMAQ model, see e.g. https://www.epa.gov/cmaq/cmaq-output/. These models use emission inventories, meteorological information, and land use to estimate average pollution levels for gridded cells (12 km^2) over successive time periods. However, it is well known that these computer models may produce biased output and, as a result, this may lead to inaccurate pollution forecasts Kang et al. (2008).

Monitoring data, on the other hand, provide much better air quality information since those are based on actual measurements and thus are free from biases in the computer model output. However, the monitoring sites are often sparsely located and irregularly spaced over large areas such as the Eastern US, which is the study region of interest in this chapter. The sparsity limits accurate air quality information for areas away from the monitoring sites. Surely, from an individual's view point the most relevant air quality information must be the one where he/she lives or works and not at or near the monitoring sites. The problem of finding accurate air quality information in space and time still remains even after obtaining data from a monitoring network. This problem is further exacerbated by the need to forecast air quality so that preventive steps can be taken to limit exposure.

The need for prediction of air quality in both space and time naturally leads to the consideration of statistical modeling as candidate solutions. The main contribution behind the current impact case study is the development of a statistical spatio-temporal model that combines information from both the numerical model (Eta-CMAQ) and real time data from the monitoring sites. The model, implemented in a Bayesian inference framework, is computationally efficient and produces instantaneous forecast maps of hourly ozone concentration levels. The space-time model lends itself to closed form analytic Bayesian posterior predictive distributions for spatial interpolation of ozone concentration level for the past hours, current hour and forecast for future hours. These predictive distributions provide instantaneous spatial interpolation maps which could be used in a real-time environment such as the U.S. EPA AIRNow system. The predictive distributions are used to obtain the eight-hour average map, which is the average of the past four hours, current hour and three hours ahead. The forecasts are evaluated by using the model fitted to a two weeks test data set.

Assume that we have observed data from n monitoring sites denoted by $\mathbf{s}_1, \ldots, \mathbf{s}_n$ where each $\mathbf{s}_i$ is described either by a latitude and longitude pair or equivalently a northing and easting pair. Observed data often have high variability which causes problems in prediction (e.g. a negative value) using

Gaussian error distribution. To address that, we model data on the square-root scale but report all predictions at the original scale for ease of interpretation. Let $Y(\mathbf{s}, t)$ denote the observed square-root ozone concentration, in parts per billion (ppb) units at location $\mathbf{s}$ and at hour t for $t = 1, \ldots, T$ where we take $T = 168$ corresponding to a seven day modeling period that captures a full weekly cycle.

The Eta-CMAQ forecasts are proposed to be used as a regressor in the model so that we can use the best estimates so far to train the model. These forecasts fill in the gaps in space where monitoring data are not available and the regression method improves the accuracy by using these in conjunction with the ground truth revealed by the observations.

There is, however, a potential problem in using the Eta-CMAQ forecasts since those correspond to an average value on a 12-kilometer square grid-cell while the monitoring data are observed at a point level, $\mathbf{s}$, described by a latitude-longitude pair. This general problem is the 'change of support problem' and the method used to solve the problem is known as 'downscaling', see e.g. Berrocal et al. (2010), and Gelfand and Sahu (2009). We follow Sahu et al. (2011) and use $x(\mathbf{s}, t)$ (in ppb units) to denote the square-root of the Eta-CMAQ ozone forecast value at the unique grid cell covering the site $\mathbf{s}$ and at time t.

Ozone concentration data often shows strong diurnal patterns and we model using a different hourly intercept for each of the 24 hours in a day. Let $\xi(t) = \beta_j$ denote the hourly intercept, where the hour $t(= 1, \ldots, T)$ corresponds to the jth hour of the day, $j = 1, \ldots, 24$. In addition, a week-day/weekend indicator, $q(t)$ taking value 1 if the hour t is within a weekday and 0 otherwise is also used as a regressor.

The model and the forecasts are validated using the root mean square error (RMSE) for the forecasts. In our illustration, we use data from $n = 694$ sites in a study region in the eastern US. We use the RMSE criterion to select the optimal values of the spatial and temporal decay parameters ϕ_s and ϕ_t. For selecting ϕ_s the candidate effective ranges ($\approx 3/\phi_s$) were taken as 3, 6, 30, 60 and 600 kilometers. For selecting the temporal decay parameter ϕ_t we searched corresponding to effective ranges of 3, 6, 9, 12 and 24 hours. The optimal selection of these two parameters is the only tuning required in the whole procedure. The optimal values of these parameters must be found for each case of model based spatial interpolation and forecasting. However, the RMSE criterion cannot be calculated when it is necessary to forecast values in the future. In such cases, we recommend to use the optimal values of ϕ_s and ϕ_t for forecasting the most recent observed values by pretending those to be as yet unobserved.

Figure 9.1, reproduced from Sahu (2015), illustrates the RMSE of the fore-casts for one hour ahead at the 694 fitting sites. Here one hour ahead forecasts are obtained for 11 hours from 6AM to 4PM for 7 days. At each forecasting occasion the data from previous seven days (i.e. 168 hours) have been used and the optimal values of the tuning parameters are found using method described above. On average, the RMSEs for the Bayesian model based forecasts are a third lower than the same for the Eta-CMAQ forecasts and are about half of the same for the forecasts based on simple linear regression method. Sahu et al. (2011) illustrate the accuracy of the forecasts in further detail. In conclusion, it is expected that forecasting using optimal Bayesian space-time model will have much better accuracy than other methods which do not explicitly take space-time correlation into account.

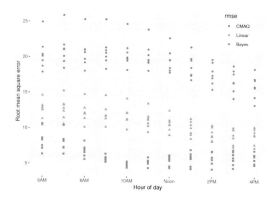

FIGURE 9.1: The RMSE's of the forecasts for the 8-hour averages at the current hour at each hour from 6AM to 4PM for three different forecasting methods.

♣ **R Code Notes 9.1. Figure 9.1** The forecasting methods used here have been coded using the C language which has been made available along with the input data. The RMSE's obtained from the C program are also made available to reproduce the plot.

9.3 Forecasting using the models implemented in spTimer

Forecasting based on general spatio-temporal models proceeds by advancing the model equations in time beyond the last time point used for model fitting. The extra parameters that emerge in the advanced model equations are integrated by sampling them from their conditional distributions by compositional sampling. In this section we discuss forecasting methods for the three models implemented in the package spTimer (Bakar and Sahu, 2015).

9.3.1 Forecasting using GP models

Recall the spatio-temporal linear regression model with full GP provided in (7.10). The model is written hierarchically as:

$$\mathbf{Y}_t = \mathbf{O}_t + \boldsymbol{\epsilon}_t, \tag{9.1}$$

$$\mathbf{O}_t = \mathbf{X}_t \boldsymbol{\beta} + \mathbf{w}_t \tag{9.2}$$

where $\boldsymbol{\epsilon}_t = (\epsilon(\mathbf{s}_1, t), ..., \epsilon(\mathbf{s}_n, t))' \sim N(\mathbf{0}, \sigma_\epsilon^2 \mathbf{I}_n)$ is the independently distributed white noise error with variance σ_ϵ^2 also known as the nugget effect, and $\mathbf{I}_n$ is the $n \times n$ identity matrix. The term $\mathbf{w}_t = (w(\mathbf{s}_1, t), ..., w(\mathbf{s}_n, t))'$ is an independent, over time, realization of a GP with zero mean and the Matérn correlation function (2.2). In effect, this implies that the smooth process, $O(\mathbf{s}, t)$ is assumed to be isotropic and stationary. Note that this does not necessarily imply the same assumptions for the un-transformed noisy data, Y since other hierarchical model components will contribute to the overall space-time correlation function. Thus, we assume that $\mathbf{w}_t \sim N(\mathbf{0}, \sigma_w^2 S_w)$ as previously in Section 7.3.

The error distributions of $\boldsymbol{\epsilon}_t$ and $\mathbf{w}_t$ are assumed to be independent of each other. For future reference, let $\boldsymbol{\theta}$ denote all the parameters, $\boldsymbol{\beta}$, σ_ϵ^2, σ_w^2, and ϕ. We assume independent normal prior distribution with zero mean and a very large variance, 10^{10}, to achieve vague prior specification, for the components of $\boldsymbol{\beta}$. The inverse of the variance components σ_ϵ^2, σ_w^2 are given independent gamma distribution with mean a/b and variance a/b^2. Although any suitable values for a and b can be chosen, following Sahu et al. (2007) we have taken $a = 2$ and $b = 1$ to have a proper prior distribution for any variance component that will guarantee a proper posterior distribution. We assume discrete uniform prior distributions for the decay parameter ϕ, although many other choices are possible in the bmstdr and spTimer packages.

To obtain the 1-step ahead forecast distribution of $Y(\mathbf{s}_0, T + 1)$ at any unobserved location $\mathbf{s}_0$ at time $T + 1$, we first note that:

$$Y(\mathbf{s}_0, T + 1) = O(\mathbf{s}_0, T + 1) + \epsilon(\mathbf{s}_0, T + 1), \tag{9.3}$$

$$O(\mathbf{s}_0, T + 1) = \mathbf{x}'(\mathbf{s}_0, T + 1)\boldsymbol{\beta} + w(\mathbf{s}_0, T + 1). \tag{9.4}$$

The 1-step ahead forecast distribution is the posterior predictive distribution of $Y(\mathbf{s}_0, T + 1)$ given $\mathbf{y}$ and is given by:

$$\pi(y(\mathbf{s}_0, T + 1)|\mathbf{y}) = \int \pi(y(\mathbf{s}_0, T + 1)|\boldsymbol{\theta}, \mathbf{o}, o(\mathbf{s}_0, T + 1), \mathbf{y})\pi(o(\mathbf{s}_0, T + 1)|\boldsymbol{\theta}, \mathbf{y})$$

$$\pi(\boldsymbol{\theta}, \mathbf{o}|\mathbf{y})do(\mathbf{s}_0, T + 1)d\mathbf{o}d\boldsymbol{\theta}, \tag{9.5}$$

where $\pi(\boldsymbol{\theta}, \mathbf{o}|\mathbf{y})$ denotes the joint posterior distribution of $\mathbf{O}$ and $\boldsymbol{\theta}$. Note that $\pi(y(\mathbf{s}_0, T+1)|\boldsymbol{\theta}, \mathbf{o}, o(\mathbf{s}_0, T+1), \mathbf{y}) = \pi(Y(\mathbf{s}_0, T+1)|\boldsymbol{\theta}, \mathbf{o}, o(\mathbf{s}_0, T+1))$ due to the conditional independence of $Y(\mathbf{s}_0, T+1)$ and $\mathbf{y}$ given $\mathbf{o}$. Similarly, $O(\mathbf{s}_0, T+1)$ does not depend on $\mathbf{y}$ given $\boldsymbol{\theta}$, hence in the following development we replace $\pi(o(\mathbf{s}_0, T + 1)|\boldsymbol{\theta}, \mathbf{y})$ by $\pi(o(\mathbf{s}_0, T + 1)|\boldsymbol{\theta})$.

Now the 1-step ahead forecast distribution (9.5) is constructed by composition sampling as follows. Assume that, at the jth MCMC iteration, we have posterior samples, $\boldsymbol{\theta}^{(j)}$ and $\mathbf{O}^{(j)}$. Then we first draw, $O^{(j)}(\mathbf{s}_0, T+1)$ from $N(\mathbf{x}'_{T+1}\boldsymbol{\beta}^{(j)}, \sigma_w^{2\,(j)})$. Finally, we draw $Y^{(j)}(\mathbf{s}_0, T+1)$ from $N(O^{(j)}(\mathbf{s}_0, T+1), \sigma_\epsilon^{2(j)})$.

Note that in the above paragraph, we use the marginal distribution instead of the conditional distribution because we have already obtained the conditional distribution given observed information up to time T at the observation locations $\mathbf{s}_1, ..., \mathbf{s}_n$, and at the future time $T+1$ there is no further new information to condition on except for the new regressor values $\mathbf{x}(\mathbf{s}_0, T+1)$ in the model. However, the conditional distribution can be used instead if it is so desired. To do this, we note that the joint distribution of $\mathbf{O}_{T+1} = (O(\mathbf{s}_1, T+1), ..., O(\mathbf{s}_n, T+1))'$ is simply given by $N(\mathbf{X}_{T+1}\boldsymbol{\beta}, \Sigma_w)$, according to (9.2). Similarly, we construct the joint distribution of $O(\mathbf{s}_0, T+1)$ and $\mathbf{O}_{T+1}$ from which we obtain the conditional distribution $\pi(O(\mathbf{s}_0, T+1)|\mathbf{o}_{T+1})$, that is Gaussian with mean

$$\mathbf{x}'(\mathbf{s}_0, T+1)\boldsymbol{\beta} + S_{w,12}S_w^{-1}(\mathbf{o}_{T+1} - \mathbf{X}_{T+1}\boldsymbol{\beta})$$

and variance

$$\sigma_w^2(1 - S_{w,12}S_w^{-1}S_{w,21}),$$

where $S'_{w,21} = S_{w,12} = e^{-\phi\,\mathbf{d}_{12}}$ and $\mathbf{d}_{12} = (||\mathbf{s}_1 - \mathbf{s}_0||, ..., ||\mathbf{s}_n - \mathbf{s}_0||)'$.

For forecasting at any observed site $\mathbf{s}_i$ for any $i = 1, ..., n$ at time $T+1$ we recall the equations (9.3) and (9.4). These two make it clear that the 1-step ahead forecast distribution of $Y(\mathbf{s}_0, T+1)$ given $\mathbf{y}$ can simply be constructed by iteratively sampling from the conditional distribution $O^{(j)}(\mathbf{s}_0, T+1) \sim N(\mathbf{x}'(\mathbf{s}_0, T+1)\boldsymbol{\beta}^{(j)}, \sigma_w^{2,(j)})$ and then $Y^{(j)}(\mathbf{s}_0, T+1)$ from the normal distribution with mean $O^{(j)}(\mathbf{s}_0, T+1)$ and variance $\sigma_\epsilon^{2(j)}$. Finally, $Y^{(j)}(\mathbf{s}_0, T+1)$ values are transformed back to the original scale if a transformation was applied in modeling.

9.3.2 Forecasting using AR models

Here we briefly describe the forecasting method based on the hierarchical AR models proposed by Sahu et al. (2007, 2009). The model equations are given by:

$$\mathbf{Y}_t = \mathbf{O}_t + \boldsymbol{\epsilon}_t, \tag{9.6}$$
$$\mathbf{O}_t = \rho\mathbf{O}_{t-1} + \mathbf{X}_t\boldsymbol{\beta} + \mathbf{w}_t \tag{9.7}$$

where $\boldsymbol{\epsilon}_t$ and $\mathbf{w}_t$ have been previously specified, and ρ is a scalar denoting site-invariant temporal correlation. These auto-regressive models also need an initialization for $\mathbf{O}_0$ which we assume to be independently normally distributed with mean $\boldsymbol{\mu}$ and the covariance matrix $\sigma_w^2 S_0$ where the correlation matrix S_0 is obtained using the exponential correlation function with a new decay parameter ϕ_0. These additional parameters and initialization random variables are added to $\boldsymbol{\theta}$ and $\mathbf{O}$ respectively.

The temporal correlation, ρ in (9.7), for the smooth process $O(\mathbf{s}, t)$, has been assumed to be site invariant given the effects of the spatially and temporally varying covariates and the spatio-temporal intercepts $w(\mathbf{s}, t)$. A site specific temporal correlation will perhaps be needed, though not pursued here, if only the last two terms are omitted from the model. We also assume, for stationarity, that $|\rho| < 1$.

We assume the same set of prior distributions for $\boldsymbol{\beta}$, the variance components σ_ϵ^2 and σ_w^2, and the correlation decay parameters ϕ as previously discussed in Section 9.3.1. For the additional ρ parameter we again provide a normal prior distribution with zero mean and a large variance (10^{10} in our implementation), but we restrict the prior distribution in the range $|\rho| < 1$.

Under the AR models the predictive distribution of $Y(\mathbf{s}_0, T + 1)$ is determined by $O(\mathbf{s}_0, T + 1)$. Following (9.7), we see that $O(\mathbf{s}_0, T + 1)$ follows the normal distribution with site invariant variance σ_w^2 and mean $\rho O(\mathbf{s}_0, T) + \mathbf{x}'(\mathbf{s}_0, T + 1)\boldsymbol{\beta}$. This depends on $O(\mathbf{s}_0, T)$ and as a result, due to this auto-regressive nature, we have to determine all the random variables $O(\mathbf{s}_0, k)$, for $k = 0, \ldots, T$. In order to simulate, all these random variables, we first simulate from the conditional distribution of $O(\mathbf{s}_0, 0)$ given $\mathbf{o}_0$, which is a univariate normal distribution. Then, at the jth MCMC iteration we sequentially simulate $O^{(j)}(\mathbf{s}_0, k)$ given $O^{(j)}(\mathbf{s}_0, k - 1)$ for $k = 1, \ldots, T + 1$ from the normal distribution with mean $\rho^{(j)} O^{(j)}(\mathbf{s}_0, k - 1) + \mathbf{x}'(\mathbf{s}_0, k)\boldsymbol{\beta}^{(j)}$ and variance $\sigma_w^{2\,(j)}$. For forecasting at any observation location $\mathbf{s}_i$ we draw $Y^{(j)}(\mathbf{s}_i, T + 1)$ from the normal distribution with mean $\rho^{(j)} O^{(j)}(\mathbf{s}_i, T) + \mathbf{x}'(\mathbf{s}_i, T + 1)\boldsymbol{\beta}^{(j)}$ and variance $\sigma_\epsilon^{2\,(j)}$. For further details regarding prediction see, Sahu *et al.* (2007). Now these Y values are transformed back to the original scale if necessary, as in the case of GP models.

9.3.3   Forecasting using the GPP models

In this section we describe forecasting methods for the models introduced in Section 7.6. Recall that the main idea here is to define the random effects $O(\mathbf{s}, t)$ at a much smaller number of locations, m say, where $m << n$, called the knots, and then use kriging to predict those random effects at the data locations. The top level model is written as (7.37)

$$\mathbf{Y}_t = \mathbf{X}_t \boldsymbol{\beta} + A\mathbf{O}_t + \boldsymbol{\epsilon}_t, \ t = 1, ..., T$$

where $A = CS_{w*}^{-1}$ and $\boldsymbol{\epsilon}_t$ has been previously specified. In the next stage of the modeling hierarchy the AR model (7.38) is assumed as:

$$\mathbf{O}_t = \rho\,\mathbf{O}_{t-1} + \mathbf{w}_t,$$

where $\mathbf{w}_t \sim N(\mathbf{0}, \sigma_w^2 S_{w*})$. Again, we assume that $\mathbf{O}_0 \sim N(\mathbf{0}, \sigma_w^2 S_0)$, where the elements of the covariance matrix S_0 are obtained using the correlation function, $\exp(-\phi_0 d_{ij})$, which is the same correlation function used previously but with a different decay parameter ϕ_0. The Bayesian model specification here is completed by assuming the same set of prior distributions as noted in the previous two sub-sections.

At an unobserved location $\mathbf{s}_0$, the 1-step ahead Bayesian forecast is given by the predictive distribution of $Y(\mathbf{s}_0, T + 1)$, that we determine from equation (7.37) replacing t with $T + 1$. At any location $\mathbf{s}_0$ we advance the model equation at the final model fitting time T by one step to obtain

$$Y(\mathbf{s}_0, T + 1) = \mathbf{x}'(\mathbf{s}_0, T + 1)\boldsymbol{\beta} + \mathbf{a}_0' S_{w*}^{-1} \mathbf{O}_{T+1} + \epsilon(\mathbf{s}_0, T + 1)$$

where $\mathbf{a}_0$ has been defined before. Thus, the 1-step ahead forecast distribution has variance σ_ϵ^2 and mean $\mathbf{x}'(\mathbf{s}_0, T + 1)\boldsymbol{\beta} + S_{w*,12} S_{w*}^{-1} \mathbf{o}_{T+1}$, where $S_{w*,12} = e^{-\phi \mathbf{d}_{12}}$ and $\mathbf{o}_{T+1}$ is obtained from (7.38) and $\mathbf{d}_{12}$ is the distance vector between the location $\mathbf{s}_0$ and m knot sites $\mathbf{s}_1^*, \ldots, \mathbf{s}_m^*$.

Thus, at each MCMC iteration, we draw a forecast value $Y^{(j)}(\mathbf{s}_0, T + 1)$ from this normal distribution. Forecasting at the observation sites $\mathbf{s}_1, \ldots, \mathbf{s}_n$ is performed by noting that, according to (7.37),

$$\mathbf{Y}_{T+1} = X_{T+1}\boldsymbol{\beta} + A\mathbf{O}_{T+1} + \boldsymbol{\epsilon}_{T+1},$$

with $\boldsymbol{\epsilon}_{T+1} \sim N(\mathbf{0}, \sigma_\epsilon^2 I_n)$. Thus, as before $\mathbf{O}_{T+1}$ is obtained from (7.38) and MCMC sampling from the forecast distribution of $\mathbf{Y}_{T+1}$ is straightforward. Again these samples are transformed back to the original scale if necessary.

9.4   Forecast calibration methods

9.4.1   Theory

Bayesian forecasting methods must be compared using suitable forecast calibration methods. Predictive Bayesian model selection methods, such as DIC, WAIC and PMCC as discussed in Chapter 4, are appropriate for comparing Bayesian models. However, the main objective of this chapter is forecasting and hence we compare the models on the basis of their forecasting performance. There is a large literature on forecast comparison and calibration methods, see e.g., Gneiting et al. (2007) and the references therein. In the Bayesian context of this chapter, we need to compare the entire forecast predictive distribution, not just summaries like the mean, since forecasting is the primary goal here.

To simplify notation, following Section 6.8, suppose that $y_\ell, \ell = 1, \ldots, m$ denote the m hold-out validation observations that have not been used in model fitting. Note that we use a single indexed notation y_ℓ, instead of the more elaborate $y(\mathbf{s}, t)$ used previously. Clearly, some of these validation observations may be future observations at the modeling sites or completely at new sites – what's important here is that those must not have been used for model fitting. Let $F_\ell(y)$ denote the model based forecast predictive distribution function of Y_ℓ, the random variable whose realized value is y_ℓ. Thus, $F_\ell(y)$ is the candidate forecast predictive distribution using a particular spatio-temporal model described previously in Chapter 7. Let $G_\ell(y)$ be the true unknown forecast predictive distribution function of the observations, which the $F_\ell(y)$ is trying to estimate. The problem here is to calibrate $F_\ell(y)$ for $G_\ell(y)$, $\ell = 1, \ldots, m$, conditional on the modeled data, $\mathbf{y}$. Let $\hat{y}_\ell$ be the intended forecast for y_ℓ, i.e., $\hat{y}_\ell$ mean or median of the forecast distribution $F_\ell(y)$, estimated using the mean or median of the MCMC samples $y_\ell^{(j)}, j = 1, \ldots, J$, where J is a large number. In our implementation in Sections 9.5 and 9.6, we have taken $J = 15,000$ after discarding the first 5000 iterations; that was deemed to be adequate to mitigate the effect of initial values. Below, we describe four additional forecast calibration and diagnostic methods and develop their computation methods using MCMC:

1. The hit and false alarm rates are considered by many authors for forecast comparison purposes, see e.g., Sahu et al. (2009). These rates are defined for a given threshold value y_0, which is often the value beyond which the pollutant is considered to be very dangerous. Hit is defined as the event where both the validation observation, y_ℓ and the forecast, $\hat{y}_\ell$, for it are either both greater or less than the threshold y_0. The false alarm, on the other hand, is defined as the event where the actual observation is less than y_0 but the forecast is greater than y_0. Thus, we define:

$$\text{Hit rate}(y_0) = \frac{1}{m} \sum_{\ell=1}^{m} \{1 \, (y_\ell > y_0 \, \& \, \hat{y}_\ell > y_0) + 1 \, (y_\ell < y_0 \, \& \, \hat{y}_\ell < y_0)\},$$

$$\text{False alarm}(y_0) = \frac{1}{m} \sum_{\ell=1}^{m} 1(y_\ell < y_0 \, \& \, \hat{y}_\ell > y_0).$$

Forecasting methods with high hit rates and low false alarm rates are preferred.

2. The concentration of the forecast distribution is compared using the sharpness diagram. A sharpness diagram plots the widths of the (m) forecast intervals as side-by-side boxplots where each boxplot is for a particular forecasting method. The forecasting method that produces narrower width forecast intervals, but with good nominal coverages, is preferred.

3. Many authors have proposed the probability integral transform (PIT) diagram as a necessary diagnostic tool for comparing forecasts. For each hold-out observation y_ℓ, the PIT value is calculated as

$$p_\ell = F_\ell(y_\ell), i = 1, \ldots, m. \tag{9.8}$$

If the forecasts are ideal, and F_ℓ is continuous, then p_ℓ has a uniform distribution. The PIT diagram is simply an histogram of the p_ℓ's, $1, \ldots, m$. Using MCMC samples, p_ℓ is estimated by:

$$\hat{p}_\ell = \frac{1}{J} \sum_{j=1}^{J} 1 \left(y_\ell^{(j)} \leq y_\ell \right), \ \ell = 1, \ldots, m.$$

4. A marginal calibration plot (MCP) is used to calibrate the equality of the forecast and the actual value, and is constructed as follows. First, take a grid, $y_k, k = 1, \ldots, K$, say, covering the domain of the forecast distribution. For each of those y_k values, calculate

$$\hat{G}(y_k) = \frac{1}{m} \sum_{\ell=1}^{m} 1 \left(y_\ell \leq y_k \right). \tag{9.9}$$

Now calculate

$$\bar{F}(y_k) = \frac{1}{m} \sum_{\ell=1}^{m} \hat{F}_\ell(y_k), \tag{9.10}$$

where

$$\hat{F}_\ell(y_k) = \frac{1}{J} \sum_{j=1}^{J} 1 \left(y_\ell^{(j)} \leq y_k \right), \ \ell = 1, \ldots, m. \tag{9.11}$$

Now, the plot of the differences $\bar{F}(y_k) - \hat{G}(y_k)$ against y_k, for $k = 1, \ldots, K$ is the desired MCP. If the forecasts are good, only minor fluctuations about 0 are expected. Thus, a forecast distribution whose MCP stays closest to 0 will be the preferred choice.

9.4.2   Illustrating the calibration plots

In this section we illustrate three forecast calibration plots using two versions of the GP model for the running New York air pollution data set `nysptime`. The first version is the full GP model discussed in Section 7.3.1 and the second version is the marginal model described in Section 7.3.2 implemented using the `Stan` package. Recall that the `nysptime` data set contains data from 28 air pollution sites for the 62 days in July and August in 2006. We fit the models using the data for the first 60 days in July and August and then forecast for the last two days in August. Hence forecast validation and calibration are performed using 56 daily data values from the 28 sites.

The commands to split the data and identify rows for forecasting are given below.

```
vs <- getvalidrows(valids=1:28, validt=61:62)
dfit <- nysptime[-vs, ]
dfore <- nysptime[vs, ]
```

The first of the above commands finds the row numbers of the data frame nysptime which contain the data for August 30 and 31. The last two commands split nysptime into a model fitting and model validation data frame. To fit the marginal model and to perform forecasting using the validation trick we issue the commands:

```
f2 <- y8hrmax ~ xmaxtemp+xwdsp+xrh
M4 <- Bsptime(package="stan",formula=f2, data=nysptime, validrows=vs
   , coordtype="utm", coords=4:5, scale.transform = "SQRT",
N=1500, burn.in=500, verbose = F)
```

Note that here we send the full data set to the model fitting program but instruct it to set aside the data with the validation rows contained in the vector vs. The model fitted object contains the forecast summaries and also the forecast samples as the component M4$valpreds which we will use to draw the forecast calibration plots.

The package spTimer can also be used to forecast by exploiting the validation trick. But it is preferable to use the more flexible dedicated forecasting methods implemented using the **predict** command tailored for spTimer model fitted objects. Here forecasting is performed by using two commands given below.

```
library(spTimer)
M3new <- Bsptime(package="spTimer", formula=f2, data=dfit,
    coordtype="utm", coords=4:5, scale.transform = "SQRT", N=5000)
nfore <- predict(M3new$fit,newdata=dfore, newcoords=~Longitude+
    Latitude, type="temporal", foreStep=2, tol=0.05)
```

The second of the two commands performs forecasting using the data provided by the dfore data frame. The argument foreStep=2 opts for two step forecasting starting after the last time point in the data set. The option tol =0.05 provides a value for the minimum distance between the forecasting and the data modeling sites. However, this is not relevant as we are forecasting at the data modeling sites, although the argument is a required argument.

The forecast output nfore from the predict command contains the forecast summaries, i.e. mean, median, sd and forecast intervals and also the forecast samples as the component nfore$fore.samples which we will use to draw the three forecast calibration plots: sharpness diagram, PIT diagram and marginal calibration plot.

The MCMC samples M4$valpreds and nfore$fore.samples are used to find the 50% and 90% forecast intervals for the 56 set aside observations. The commands used are

```
stanpreds <- M4$valpreds
dim(stanpreds)
sptpreds <- nfore$fore.samples
ints50stan <- get_parameter_estimates(t(stanpreds), level=50)
ints50spt <- get_parameter_estimates(t(sptpreds), level=50)
```

```
ints90stan <- get_parameter_estimates(t(stanpreds), level=90)
ints90spt <- get_parameter_estimates(t(sptpreds), level=90)
```

From the last four objects the lengths of the intervals are calculated and those are collected in a data frame suitable for plotting the side by side box plots in Figure 9.2. Clearly the forecast intervals are shorter for the spTimer package.

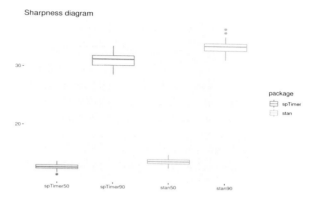

FIGURE 9.2: Sharpness diagram for the forecasts based on Stan and spTimer.

To obtain the PIT diagram we first write a function, findprop which obtains the proportion of a vector of sample values which are less than a given value. The given value is the first element of the input vector of the function and the remaining values of the vector are the sample values. This is constructed so that we can estimate $\hat{p}_\ell$ given in (9.8) using vectorised calculations in R. Example code for the Stan package output is given below:

```
findprop <- function(x) {
  value <- x[1]
  samples <- x[-1]
  prop <- length(samples[samples<value])/length(samples)
  prop
}
yobs <- dfore$y8hrmax
x <- cbind(yobs, stanpreds)
y <- na.omit(x)
stanpitvals <- apply(y, 1, findprop)
```

Note that the na.omit command removes the missing observations for forecast validation. We use similar code to find the values for the spTimer forecasts. Finally we collect the output in a data frame suitable for plotting. Left panel of Figure 9.3 provides the resulting PIT diagram. It does not show a uniform distribution for either of the two forecasting methods.

In order to obtain the marginal calibration plot we use the function `findprop` to obtain the estimates of $\hat{G}(y_k)$ in (9.9) and $\hat{F}_\ell(y_k)$ for $\ell = 1, \ldots, m(= 56)$ in (9.11) for each value of y_k. Here the y_k values are taken as a regular grid of 100 values in the interval spanning the range of observed values of the ozone levels. To calculate $\bar{F}(y_k)$ in (9.10) we loop through the 56 forecast observations. The resulting differences are collected in a data frame suitable for plotting. Right panel of figure 9.3 provides the plots for the two forecasting methods. The `spTimer` package is performing slightly better here.

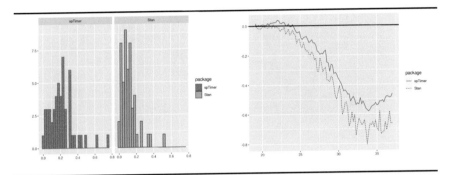

FIGURE 9.3: PIT diagram (left) and marginal calibration plot for the forecasts based on `Stan` and `spTimer`.

9.5 Example comparing GP, AR and GPP models

The computation of all the forecast calibration methods for the whole eastern US data set is prohibitive because of the big-n problem as mentioned in the Introduction, see also the next section. Due to this reason, we compare all three forecasting methods using a subset of the whole eastern US data, consisting of four states: Illinois, Indiana, Ohio and Kentucky. There are 156 ozone monitoring sites in these states, see Figure 9.4. We illustrate the 1-step ahead forecasts for a running window of seven days as follows. We use seven days data to fit model and obtain forecasts for the next day at all 156 model fitting locations. The running window of fitting data consists of data from day 1 to day 13 in July 2010 and the forecasting days are from day 8 to day 14 in July 2010.

For the GPP model the knot size is taken as 108, that has been chosen from a sensitivity analysis similar to the ones reported in Sahu and Bakar (2012b). We also have performed a number of different sensitivity analysis with respect to the choice of the hyper-parameter values in the prior distribution, tuning

156 data and 108 knot locations

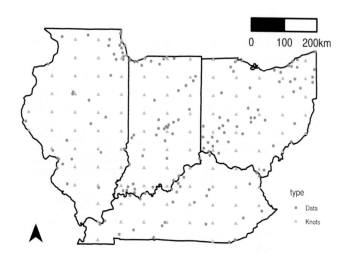

FIGURE 9.4: Map of the four states: Ohio, Indiana, Illinois, and Kentucky.

of the MCMC algorithms and also have monitored convergence using trace plots and the package CODA (Plummer et al., 2006).

We fit each of the three models: GP, AR and GPP for all seven days of fitting and forecasting windows. Thus, there are 21 models to fit and then each model fitting occasion is followed by a prediction step as detailed in Section 9.4.2. The repetitive code for these are not included here for brevity but is provided online from github[1]. We then develop code lines similar to the ones reported in Section 9.4.2 to obtain various forecast evaluation and calibration summaries and plots reported below.

The RMSE, MAE and CRPS for the seven validation days are plotted in Figure 9.5. As expected, the RMSE and the MAE are very similar, see the second row. Coverage values are closer to the 95% for the AR model compared to the other two. According to CRPS, the GP model seems to be the best but it is the worst according to the coverage criterion. Thus, this investigation does not nominate a best model for forecasting. We continue to explore more using the forecast calibration methods detailed in Section 9.4.

The hit and false alarm rates using all seven validation days data are provided in Table 9.1 for two values of threshold ozone levels. The AR and GPP models perform slightly better than the GP model.

The PIT diagrams for all three forecasting models for the days data modeling case are provided in Figure 9.6. Here also the GP model is worse than the other two. However, the PIT diagram does not discriminate a great deal

	65 ppb			75 ppb		
Model	GP	AR	GPP	GP	AR	GPP
Hit rate	83.00	90.61	90.52	95.77	96.34	96.53
False alarm	14.08	5.26	0.09	1.78	1.60	0.00

TABLE 9.1: False alarm and hit rates for ozone threshold values of 65 and 75 for the four states data set.

between the other two models: AR and GPP. Figure 9.7 provides the sharpness diagrams and the marginal calibration plots. The forecast intervals are shorter for the GP model but it does not have adequate coverage properties as seen in Figure 9.5. The AR model seems to be the best according to the marginal calibration plot, see right panel in Figure 9.7. Thus, it seems that the AR model is the best one among the three models considered here.

9.6   Example: Forecasting ozone levels in the Eastern US

Here we consider the problem of forecasting ozone levels for the whole of the eastern US a day in advance. The data set, we use to illustrate, has daily data from 694 monitoring sites for the 14 days in July 2010 as in the previous section. Figure 1.3 provides a plot of the 691 of these monitoring sites. The GP and AR models of the previous section are much more computationally intensive than the GPP model. For the model fitting (a data set with 14 days data) and forecasting using 20,000 iterations, using a personal computer, we have estimated that the GP model will take about 40 hours, while the AR model will take about 66 hours to run. This excludes the use of GP and AR models for forecasting next day ozone levels, which must be produced within 24 hours of computing time. This computing problem gets exacerbated by the additional intention that here we want to produce a forecast map by forecasting at a grid of 1451 locations covering the study region. The GPP model, on the other hand, takes only about 50 minutes to run the same experiment on the same personal computer and is the only feasible method that we henceforth adopt.

The implementation of the GPP model requires the selection of the number of knots. Using a similar sensitivity study that we have used in Sahu and Bakar (2012b), but with the forecast RMSE, as the criterion we compare the GPP model with 68, 105, 158 and 269 knots which were all inside the land boundary of the United States. The forecast RMSE improved with the increasing knot sizes, but only slightly when the size increased to 269 from 158. Henceforth, we adopt 158 as the knot size that implies a much smaller computational burden.

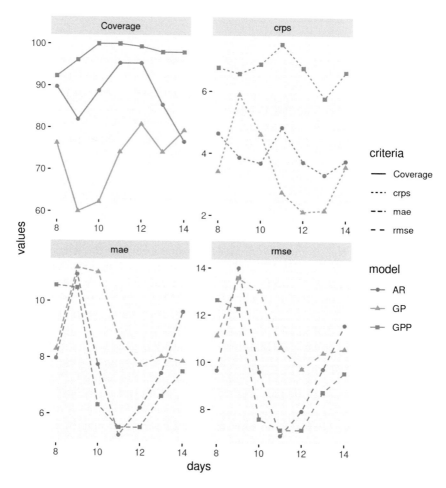

FIGURE 9.5: Plots of RMSE, MAE, CRPS and coverage based on modeling
7 days data.

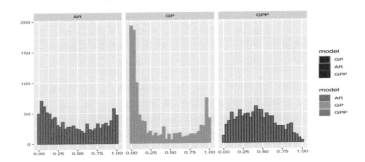

FIGURE 9.6: PIT diagrams.

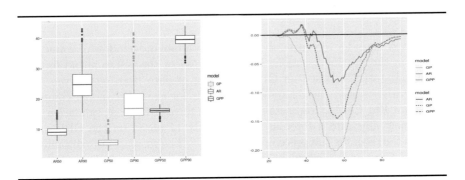

FIGURE 9.7: Sharpness diagram (left) and marginal calibration plot.

The regression model is based on the square-root scale of the data having a single covariate given by the square root of the output of the CMAQ model. The model is fitted by issuing the following commands:

```
f2 <- obs08hrmax ~ sqrt(cmaq08hrmax)
M2 <- Bsptime(formula=f2, data=dailyobs14day, package="spTimer",
    model="GPP", coordtype = "lonlat", coords=2:3, prior.phi="Gamm",
    scale.transform = "SQRT", knots.coords = knots.coords.158, n.report
    =20, N=5000)
```

The above command assumes that the data set is `dailyobs14day` and the knot locations are available as the matrix `knots.coords.158`. The parameter estimates for this model are provided in Table 9.2. The Eta CMAQ output is a significant predictor with very small standard deviation relative to the mean. The temporal correlation, estimated to be 0.405, is moderate. The random effect variance σ_w^2 is estimated to be larger than the nugget effect σ_ϵ^2. The estimate of the spatial decay parameter is 0.005, that corresponds to an effective range of about 600 kilometers.

	mean	sd	2.5%	97.5%
Intercept	4.749	0.233	4.299	5.246
sqrt(cmaq08hrmax)	0.268	0.031	0.201	0.328
ρ	0.403	0.032	0.339	0.464
σ_ϵ^2	0.306	0.005	0.296	0.317
σ_w^2	0.617	0.038	0.545	0.694
ϕ	0.005	0.000	0.004	0.006

TABLE 9.2: Estimates of parameters of the GPP model fitted to the full 14 days data set.

We now consider forecast validation using a moving window of seven days data as in the previous section. Table 9.3 presents the 1-step ahead forecast validation statistics based on GPP modeling of data from past seven days. These statistics show better forecasting skills for the GPP model compared to the raw Eta CMAQ forecasts for all seven days of forecast validation.

We now illustrate a forecast map for July 14 based on the GPP model fitted to the data for the preceding seven days. The forecasts are obtained by using the `predict` command in the `spTimer` package as noted previously in this chapter. Note that this `predict` command requires the values of the Eta CMAQ output at the grid of 1451 locations where we forecast using the fitted model.

Figure 9.8 provides the forecast map and also a map of the standard deviations of the forecasts. The map of standard deviations reveals that higher ozone levels are associated with higher uncertainty levels, which is a common phenomenon in ozone concentration modeling.

Start day	Forecast day	CMAQ RMSE	GPP			
			RMSE	MAE	CRPS	CVG
1	8	17.81	12.91	10.24	7.39	97.68
2	9	16.52	10.78	8.29	7.59	98.12
3	10	15.82	9.24	7.05	7.37	98.09
4	11	13.75	9.13	7.33	7.33	98.97
5	12	13.91	9.59	7.28	7.64	98.24
6	13	16.97	10.61	8.00	7.02	95.47
7	14	18.38	11.70	8.56	6.91	96.76

TABLE 9.3: Forecast validation using the four validation criteria for the GPP model.

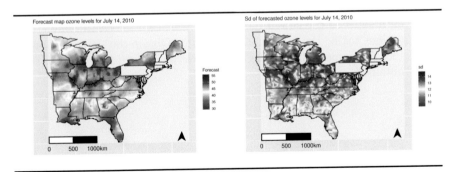

FIGURE 9.8: Maps showing the forecasts and their standard deviations for July 14 in 2010.

9.7 Conclusion

This chapter has discussed Bayesian forecasting methods using many different Bayesian hierarchical models for spatio-temporal data. MCMC methods have been developed to compute the Bayesian forecast distributions based on large space-time data. These methodological developments have enabled us to add the suite of forecasting routines in the contributed R software package, spTimer (Bakar and Sahu, 2015).

The contribution of the chapter also includes development of methods for estimating several forecast calibration measures using output from the implemented Markov chain Monte Carlo algorithms. We have demonstrated that these measures are able to compare different Bayesian forecasting methods rigorously and conclusively. A forecasting method based on a space-time model developed using the GPP model has been shown to be fast and the best for the illustrative ozone concentration forecasting problem of this chapter.

9.8 Exercises

1. Write computer code to perform forecasting using the exact method in Section 9.2. Test your code using the data from the four states used in Section 9.5. Advanced users can try to extend the methods to handle missing observations in the training data.

2. Pick four different adjacent states from the Eastern US air pollution data set. Use the subset data to run your own forecasting investigations taking the Section 9.5 as a guide. Obtain the sharpness and PIT diagrams and the marginal calibration plot for the forecasts so obtained.

10

Bayesian modeling for areal unit data

10.1 Introduction

Response variables that are observed on a discrete set of spatially referenced entities such as administrative areas in a country require different modes of analysis and modeling compared to the point referenced spatial data of the previous chapters. The first obvious key difference lies in the discrete and continuous natures of the respective spatial domain of the areal and point referenced data. This changes the nature and definition of spatial association as has been discussed previously in Chapter 2. The second most important difference, especially for the purposes of modeling, comes from the reality that often the observed response variable is recorded as a count, which is a discrete number. This trumps out our assumption of the Gaussian distribution we have been using in all the previous chapters. Hence there is the need to define what are called the generalized linear models (GLM) (McCullagh and Nelder, 1989), which allow us to model not only using the most common discrete distributions such as Bernoulli, binomial and Poisson but also the Gaussian distribution. Section 10.2 in this chapter defines the key concepts we need to understand for modeling spatial and spatio-temporal data using the GLMs.

Having specified a top level model for the response variable we turn our attention to modeling the linear predictor (defined in Section 10.2). This is modeled with at least two components similar to the nugget effect model (6.12) in Section 6.5. The regression part takes care of the effects of the covariates. It is then customary to split the random effects part in two pieces: one is spatially structured and the other is a hierarchical error akin to the nugget effect in the model (6.12). This decomposition into two parts for the random effects is the celebrated BYM (Besag, York and Mollié) model due to Besag et al. (1991). Here the model with spatial effects only, where the hierarchical error is assumed to be zero, is rarely used in practice and will not be discussed in this book at all. Section 10.4 describes the models with a running example. The basic structure of the decomposed model is creatively adapted for spatio-temporal data by bringing in auto-regressive and other models for time series data, see Section 10.6.

The GLMs assumed at the top level do not, in general, render a closed form posterior distribution for the regression parameters introduced in the linear predictor. Hence, we do not attempt to calculate the exact posterior

DOI: 10.1201/9780429318443-10

distributions as we have done in Section 6.3 and also later on in Section 7.2.1. Bayesian computation for these models then proceeds either with MCMC methods or approximation techniques such as INLA. Indeed, we have developed a new function Bcartime in the accompanying package bmstdr to fit both spatial and spatio-temporal models using three packages, CARBayes, CARBayesST and INLA. Like the Bsptime function, Bcartime can fit and validate the models with a top level assumption of a GLM with its associated components. We illustrate model fitting and validation with the running example in Sections 10.4 and 10.6. The subsequent Chapter 11 showcases many further practical examples.

10.2   Generalized linear models

10.2.1   Exponential family of distributions

The generalized linear models are assumed for random variables Y_i, $i = 1, \ldots, n$ which may be spatially or spatio-temporally referencedd as well. In this section, we describe the key concepts that we need for our practical modeling. Dropping the subscript i, suppose that we would like to model the distribution of the random variable Y based on q (notation changed to avoid clashing with p for the binomial distribution which we encounter below) observed covariates which we continue to denote by $\mathbf{x}$ and consequently $\mathbf{x}'\boldsymbol{\beta}$ denotes the regression part of the model specification. Of course, with observations y_i and covariates $\mathbf{x}_i$, we will use the joint distribution of Y_i, $i = 1, \ldots, n$ to obtain the likelihood function, and hence the posterior distribution, of the parameters of interest $\boldsymbol{\beta}$.

The starting assumption of a GLM is that the random variable Y follows a probability distribution, which is a member of the exponential family having the probability density function (or probability function, if discrete):

$$f(y; \theta, \phi) = \exp \left(\frac{y\theta - b(\theta)}{a(\phi)} + c(y, \phi) \right), \tag{10.1}$$

for suitable functions $a(\cdot), b(\cdot)$ and $c(\cdot, \cdot)$, where $b(\theta)$ is free of y and $c(y, \phi)$ is free of $\theta \in \mathbb{R}$. The parameter θ is called the *natural* or *canonical* parameter. The parameter ϕ is usually assumed known. If it is unknown then it is often called the *nuisance* parameter as in the Gaussian distribution case. By using well-known theory of score functions, see e.g.McCullagh and Nelder (1989), it can be shown that

$$E(Y) = b'(\theta), \quad \text{and} \quad \text{Var}(Y) = a(\phi)b''(\theta),$$

where $b'(\theta)$ and $b''(\theta)$ are respectively the first and second partial derivatives of the function $b(\theta)$ with respect to θ. Thus, the variance is the product of two functions: (i) $b''(\theta)$ which depends on the canonical parameter θ (and hence μ) only and is called the *variance function*, denoted by $V(\mu)$; (ii) $a(\phi)$ which is sometimes of the form $a(\phi) = \sigma^2/r$ where r is a known *weight* and σ^2 is called the *dispersion parameter* or *scale parameter*.

♡ **Example 10.1. Bernoulli distribution.** Suppose $Y \sim$ Bernoulli(p). Then

$$f(y; p) = p^y(1 - p)^{1-y}, \qquad y \in \{0, 1\}; \quad p \in (0, 1)$$

$$= \exp\left(y \log \frac{p}{1 - p} + \log(1 - p)\right)$$

This is in the form (10.1), with $\theta = \log \frac{p}{1-p}$, $b(\theta) = \log(1 + \exp\theta)$, $a(\phi) = 1$ and $c(y, \phi) = 0$. Therefore,

$$E(Y) = b'(\theta) = \frac{\exp\theta}{1 + \exp\theta} = p,$$

$$\mathrm{Var}(Y) = a(\phi)b''(\theta) = \frac{\exp\theta}{(1 + \exp\theta)^2} = p(1 - p),$$

and the variance function is $V(\mu) = \mu(1 - \mu)$. □

♡ **Example 10.2. Binomial distribution.** Suppose $Y^* \sim$ Binomial(m, p). Here, m is assumed known (as usual) and the random variable $Y = Y^*/m$ is taken as the *proportion* of successes, so

$$f(y; p) = \binom{m}{my} p^{my}(1 - p)^{m(1-y)}, \qquad y \in \left\{0, \frac{1}{m}, \frac{2}{m}, \dots, 1\right\}; \quad p \in (0, 1)$$

$$= \exp\left(\frac{y \log \frac{p}{1-p} + \log(1 - p)}{\frac{1}{m}} + \log\binom{m}{my}\right).$$

This is in the form (10.1), with $\theta = \log \frac{p}{1-p}$, $b(\theta) = \log(1+\exp\theta)$, $a(\phi) = \frac{1}{m}$ and $c(y, \phi) = \log\binom{m}{ny}$. Therefore,

$$E(Y) = b'(\theta) = \frac{\exp\theta}{1 + \exp\theta} = p,$$

$$\mathrm{Var}(Y) = a(\phi)b''(\theta) = \frac{1}{m}\frac{\exp\theta}{(1 + \exp\theta)^2} = \frac{p(1 - p)}{m}$$

and the variance function is $V(\mu) = \mu(1 - \mu)$. Here, we can write $a(\phi) \equiv \sigma^2/r$ where the scale parameter $\sigma^2 = 1$ and the weight r is m, the binomial denominator. □

♡ **Example 10.3. Poisson distribution.** Suppose $Y \sim \text{Poisson}(\lambda)$. Then

$$f(y; \lambda) = \frac{\exp(-\lambda)\lambda^y}{y!}, \qquad y \in \{0, 1, \ldots\}; \quad \lambda \in \mathcal{R}_+$$

$$= \exp\left(y \log \lambda - \lambda - \log y!\right).$$

This is in the form (10.1), with $\theta = \log \lambda$, $b(\theta) = \exp \theta$, $a(\phi) = 1$ and $c(y, \phi) = -\log y!$. Therefore, $E(Y) = b'(\theta) = \exp \theta = \lambda$, $\text{Var}(Y) = a(\phi)b''(\theta) = \exp \theta = \lambda$ and the variance function is $V(\mu) = \mu$. ☐

♡ **Example 10.4. Normal distributions.** Suppose $Y \sim N(\mu, \sigma^2)$. Then

$$f(y; \mu, \sigma^2) = \frac{1}{\sqrt{2\pi\sigma^2}} \exp\left(-\frac{1}{2\sigma^2}(y - \mu)^2\right), \qquad y \in \mathbb{R}; \quad \mu \in \mathbb{R}$$

$$= \exp\left(\frac{y\mu - \frac{1}{2}\mu^2}{\sigma^2} - \frac{1}{2}\left[\frac{y^2}{\sigma^2} + \log(2\pi\sigma^2)\right]\right).$$

This is in the form (10.1), with $\theta = \mu$, $b(\theta) = \frac{1}{2}\theta^2$, $a(\phi) = \sigma^2$ and

$$c(y, \phi) = -\frac{1}{2}\left[\frac{y^2}{a(\phi)} + \log(2\pi a[\phi])\right].$$

Therefore, $E(Y) = b'(\theta) = \theta = \mu$, $\text{Var}(Y) = a(\phi)b''(\theta) = \sigma^2$ and the variance function is $V(\mu) = 1$. ☐

10.2.2 The link function

For specifying the pattern of dependence of the response variable Y on the explanatory variables $\mathbf{x}$, to avoid the kind of problems illustrated below, we do not simply write $E(Y) = \eta$ where η is the linear predictor, which is sum of the regression part $\mathbf{x}'\boldsymbol{\beta}$ and the spatial and spatio-temporal random effects as we have done so far starting with Chapter 6. Here the linear predictor, η, is allowed to take any value on the real line, i.e. $-\infty < \eta < \infty$. But note that if for example, $Y \sim \text{Poisson}(\lambda)$ where $\lambda > 0$, then $E(Y) = \lambda$ and the equation $\lambda = \eta$ is problematic since the domains of the two sides of this equation do not match. Similarly we do not simply write $p = \eta$ in the binomial distribution case where $0 < p < 1$ but again $-\infty < \eta < \infty$.

The link between the distribution of Y and the linear predictor η is provided by the concept of the *link function* which states that

$$g(\mu) = \eta$$

where $\mu = E(Y)$ and $g(\cdot)$ is a carefully chosen one-to-one and differentiable function of its argument which avoids the above mentioned difficulties. If $g(\cdot)$ is not chosen carefully, then there may exist a possible pair of values $\mathbf{x}$ and $\boldsymbol{\beta}$ such that $\eta \neq g(\mu)$ for any possible value of μ. Therefore, "sensible" choices of link function map the set of allowed values for μ onto $\mathbb{R}$, the real line.

The particular choice of $g(\cdot)$ for which

$$\eta = g(\mu) = \theta$$

is called the *canonical link function* since for this choice the canonical parameter, θ, is exactly equal to the linear predictor η. In practice, the canonical link function is the one that is used most often. For Poisson distribution with parameter λ, recall that

$$\theta = \log(\lambda) = \log(\mu).$$

Hence the canonical link, corresponding to $\eta = \theta$, is the log function, and this link function is called the "log" link. The GLM in this case is sometimes called a *log*-linear model.

For both the Bernoulli and binomial distributions, recall that,

$$\theta = \log\left(\frac{p}{1-p}\right) \quad \text{or equivalently,} \, p = \frac{e^\theta}{1+e^\theta}.$$

Hence the canonical link, corresponding to $\eta = \theta$, is the logit function, $\log(p/(1-p))$, and this is popularly known as logistic regression.

An alternative to the logit is the probit link, which is defined by the inverse of the cdf of the standard normal distribution, $\Phi^{-1}(\cdot)$. For probit link, $p = \Phi(\eta)$. Another alternative link function is the complementary log link function which sets

$$\eta = \log(-\log(1-p)), \quad \text{or equivalently,} \quad p = 1 - e^{-e^\eta}.$$

There are large differences in the shape of these three alternative link functions, see Figure 10.1. Both the probit and logit links are symmetric in the tails but the complementary log link is not symmetric. The probit link rises to probability one on the right or decreases to zero probability on the left much faster than the logit link. However, it is difficult to objectively choose the most appropriate link function for small sample sizes.

For the normal distribution example discussed above, we have seen that $\theta = \mu$ and hence the requirement that $\eta = \theta$ for the canonical link function implies that the canonical link is the identity function, i.e. $g(\mu) = \mu$. Hence, it is claimed that the Gaussian linear model is also a GLM with the identity link function.

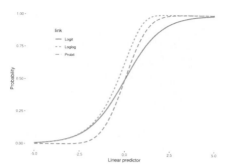

FIGURE 10.1: Three different link functions.

10.2.3   Offset

In defining the linear predictor η_i, for the ith observation y_i, we sometimes add another term without an associated regression co-efficient β, e.g.

$$\eta_i = o_i + \sum_{j=1}^{q} x_{ij}\beta_j, i = 1,\ldots,n \qquad (10.2)$$

where we assume the o_i's to be known. Thus, we do not want to estimate any regression coefficient for the values contained in the variable o. Such a variable is called an *offset* in the GLM. There can be several offsets in a GLM. An offset variable may serve several different purposes depending on the modeling context. For example, in a longitudinal trial measuring certain characteristics such as blood pressure the baseline measure obtained before administering any treatment can be treated as an offset.

    The concept of offset plays a very important role when we model areal count data. For example, let Y_i denote a count of the number of events which occur in a given region i, for instance the number of times a particular drug is prescribed on a given day, in a district i of a country. We might want to model the prescription rate **per patient** in the district λ_i^*. Suppose N_i is the number of patients registered in district i, often called the **exposure** of observation i. We model $Y_i \sim \text{Poisson}(N_i\lambda_i^*)$, where

$$\log \lambda_i^* = \mathbf{x}_i^T \boldsymbol{\beta}.$$

Equivalently, we may write the model as $Y_i \sim \text{Poisson}(\lambda_i)$, where

$$\log \lambda_i = \log N_i + \mathbf{x}_i^T \boldsymbol{\beta},$$

since $\lambda_i = N_i\lambda_i^*$, so $\log \lambda_i = \log N_i + \log \lambda_i^*$. The log-exposure $\log N_i$ appears as a fixed term in the linear predictor, without any associated parameter. In this modeling context $\log N_i$ is the offset term.

The concept of offset is directly related to the concepts of standardization discussed in Section 2.12. The standardized values of the counts are used as offsets and if a Poisson log linear model is used then the offsets are sent to the model on the log-scale as well.

10.2.4 The implied likelihood function

Returning to the general discussion, recall that for a random variable Y with a distribution from the exponential family, $\mu = E(Y) = b'(\theta)$. Hence, $\theta = b'^{-1}(\mu)$ and $g(\mu) = \eta = \mathbf{x}'\boldsymbol{\beta}$ implies,

$$\theta = b'^{-1}(g^{-1}[\eta]). \tag{10.3}$$

The canonical link function is the function $g = b'^{-1}$ so that $\theta = \eta$ as discussed above. Under this canonical link function it is straightforward to write down the implied likelihood function of $\boldsymbol{\beta}$. Note that we can express the probability function (10.1) as:

$$\exp\left(\frac{y\,\mathbf{x}^T\boldsymbol{\beta} - b(\mathbf{x}'\boldsymbol{\beta})}{a(\phi)} + c(y, \phi)\right),$$

which can be evaluated for a given value of $\boldsymbol{\beta}$ and the other parameters in ϕ. Hence, this last expression, written with each y_i and $\mathbf{x}_i$ and then multiplied for $i = 1, \ldots, n$ gives the following likelihood function of $\boldsymbol{\beta}$ and the other parameters involved in ϕ:

$$L(\boldsymbol{\beta}, \phi) = \prod_{i=1}^{n} \exp\left(\frac{y_i\,\mathbf{x}_i'\boldsymbol{\beta} - b(\mathbf{x}_i'\boldsymbol{\beta})}{a(\phi_i)} + c(y_i, \phi_i)\right). \tag{10.4}$$

The expression for the likelihood function for arbitrary (non-canonical) link functions are more complicated but can be written down by involving the inverse link function $g^{-1}(\cdot)$.

10.2.5 Model specification using a GLM

To recap, for a GLM we assume a particular distribution from binomial, Poisson, or Gaussian as the distribution for the response Y. This is the "family" argument in the model fitting routines. We also need to specify the link function, as the link argument, that we want to use for modeling. The log link, which is the canonical link, is most often used when the distribution is Poisson. The link functions "logit", "probit" and "complementary log" are meaningful in the case of Bernoulli and binomial distributions only. For the binomial distribution we also need to specify the number of trials for each data point. In the case of Gaussian distribution the canonical link is the identity link.

It now remains to specify the linear predictor, which is passed to the model fitting functions on as a formula argument. This formula argument can take an offset argument if one such data column is available. The fixed effects regression part is written down in this formula argument. The specification for the linear predictor is completed by assuming different distributions for the random effects. The specification of the random effects depends on the data modeling settings and is discussed in Section 10.4.

10.3   Example: Bayesian generalized linear model

This section illustrates Bayesian fitting of the GLMs described in the preceding Section 10.2 without any spatial or iid random effects. This is an important warming-up act before introducing the more complicated random effects varying in space and/or time. These models serve as base models which must be improved by including spatially structured random effects in the developments below. This is similar in spirit to the non-spatial modeling in Section 6.3 but with the GLM as the data model accommodating both continuous and discrete random variables. The non-linearity of the likelihood function (10.4), in general, does not allow closed form solutions for the Bayes estimates with any prior distributions except for some special cases such as the normal distribution with conjugate priors as in Section 6.3. Bayesian model fitting here proceeds by assuming suitable prior distributions for the parameters β and ϕ, if these are unknown.

As mentioned in the introduction to this chapter, Section 10.1, we will use the model fitting and validation function `Bcartime` in the accompanying package `bmstdr`. To fit the Bayesian GLMs without any random effects `Bcartime` employs the `S.glm` function of the `CARBayes` package to perform sampling from the posterior distribution. Deploying the `Bcartime` function requires the following three essential arguments:

- **formula** specifying the response and the covariates for forming the linear predictor η in Section 10.2;

- **data** containing the data set to be used;

- **family** being one of either "binomial", "gaussian", "multinomial", "poisson" or "zip", which respectively specifies a binomial likelihood model with a logistic link function, a Gaussian likelihood model with an identity link function, a multinomial likelihood model with a logistic link function, a Poisson likelihood model with a log link function, or a zero-inflated Poisson model with a log link function.

In addition, if the binomial family is chosen then the `trials` argument, a numeric vector containing the number of trials for each row of data, must be provided.

Like the `Bsptime` function, model validation is performed automatically by specifying the optional vector valued `validrows` argument containing the row numbers of the supplied data frame that should be used for model validation. As before, the user does not need to split the data set into training and test sets for facilitating validation. This task is done by the `Bcartime` function.

The function `Bcartime` automatically chooses the default prior distributions which can be modified by the many optional arguments, see the documentation of this function and also the `S.glm` function from `CARBayes`. Three MCMC control parameters `N`, `burn.in` and `thin` determine the number of iterations, burn-in and thinning interval. The default values of these are 2000, 1000 and 10 respectively. In all of our analysis in this chapter, unless otherwise mentioned, we take these to be 120,000, 20,000 and 10 respectively. These choices result in 10,000 MCMC samples being available for inference and prediction purposes.

In the remainder of this section we will only illustrate the "binomial","poisson" and "gaussian" distribution based model fitting with the Covid-19 mortality data set `engtotals` from England introduced in Section 1.4.1. See the documentation `?engtotals` for definitions of the various columns in this data set.

10.3.1   GLM fitting with binomial distribution

For the binomial model the response variable is `noofhighweeks`, which is the number of weeks out of 20 when the SMR (Standardized Mortality Rate, see Section 2.12) for the number of positive Covid-19 cases was greater than 1. In this case, the `trials` argument is a vector of constant values supplied as the column `nweek`. Three relevant covariates in this GLM are `jsa`, `houseprice` on the log10 scale, `popdensity` on the log scale and NO_2 on the square-root scale, see Section 3.4.2. Thus, the formula argument is:

```
f1 <- noofhighweeks ~jsa + log10(houseprice) + log(popdensity) +
    sqrt(no2)
```

We obtain the number of trials as `ntrails <-engtotals$nweek`. Essentially the command for fitting the model using the full data set is:

```
M1 <- Bcartime(formula=f1, data=engtotals, family="binomial",
    trials=ntrails)
```

The summary command `summary(M1)` on the fitted **bmstdr** model produces the estimates in Table 10.1. The last four columns in Table 10.1 show the convergence behavior of the MCMC used to fit the model. Allowing for 20,000 burn-in iterations and a thinning value of 10, 120,000 iterations yield 10,000 samples noted in the n.s column. The effective number of samples (see Section 5.11.1) for each of the parameters is shown in the column n.e in this table. The percentage of acceptance, see Section 5.7, shows the percentage of MCMC proposals that has been accepted. The Diag column shows the MCMC convergence diagnostics due to Geweke (1992). This statistic should

	Median	2.5%	97.5%	n.s	% accept	n.e	Diag
Intercept	−1.220	−2.995	0.557	10000	43.1	2759.3	−0.9
jsa	0.207	−0.087	0.495	10000	43.1	2575.7	1.3
log10(houseprice)	−0.100	−0.450	0.249	10000	43.1	2555.4	1
log(popdensity)	0.082	0.036	0.131	10000	43.1	2643.1	0.5
sqrt(no2)	0.126	0.024	0.232	10000	43.1	3184.3	−1.7

Criteria	Penalty	Value
DIC	5.0	1504.0
WAIC	6.2	1505.4

TABLE 10.1: Parameter estimates and the two model choice criteria for the binomial GLM. Here n.s and n.e stand for n.sample and n.effective respectively while Diag stands for MCMC convergence diagnostics due to Geweke (1992).

be referred to a standard normal distribution, i.e. an absolute value higher than 1.96 (5% cut-off point of the standard normal distribution) would indicate lack of convergence. The results in the above parameter estimates table do not show any problem in convergence. In fact, our choice of values for N, burn.in and thin always ensure adequate convergence in all the model fitting illustrations of this chapter. Henceforth, we do not report the last four columns in the parameter estimates table.

The parameter estimates in Table 10.1 show that population density and NO_2 are significant predictors while job seekers allowance and house price are not significant in the model. We note the two model choice criteria values followed by the penalty parameters in the table for comparison with spatial models that will follow.

To validate the model using 10% of the data set (31 observations selected randomly) we obtain:

```
vs <- sample(nrow(engtotals), 0.1*nrow(engtotals))
```

Then the model fitting command above is modified to include the additional argument validrows = vs. The summary command on the modified model fitted object produces RMSE=2.47, MAE=1.71, CRPS =1.22, Coverage=93.5% for the 95% prediction intervals. These results will be revisited when we introduce an additional spatial random effect parameter in Section 10.4.

10.3.2   GLM fitting with Poisson distribution

The number of Covid-19 deaths in the data set engtotals is now modeled by the Poisson distribution. The Poisson GLM uses the default canonical log link and the model formula argument in Bcartime now includes an offset term, which is log of the expected number of deaths. Note that the data frame engtotals already contains the logEdeaths column holding these offset values. Now the earlier model formula, f1, for the binomial GLM is changed to:

```
f2 <- covid ~ offset(logEdeaths) + jsa + log10(houseprice) +
log(popdensity) + sqrt(no2)
```

The model fitting command is:

```
M2 <- Bcartime(formula=f2, data=engtotals, family="poisson")
```

The MCMC run with `N=120,000`; `burn.in=20,000` and `thin=10` produces 10,000 samples from the posterior distribution and the summary command summary (`M2`) on the fitted `bmstdr` model produces the estimates given in Table 10.2.

	Median	2.5%	97.5%
Intercept	−2.848	−3.133	−2.569
jsa	0.204	0.161	0.248
log10(houseprice)	0.325	0.269	0.381
log(popdensity)	0.100	0.091	0.108
sqrt(no2)	0.086	0.070	0.102
Criteria	Penalty	Value	
DIC	4.9	5430.2	
WAIC	57.9	5485.2	

TABLE 10.2: Parameter estimates and model choice criteria values for the Poisson GLM.

Note that we do not report the MCMC convergence diagnostics as commented while discussing Table 10.1. The parameter estimates show that all four covariates are significant in the model for Covid-19 deaths. We note the two model choice criteria values followed by the penalty parameters in the table for comparison with spatial models that will follow. A much larger penalty for the WAIC highlights the differences between the DIC and the WAIC as alluded to in Section 4.16.2.

Model validation proceeds as previously in Section 10.3.1 with the additional argument `validrows = vs`. The summary command on the modified model fitted object produces RMSE=58.1, MAE=34.3, CRPS=5.9 and Coverage=54.8% for the 95% prediction intervals. These results will be revisited when we introduce an additional spatial random effect parameter in Section 10.4.

10.3.3   GLM fitting with normal distribution

For the sake of illustration in this section we model the NO_2 values present in the data set `engtotals` on the square-root scale. We continue to use the three earlier covariates: jsa, house price and population density. The model formula is now changed to:

```
f3 <- sqrt(no2) ~ jsa + log10(houseprice) + log(popdensity)
```

The model fitting command is:

```
M3 <- Bcartime(formula=f3, data=engtotals, family="gaussian")
```

The unknown variance parameter in the Gaussian model is by default given an inverse gamma prior distribution with shape parameter=1 and scale parameter=0.01 in the CARBayes package. The default values of the parameters can be changed by the prior.nu2 argument in the calling function Bcartime.

The MCMC run with N=120,000, burn.in=20,000 and thin=10 produces 10,000 samples from the posterior distribution and the summary command summary(M3) on the fitted bmstdr model produces the following estimates:

	Median	2.5%	97.5%
Intercept	−2.902	−4.839	−0.986
jsa	0.043	−0.283	0.365
log10(houseprice)	1.017	0.651	1.389
log(popdensity)	0.168	0.119	0.216
ν^2	0.260	0.223	0.305
Criteria	Penalty	Value	
DIC	5.0	473.4	
WAIC	6.0	474.6	

TABLE 10.3: Parameter estimates and the two model choice criteria for the Gaussian GLM fitted to the NO_2 data.

Here the higher levels of NO_2 are associated significantly with house price and population density. Model validation is performed by supplying the additional argument validrows = vs produces RMSE=0.58, MAE=0.44, CRPS=0.34 and 90.3% coverage for the 95% prediction intervals. These model fitting results will be compared with the results from spatial model fitting discussed below.

10.4 Spatial random effects for areal unit data

At the top level of modeling, the data random variables Y_i's are assumed to independently follow the exponential family of distributions discussed in Section 10.2 in this chapter. The objective now is to introduce a spatially structured random effects into the linear predictor (10.2). These random effects are designed to capture the omni-present dependence in the response variable observed in the neighboring areal units.

The most popular parametric method to specify those random effects is through the use of the CAR models discussed in Section 2.14. The model specification (2.6) is re-written for the random effects, $\psi = (\psi_1, \ldots, \psi_n)$

$$\psi_i | \tau^2, \psi_j, j \neq i \sim N \left(\frac{1}{w_{i+}} \sum_j w_{ij} \psi_j, \frac{\tau^2}{w_{i+}} \right) \tag{10.5}$$

where the $w_{i+} = \sum_{j=1}^n w_{ij}$ and w_{ij} is the ijth entry of the n-dimensional adjacency matrix W. The spatial variance of the random effects is τ^2 which will be assigned an inverse gamma prior distribution. Note that, here we specify the CAR model for the random effects denoted ψ_i – not for the data Y_i for $i = 1, \ldots, n$.

From the discussion of the CAR models in Section 2.14, note that this model implies the joint distribution:

$$f(\psi | \tau^2) \propto e^{-\frac{1}{2\tau^2} \psi'(D_w - W)\psi}, \tag{10.6}$$

for ψ similar to (2.7), where D_w is a diagonal matrix with w_{i+} as the ith diagonal entry. The two equivalent specifications (10.5) and (10.6) are often referred to as an *intrinsically* autoregressive model. Note that $(D_w - W)\mathbf{1} = \mathbf{0}$ where $\mathbf{1}$ is the vector of one's and $\mathbf{0}$ is the vector of all zeros. This linear constraint implies that the matrix $D_w - W$ is not of full rank. Hence (10.6) does not define a proper prior distribution. This is not a problem in Bayesian modeling as long as the joint posterior distribution is proper. However, several modifications have been suggested to overcome the singularity.

One notable modification to avoid the singularity has been suggested by Leroux et al. (2000). The suggestion is to replace the matrix $D_w - W$ in (10.6) by

$$Q(W, \rho) = \rho(D_w - W) + (1 - \rho)I \tag{10.7}$$

where I is the identity matrix and $0 \leq \rho \leq 1$ is a parameter dictating the strength of spatial correlation. Clearly, if $\rho = 1$ then the modification reduces to the original specification (10.6). In Bayesian modeling ρ is afforded a prior distribution and is then estimated along with the other parameters. With this substitution, the density (10.6) is re-written:

$$f(\psi | \rho, \tau^2) \propto e^{-\frac{1}{2\tau^2} \psi' Q(W, \rho)\psi}. \tag{10.8}$$

This, in turn, modifies the CAR distribution (10.5) to

$$\psi_i | \rho, \tau^2, \psi_j, j \neq i \sim N \left(\frac{\rho \sum_{j=1}^n w_{ij} \psi_j}{\rho \sum_{j=1}^n w_{ij} + 1 - \rho}, \frac{\tau^2}{\rho \sum_{j=1}^n w_{ij} + 1 - \rho} \right). \tag{10.9}$$

We denote this CAR prior distribution as $\text{NCAR}(\psi | \rho, \tau^2, W)$ for future reference.

Introduction of the spatial random effects, ψ, modifies the linear predictor in (10.2) to

$$\eta_i = o_i + \sum_{j=1}^q x_{ij} \beta_j + \psi_i, i = 1, \ldots, n. \tag{10.10}$$

This GLM with the NCAR($\boldsymbol{\psi}|\rho, \tau^2, W$) for $\boldsymbol{\psi}$ is denoted as the Leroux model in the `bmstdr` package.

The model (10.10) is further modified to include a 'nugget' effect like term ϵ_i, which is independently assumed to follow $N(0, \sigma^2)$ distribution but the assumption for the CAR model is also modified to have $\rho = 1$. Thus, the CAR prior distribution here is NCAR($\boldsymbol{\psi}|\rho = 1, \tau^2, W$). This model is the famous Besag, York and Mollié (BYM) CAR model (Besag et al., 1991). Thus, the final specification of the linear predictor is given by:

$$\eta_i = o_i + \sum_{j=1}^{q} x_{ij}\beta_j + \psi_i + \epsilon_i, i = 1, \ldots, n. \tag{10.11}$$

As in the case of the nugget effect model in Section 6.5 for point referenced data, here also there are identifiability concerns regarding the spatial random effect ψ_i and independent error term ϵ_i. However, the spatial structure in the ψ_i helps in their identification especially when that structure is strong. There are various remedies such as "hierarchical centering" and "centering in the fly" to recenter the random effects during MCMC implementations for such models. Primitive articles in this area include the papers such as Gelfand et al. (1995b), Gelfand et al. (1996), and Gelfand and Sahu (1999). The current literature in this area is huge.

10.5 Revisited example: Bayesian spatial generalized linear model

We now revisit the Bayesian GLM fitting in Section 10.3 but now including the spatial random effects in the linear predictor as discussed in the preceding Section 10.4.

As in Section 10.3 we continue to use the model fitting and validation function `Bcartime` in the `bmstdr` package but with three additional crucial arguments. The first of these is the `scol` argument which identifies the spatial indices of the areal units. This argument must be present as a column in the supplied data frame for model fitting and the argument itself can be supplied by the name or number of the column in the data frame. For example, the `engtotals` data frame has a column called 'spaceid' which numbers the spatial units and for model fitting we need to supply the argument as `scol="spaceid"`. Note the quotation marks since this is supplied as a column name. An alternative will be `scol=3` since the 3rd column of the `engtotals` data frame is the "spaceid" column. The function `Bcartime` will attempt to evaluate `data[, scol]` if the supplied data frame is `data`.

The second of the three required arguments is the proximity matrix `W` which should be a 0-1 matrix identifying the neighbors of each spatial unit as discussed in the above section. For the `engtotals` data example this will be `W=Weng` which will use the built-in `Weng` matrix already present in the `bmstdr` package.

The third argument is the model identifier, model, which should be either `"bym"` for the BYM model or `"leroux"` for the Leroux model, see Section 10.4. With these three additional arguments we are ready to fit and validate spatial models for the GLMs using the default `CARBayes` package. The two models use the `CARBayes` functions `S.CARbym` and `S.CARleroux` respectively. The other three models: localised, multilevel and diss-similarity using corresponding `CARBayes` functions `S.CARlocalised`, `S.CARmultilevel` and `S.CARdissimilarity` can be fitted by supplying the model argument as model `="localised"`, model=`"multilevel"`, and model=`"dissimilarity"` respectively. However, those advanced models are not considered here. Instead, we illustrate some of those models in the next Chapter 11.

10.5.1 Spatial GLM fitting with binomial distribution

For the binomial model we use the same formula `f1` as before in Section 10.3.1. The model fitting and summary commands for the BYM and Leroux models are:

```
M1.leroux <- Bcartime(formula=f1, data=engtotals, scol="spaceid",
    W=Weng, family= "binomial", trials=nweek, model="leroux", N=N,
    burn.in = burn.in, thin=thin)


M1.bym <- Bcartime(formula=f1, data=engtotals, scol="spaceid", W=
    Weng, family= "binomial", trials=nweek, model="bym", N=N,
    burn.in = burn.in, thin=thin)
```

	Leroux			BYM		
	Median	2.5%	97.5%	Median	2.5%	97.5%
Intercept	-2.410	-6.454	1.674	-2.550	-6.512	1.354
jsa	0.052	-0.335	0.441	0.034	-0.353	0.432
log10(houseprice)	0.187	-0.562	0.929	0.220	-0.488	0.943
log(popdensity)	0.074	0.007	0.143	0.075	0.007	0.141
sqrt(no2)	0.043	-0.096	0.185	0.039	-0.101	0.176
τ^2	0.269	0.182	0.393	0.259	0.175	0.381
ρ	0.977	0.909	0.998	$-$	$-$	$-$
σ^2	$-$	$-$	$-$	0.004	0.002	0.012
Criteria	Penalty	Value		Penalty	Value	
DIC	85.1	1352.1		86.2	1352.5	
WAIC	52.2	1329.6		52.9	1329.8	

TABLE 10.4: Parameter estimates from the Leroux and BYM models for the spatial binomial GLM.

The parameter estimates from the above two models are presented in Table 10.4. The parameter estimates are very similar since ρ has been estimated to be near one in the Leroux model and σ^2 has been estimated to be near zero in the BYM model. Here we do not report the MCMC convergence diagnostics as commented while discussing Table 10.1. Note that the overall DIC and WAIC criteria values are much smaller than the same for the independent error model shown in Table 10.1, although the penalties for the spatial models are much higher as expected.

To perform model validation, the model fitting command above for fitting the Leroux model is modified to include the additional argument `validrows` = `vs`. The summary command on the modified model fitted object produces RMSE=2.06, MAE=1.52, CRPS=1.48 and Coverage=96.8% for the 95% prediction intervals. These results, compared to the same for the non-spatial GLM in Section 10.3.1 are much better as can be hoped for. Thus, indeed both the model choice criteria and all the model validation statistics emphatically support in favor of spatial modeling.

10.5.2   Spatial GLM fitting with Poisson distribution

We now include spatial random effects in the independent error Poisson GLM illustrations in Section 10.3.2. The model fitting commands from Section 10.3.2 are modified in the same way by including the three additional arguments `scol`, `W` and model as in the case of spatial binomial GLM in Section 10.5.1. The modified commands are:

```
M2.leroux <- Bcartime(formula=f2, data=engtotals, scol="spaceid",
    W=Weng, family="poisson", model="leroux", N=N, burn.in = burn.
    in, thin=thin)

M2.bym <- Bcartime(formula=f2, data=engtotals, scol="spaceid", W=
    Weng, family="poisson", model="bym", N=N, burn.in = burn.in,
    thin=thin)
```

Compared to the parameter estimates reported in Table 10.2 we can see that the jsa and NO_2 are no longer significant in the spatial model in Table 10.5, which is a usual phenomenon in spatial modeling as the spatial random effects attempt to explain more variability than the fixed effects covariates. However, the other two covariates: house price and population density remain significant which shows considerable strength in those effects. The estimate of ρ in the Leroux model is high showing the presence of spatial correlation. Lastly, note that the model choice criteria values are almost half of the values in the non-spatial GLM in Table 10.2 pointing the need to fit spatial models. As previously for the spatial binomial GLMs the penalties are much higher for the spatial models. The two criteria DIC and WAIC do not seem to find much difference at all between the Leroux and BYM models.

	Leroux			BYM		
	Median	2.5%	97.5%	Median	2.5%	97.5%
Intercept	−2.704	−4.625	−1.025	−2.710	−4.676	−0.731
jsa	0.094	−0.066	0.263	0.093	−0.068	0.254
log10(houseprice)	0.386	0.092	0.726	0.387	0.034	0.737
log(popdensity)	0.068	0.041	0.095	0.068	0.042	0.098
sqrt(no2)	0.010	−0.041	0.060	0.007	−0.047	0.060
τ^2	0.145	0.119	0.178	0.135	0.108	0.169
ρ	0.971	0.899	0.997	−	−	−
σ^2	−	−	−	0.003	0.001	0.007
Criteria	Penalty	Value	−	Penalty	Value	
DIC	245.0	2640.4		246.6	2639.4	
WAIC	147.8	2596.8		146.9	2593.0	

TABLE 10.5: Parameter estimates and model choice criteria values for the spatial Poisson GLM.

Model validation proceeds with the additional argument `validrows = vs`. The summary command on the modified model fitted object for the Leroux model produces RMSE=44.9, MAE=27.8, CRPS=19.8, and Coverage=96.8% for the 95% prediction intervals. These validation results for the same validation data are overwhelmingly better than the same results for the independent Poisson GLMs of Section 10.3.2.

10.5.3 Spatial GLM fitting with normal distribution

We now illustrate spatial random effects fitting of the model for NO_2 in Section 10.3.3. The CARBayes package does not allow the BYM model for Gaussian observations. Hence we fit the Leroux model using the modified command:

```
M3.leroux <- Bcartime(formula=f3, data=engtotals, scol="spaceid",
    W=Weng, family="gaussian", model="leroux")
```

The parameter estimates and the model choice statistics are given in Table 10.6. The DIC and WAIC criteria are both lower than the ones for the independent error regression model in Table 10.3. The parameter estimates for the spatial models are comparable. Note that the estimate of the error variance ν^2 is smaller due to the presence of the spatial variance τ^2. Model validation can be performed by supplying the additional `validrows` argument. However, we do not pursue that in this illustration.

	Median	2.5%	97.5%
Intercept	-1.498	-5.370	2.279
jsa	0.063	-0.221	0.373
log10(houseprice)	0.799	0.139	1.543
log(popdensity)	0.119	0.073	0.163
ν^2	0.114	0.079	0.147
τ^2	0.318	0.209	0.445
ρ	0.882	0.622	0.980

Criteria	Penalty	Value
DIC	138.1	329.6
WAIC	109.1	333.3

TABLE 10.6: Parameter estimates from the Leroux model for Gaussian observations.

10.6 Spatio-temporal random effects for areal unit data

The random effects models in Section 10.4 are now extended for areal spatio-temporal data. Here the data random variables are denoted by Y_{it} for $i = 1, \ldots, n$ and $t = 1, \ldots, T$. Again the top-level distribution is assumed to be a member of the Exponential family (10.1) and we now have the task of defining spatio-temporal random effects ψ_{it} to be included in the linear predictor η_{it}. We consider several models introducing spatio-temporal interactions below.

10.6.1 Linear model of trend

Often, it is of interest to see if there is any simple linear trend in the random effects ψ_{it} for each region i. For such investigations the appropriate model is the linear trend model given by

$$\psi_{it} = \beta_1 + \phi_i + (\beta_2 + \delta_i)\frac{t - \bar{t}}{T} \tag{10.12}$$

where $\bar{t} = \frac{T+1}{2}$. The other parameters are described below. Here β_1 and β_2 are overall intercept and slope (trend) parameters which will be given a flat prior distribution in a Bayesian model. The parameters ϕ_i and δ_i are incremental intercept and slope parameters for the ith region for $i = 1, \ldots, n$. These two sets of parameters are assigned the modified CAR prior distribution (10.9) with different values of ρ and τ^2. The parameters $\boldsymbol{\phi} = (\phi_1, \ldots, \phi_n)$ are assigned the

$$\text{NCAR}(\boldsymbol{\phi}|\rho_{\text{int}}, \tau^2_{\text{int}}, W)$$

as defined in (10.9) above. Similarly, the parameters $\boldsymbol{\delta} = (\delta_1, \ldots, \delta_n)$ are assigned the

$$\text{NCAR}(\boldsymbol{\delta}|\rho_{\text{slo}}, \tau^2_{\text{slo}}, W)$$

distribution as defined in (10.9) above. Here $\rho_{\text{int}}, \rho_{\text{slo}}, \tau^2_{\text{int}}, \tau^2_{\text{slo}}$ are auto-regression and variance parameters for the intercept (ϕ_i) and slope (δ_i) processes. The parameters ρ_{int} and ρ_{slo} are given independent uniform prior distributions in the unit interval $(0, 1)$ and the variance parameters τ^2_{int} and τ^2_{slo} are given the inverse gamma prior distributions. This model has been implemented as the ST.CARlinear model in the CarBayesST (Lee et al., 2018) package. The argument model="linear" when package ="CarBayesST" chooses this model in the Bcartime function in the bmstdr package.

10.6.2 Anova model

An analysis of variance type model with or without for space time interaction has been suggested by Knorr-Held (2000). The model is given by:

$$\psi_{it} = \phi_i + \delta_t + \gamma_{it}, i = 1, \ldots, n, \ t = 1, \ldots, T. \tag{10.13}$$

All three sets of parameters denoted by ϕ_i, δ_t and γ_{it} in the Anova model (10.13) are random effects with the distributions given by:

$$\phi|\rho_S, \tau^2_S, W \ \sim \ \text{NCAR}\left(\phi|\rho_S, \tau^2_S, W\right),$$
$$\delta|\rho_T, \tau^2_T, D \ \sim \ \text{NCAR}\left(\delta|\rho_T, \tau^2_T, D\right),$$
$$\gamma_{it} \ \sim \ N(0, \tau^2_I), i = 1, \ldots, n, \ t = 1, \ldots, T,$$

where D denotes the $T \times T$ temporal adjacency matrix with $d_{ij} = 1$ if $|i-j| = 1$ and 0 otherwise. The interaction effect, γ_{it}, is assumed to be independent for all values of i and t. As before, the parameters ρ_S and ρ_T are given independent uniform prior distributions in the unit interval $(0, 1)$ and the variance parameters τ^2_S, τ^2_T and τ^2_I are given the inverse gamma prior distributions. This model has been implemented as the ST.CARanova model in the CarBayesST (Lee et al., 2018) package. The argument model="anova" when package =" CarBayesST" chooses this model in the Bcartime function in the bmstdr package. Here the additional argument interaction can be set to FALSE to suppress the interaction term if desired. The default for this is TRUE.

10.6.3 Separable model

An alternative to the Anova model is the separable model given by:

$$\psi_{it} = \phi_{it} + \delta_t, i = 1, \ldots, n, \ t = 1, \ldots, T, \tag{10.14}$$

where independent NCAR models, see (10.9), are specified for $\phi_t = (\phi_{1t}, \ldots, \phi_{nt})$ for each $t = 1, \ldots, T$ and also another one for $\delta = (\delta_1, \ldots, \delta_T)$. Here are the details:

$$\phi_t|\rho_S, \tau^2_t, W \ \sim \ \text{NCAR}\left(\phi|\rho_S, \tau^2_t, W\right), \ t = 1, \ldots, T$$
$$\delta|\rho_T, \tau^2, D \ \sim \ \text{NCAR}\left(\delta|\rho_T, \tau^2, D\right),$$

where D has been defined above. As before, the parameters ρ_S and ρ_T are given independent uniform prior distributions in the unit interval $(0, 1)$ and the variance parameters $\tau_t^2, t = 1, \ldots, T$ and τ^2 are given the inverse gamma prior distributions. This model has been implemented as the `ST.CARsepspatial` model in the `CarBayesST` (Lee et al., 2018) package. The argument `model="sepspatial"` when `package ="CARBayesST"` chooses this model in the `Bcartime` function in the `bmstdr` package.

10.6.4   Temporal autoregressive model

A temporal autoregressive model of order one is assumed as a special case of the separable model (10.14) where $\delta_t = 0$ for all t and

$$\phi_t|\phi_{t-1}, W \quad \sim \quad N\left(\rho_{T,1}\phi_{t-1}, \tau^2 Q(W, \rho_S)^{-1}\right), \quad t = 2, \ldots, T \quad (10.15)$$
$$\phi_1|W \quad \sim \quad N\left(\mathbf{0}, \tau^2 Q(W, \rho_S)^{-1}\right), \quad (10.16)$$

where $Q(W, \rho_S)$ is the spatially dependent precision matrix defined in (10.7). Here the temporal auto-correlation is induced by the mean $\rho_T\phi_{t-1}$. Prior distributions are assumed as before. This model has been implemented as the `ST.CARar` model in the `CarBayesST` (Lee et al., 2018) package. The argument `model="ar"` when `package ="CARBayesST"` chooses this model in the `Bcartime` function in the `bmstdr` package. A second order auto-regressive model is also used where (10.15) is allowed to have the additional term $\rho_{T,2}\phi_{t-2}$.

10.7   Example: Bayesian spatio-temporal generalized linear model

The data set used in this example is the `engdeaths` data frame available in the `bmstdr` package. Some exploratory data analysis plots have been presented in Section 3.4. Documentation for this data set can be read by issuing the command `?engdeaths`.

In this section we will modify the `Bcartime` commands presented in Section 10.5 to fit all the spatio-temporal models discussed in Section 10.6. We will illustrate model fitting, choice and validation using the binomial, Poisson and normal distribution based models as in Section 10.5. The user does not need to write any direct code for fitting the models using the `CARBayesST` package. The `Bcartime` function does this automatically and returns the fitted model object in its entirety and in addition, performs model validation for the named rows of the supplied data frame as passed on by the `validrows` argument.

The previously documented arguments of `Bcartime` for spatial model fitting remain the same for the corresponding spatio-temporal models. For example, the arguments **formula**, **family**, `trials`, `scol` and `W` are unchanged in spatial-temporal model fitting. The `data` argument is changed to the spatio-temporal data set `data=engdeaths`. We keep the MCMC control parameters `N`, `burn.in` and `thin` to be same as before.

The most important additional argument is `tcol`, similar to `scol`, which identifies the temporal indices. Like the `scol` argument this may be specified as a column name or number in the supplied data frame. The `package` argument must be specified as `package="CARBayesST"` to change the default `CARBayes` package. The model argument should be changed to one of four models, `"linear"`, `"anova"`, `"sepspatial"` and `"ar"`, as discussed in Section 10.6. Other possibilities for this argument are `"localised"`, `"multilevel"` and `"dissimilarity"`, but those are not illustrated here. For the sake of brevity it is undesirable to report parameter estimates of all the models. Instead, below we report the two model choice criteria, DIC and WAIC, along with the values of the penalty parameters. After model choice we only report the parameter estimates of the chosen model in each case of binomial, Poisson and normal model fitting.

10.7.1 Spatio-temporal GLM fitting with binomial distribution

For the binomial model the response variable is `highdeathsmr`, which is a binary variable taking the value 1 if the SMR for death is larger than 1 in that week and in that local authority. Consequently, the number of trials is set at the constant value 1 by setting `nweek <-rep(1, nrow(engdeaths))`. The right hand side of the formula is same as before in Sections 10.3.1 and 10.5.3. We now execute the following commands.

```
f1 <- highdeathsmr ~ jsa + log10(houseprice) + log(popdensity)
scol <- "spaceid"; tcol <- "Weeknumber"
N <- 120000; burn.in <- 20000; thin <- 10
vs <- sample(nrow(engdeaths), 0.1*nrow(engdeaths))
```

The basic model fitting command for fitting the linear trend model is:

```
M1st <- Bcartime(formula=f1, data=engdeaths, scol=scol, tcol=tcol,
       trials=nweek, W=Weng, model="linear", family="binomial",
       package="CARBayesST", N=N, burn.in=burn.in, thin=thin)
```

To fit the other models we simply change the `model` argument to one of `"anova"`, `"sepspatial"` and `"ar"`. For the choice `"anova"` an additional argument `interaction=F` may be supplied to suppress the interaction term.

Clearly, the AR model of Section 10.6.4 is chosen by both the DIC and WAIC. The parameter estimates of the chosen AR model are provided in Table 10.8.

The population density still remains significant but jsa and house price are no longer significant. The spatial correlation parameter ρ_S is estimated to be much larger than the temporal auto-correlation parameter ρ_T. Model

	p.dic	DIC	p.waic	WAIC
Linear	56.5	8035.8	56.9	8036.7
Anova (no interaction)	59.1	8015.8	59.3	8016.6
Separable	767.5	7804.3	598.6	7704.7
AR (1)	1800.7	7474.3	1343.3	7337.0

TABLE 10.7: Model choice criteria values for four spatio-temporal binomial models fitted to the `engdeaths` data set. Here p.dic and p.waic denote the penalties for the DIC and WAIC, respectively.

	Median	2.5%	97.5%
Intercept	-3.0803	-6.6223	0.3979
jsa	0.3673	-0.1281	0.8888
log10(houseprice)	0.1742	-0.4804	0.8126
log(popdensity)	0.1506	0.0707	0.2416
τ^2	6.9696	4.0990	12.8172
ρ_S	0.6561	0.4809	0.8013
ρ_T	0.0016	0.0001	0.0085

TABLE 10.8: Parameter estimates from the spatio-temporal AR binomial GLM.

validation by supplying the `validrows` = `vs` is straightforward and it produces RMSE=0.6, MAE=0.4 and 100% coverage for the binary data.

10.7.2 Spatio-temporal GLM fitting with Poisson distribution

For fitting the Poisson distribution based model we take the response variable as the column `covid`, which records the number of Covid-19 deaths, of the `engdeaths` data set. The column `logEdeaths` is used as an offset in the model with the default log link function.

The formula argument for the regression part of the linear predictor is chosen to be the same as the one used by Sahu and Böhning (2021) for a similar data set. The formula contains, in addition to the thre socio-economic variables, the log of the SMR for the number cases in the current week and three previous weeks denoted by `n0`, `n1`, `n2` and `n3`. The formula is given below:

```
f2 <- covid ~ offset(logEdeaths) + jsa + log10(houseprice) + log(
    popdensity) + n0 + n1 + n2 + n3
```

We now fit the Poisson model by keeping the other arguments same as before in Section 10.7.1. The command for fitting the temporal auto-regressive model is:

```
M2st <- Bcartime(formula=f2, data=engdeaths, scol=scol, tcol=tcol,
     W=Weng, model="ar", family="poisson", package="CARBayesST", N
     =N, burn.in=burn.in, thin=thin)
```

The model argument can be changed to fit the other models. The resulting
model fits are used to record the model choice statistics reported in Table 10.9.
According to both DIC and WAIC the separable model is the best model. The
parameter estimates of this model are presented in Table 10.10. All the covariate effects are significant. The significance of jsa shows higher death rate in
the more deprived areas. Also, higher house price areas in more densely populated cities experience significantly higher death rates even after accounting
for spatio-temporal correlations in the data as estimated by ρ_S and ρ_T. The
temporal auto-correlation ρ_T is estimated to be larger than the spatial correlation ρ_S. The weekly variances τ_t^2 for $t = 1, \ldots, 20$ are quite similar in
value.

	p.dic	DIC	p.waic	WAIC
Linear	434.1	28965.2	957.9	29670.7
Anova (with interaction)	2148.9	26182.9	1616.5	26092.7
Anova (no interaction)	237.4	28872.0	566.7	29270.6
Separable	2027.9	26114.1	1546.9	26041.9
AR (1)	1958.3	26153.1	1536.4	26137.3
AR (2)	1965.1	26150.4	1538.8	26131.2

TABLE 10.9: Model choice criteria values for six spatio-temporal Poisson models fitted to the engdeaths data set. Here p.dic and p.waic denote the penalties
for the DIC and WAIC, respectively.

To investigate model validation we consider the Anova (with interaction)
and the AR (2) models only since fitting of the separable model does not
currently allow missing values, see the documentation ?ST.CARsepspatial for
this model in the CARBayesST package. We now re-fit the two models by setting
aside 10% randomly selected data rows. This is done by passing the additional
argument validrows = vs where

```
vs <- sample(nrow(engdeaths), 0.1 * nrow(engdeaths))
```

Table 10.11 provides the model validation statistics, and it shows that the AR
model is slightly more accurate than the Anova model.

10.7.3   Examining the model fit

The fitted model object can be graphically scrutinized in many different
ways as the investigator may wish. For example, see the illustrations of the
CarBayesST package by Lee et al. (2018). To illustrate we examine the fitted
AR (2) model of Section 10.7.2. We perform two analyses: one to visualize a

	Median	2.5%	97.5%
Intercept	-3.469	-4.423	-2.540
jsa	0.136	0.026	0.244
log10(houseprice)	0.567	0.395	0.739
log(popdensity)	0.054	0.033	0.076
n0	0.491	0.469	0.512
n1	0.209	0.185	0.231
n2	0.084	0.059	0.108
n3	0.059	0.043	0.075
τ_1^2	0.296	0.208	0.417
τ_2^2	0.247	0.173	0.356
τ_3^2	0.221	0.152	0.320
τ_4^2	0.388	0.279	0.539
τ_5^2	0.286	0.179	0.448
τ_6^2	0.316	0.198	0.489
τ_7^2	0.267	0.171	0.420
τ_8^2	0.202	0.120	0.327
τ_9^2	0.192	0.120	0.297
τ_{10}^2	0.217	0.131	0.346
τ_{11}^2	0.329	0.202	0.514
τ_{12}^2	0.216	0.134	0.344
τ_{13}^2	0.147	0.092	0.234
τ_{14}^2	0.160	0.097	0.262
τ_{15}^2	0.214	0.135	0.328
τ_{16}^2	0.277	0.192	0.398
τ_{17}^2	0.203	0.144	0.286
τ_{18}^2	0.344	0.253	0.468
τ_{19}^2	0.275	0.188	0.398
τ_{20}^2	0.305	0.209	0.448
τ_T^2	0.102	0.061	0.192
ρ_S	0.377	0.282	0.484
ρ_T	0.591	0.126	0.922

TABLE 10.10: Parameter estimates and model choice criteria values for the spatio-temporal Poisson GLM (separable).

spatial map of the residuals and the other to compare a time series plot of observed and fitted values. The spatial plot enables us to look at the residuals to find any detectable spatial pattern and the temporal plot is designed to assess a goodness-of-fit of the means over time. The code lines for these calculations are not included in the text here but are provided online on github[1].

To obtain the spatial residual plot we obtain the so called response residuals (observed – fitted) for each observed data point y_{it} at each MCMC iteration

---

[1]https://github.com/sujit-sahu/bookbmstdr.git

	RMSE	MAE	CRPS	CVG
Anova	7.08	2.80	1.81	97.92
AR	5.42	2.60	1.80	96.96

TABLE 10.11: Model validation statistics for the two models.

j, i.e.

$$r_{it}^{(j)} = y_{it} - \hat{y}_{it}^{(j)}, \ j = 1, \ldots, J$$

where J is the total number of saved MCMC iterations for each $i = 1, \ldots, n$ and $t = 1, \ldots, T$. To obtain the spatial residual we first obtain

$$r_{i.}^{(j)} = \frac{1}{T} \sum_{t=1}^{T} r_{it}^{(j)},$$

for i and j. These MCMC replicates are then used to estimate the residual value and its standard deviation for each spatial unit i. We then simply plot these two surfaces as in Figure 10.2 for the spatio-temporal Poisson Anova model fitted to the Covid-19 deaths in the engdeaths data set. The residual plot does not show any overwhelming spatial pattern that requires further investigation.

For the second plot of temporally varying values, aggregated over all the spatial units, we obtain

$$\hat{y}_{.t}^{(j)} = 100000 \frac{1}{n} \sum_{i=1}^{n} \frac{\hat{y}_{it}^{(j)}}{p_{it}}, \ j = 1, \ldots, J, \ t = 1, \ldots, T$$

where p_{it} is the population size of the spatial unit i. Division by p_{it} and then multiplication by 100,000 in the above enables us to get the fitted values for the number of deaths per 100,000 residents. The MCMC replicates, $j = 1, \ldots, J$ are then used to estimate the fitted death rate at time t. We apply the same adjustment to the observed Covid-19 death numbers. The resulting plot is provided in Figure 10.3. The plot shows slight under estimation at the peak of the curve at week numbers 15 and 16, and slight over estimation at the beginning and the end of the time domain when death rates are low. Other than these extremes we see a very high degree of agreement between the plots of the observed and fitted weekly summaries. Moreover, the upper and lower limits for the fitted values are very close to the fitted values which indicates a very high degree of accuracy due to the large sample size of the spatio-temporal data set.

10.7.4 Spatio-temporal GLM fitting with normal distribution

We now illustrate spatio-temporal random effects fitting of the model f3 for NO_2 in Section 10.5.3. We fit the "gaussian" family model but keep the other

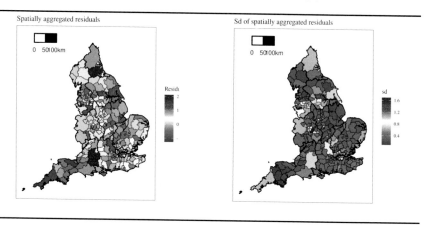

FIGURE 10.2: Map of spatially aggregated residuals and their standard deviations.

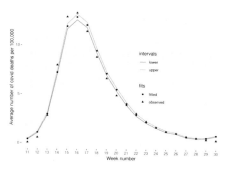

FIGURE 10.3: Fitted and observed weekly death rates per 100,000 people. The 95% limits are also superimposed.

arguments same as before in the previous two sections for fitting binomial and Poisson models. The command for fitting the temporal auto-regressive model is:

```
M3st <- Bcartime(formula=f3, data=engdeaths, scol=scol, tcol=tcol,
    W=Weng, model="ar", family="gaussian", package="CARBayesST",
    N=N, burn.in=burn.in, thin=thin)
```

The model choice statistics are presented in Table 10.12. The AR (2) model is the best according to both DIC and WAIC, although it receives much higher penalty. The parameter estimates of the chosen model are presented in Table 10.13. Significance of house price and population density show higher level of NO_2 concentration in the densely populated cities. Both the AR parameters show negative lagged correlation. Model validation, not done here, can be performed by supplying the `validrows` argument.

	p.dic	DIC	p.waic	WAIC
Linear	151.6	12899.9	152.2	12904.9
Anova	118.3	12707.2	116.7	12708.0
AR (1)	1789.8	11530.0	1449.0	11491.9
AR (2)	1766.7	11504.8	1433.6	11468.8

TABLE 10.12: Model choice criteria values for five spatio-temporal normal models fitted to the `engdeaths` data set. Here p.dic and p.waic denote the penalties for the DIC and WAIC, respectively.

	Median	2.5%	97.5%
Intercept	−1.041	−1.909	−0.151
jsa	0.007	−0.103	0.119
log10(houseprice)	0.733	0.571	0.892
log(popdensity)	0.116	0.097	0.135
τ^2	0.351	0.292	0.416
ν^2	0.278	0.257	0.298
ρ_S	0.944	0.900	0.972
ρ_1	−0.174	−0.271	−0.088
ρ_2	−0.065	−0.158	0.019

TABLE 10.13: Parameter estimates from the spatio-temporal AR (2) model for NO_2.

10.8 Using INLA for model fitting and validation

The INLA package is very popular for fitting GLMs for areal spatial and spatio-temporal data. Hence we devote this section for illustrating model fitting with the package. The commands for fitting and interpreting INLA output are similar to those described in Section 6.6.3 with the exception that we do not need to setup the mesh for spde with Matèrn covariance function. Model fitting for areal unit data using INLA is much more straightforward. Instead of the supplying mesh we need to send the proximity matrix W or equivalently an INLA graph object for obtaining the proximity matrix. The bmstdr function Bcartime has been setup to work with either of the two forms of specifications: (i) the W matrix or (ii) a file name containing the adjacency graph.

The general model structure implemented for INLA based fitting is given by:

$$
\begin{aligned}
\psi_{it} &= \psi_i + \delta_t, \quad i = 1, \ldots, n, \ t = 1, \ldots, T, \\
\psi &\sim \mathrm{NCAR}\left(\phi | W, 1, \tau^2\right), \\
\delta_t &\sim N(\rho_T \delta_{t-1}, \tau^2), \\
\delta_0 &\sim N(0, \tau^2).
\end{aligned}
$$

The is denoted by the BYM-AR1 model. It is also possible to fit an iid random effect model for δ_t or take $\delta_t = 0$.

The `Bcartime` function can fit and validate using the INLA package when the option is set to `package="inla"` instead of the default CARBayes package. For fitting areal unit data only with no temporal variation the model argument can be any valid model that has been implemented in the INLA package. The `Bcartime` function has been tested with the model specifications `"bym"` and `"iid"` which correspond to fitting the BYM model or the iid model for the spatial random effects respectively.

For fitting spatio-temporal data the model argument should be specified as a two element vector where the first element specifies the model for the spatial random effect and the second element specifies the model to be fitted for the temporal component. The second element, model for the temporal component, can be either `"ar1"`, `"iid"` or `"none"` which corresponds to fitting the first order auto-regressive model, the iid model or no model at all for the temporal random effects respectively. That is, the option `"none"` will not include any random effects for the temporal component. If indeed it is desired to have no random effects for the temporal component then the model argument for spatio-temporal data may simply be specified as a singleton specifying the spatial model.

The argument link may be supplied to the `Bcartime` function for INLA based model fitting. Offsets in GLM fitting with the INLA package option can be supplied by the additional `offsetcol` argument which should be a name or number of a column in the supplied data frame argument where the column contains the offsets in the transformed scale as dictated by the link function which may have been supplied by the link argument. For example, for modeling the number of Covid-19 deaths in the `engdeaths` data set using the Poisson GLM with log-link we specify `offsetcol="logEdeaths"` and do not include the offset in the formula argument for specifying the fixed effects regression part of the model.

As before, the `Bcartime` function can validate if the additional argument `validrows` containing the row numbers to be validated has been passed on as well. The output of the `Bcartime` contains the parameter estimates, model choice statistics and also the validation statistics if validation has been requested for. In addition, the component `fit` contains the fully fitted INLA model object from which all the marginal distributions can be extracted. The documentation for the INLA package, see e.g. https://www.r-inla.org/ should be consulted to know more.

We illustrate INLA based model fitting for the built-in spatio-temporal data set `engdeaths` in the `bmstdr` package. The code lines for fitting the Poisson distribution based model are given below.

```
f2inla <- covid ~ jsa + log10(houseprice) + log(popdensity) + n0
    + n1 + n2 + n3
scol <- "spaceid"; tcol <- "Weeknumber";
model <- c("bym", "ar1")
```

```
M2stinla <- Bcartime(data=engdeaths, formula=f2inla, W=Weng,
    scol =scol, tcol=tcol, offsetcol="logEdeaths", model=model,
    link="log",family="poisson", package="inla")
```

	mean	sd	2.5%	97.5%
Intercept	−5.531	0.839	−7.160	−3.861
jsa	−0.122	0.071	−0.263	0.020
log10(houseprice)	0.584	0.151	0.299	0.881
log(popdensity)	0.096	0.013	0.070	0.121
n0	0.157	0.009	0.138	0.175
n1	0.520	0.039	0.443	0.594
n2	0.253	0.031	0.193	0.313
n3	0.085	0.011	0.064	0.109
τ^2	93.856	22.449	56.656	144.549
σ^2	0.001	0.003	0.000	0.007
ρ	0.613	0.135	0.338	0.841
$\tau2_T$	0.485	0.215	0.213	1.035
Model Choice	p.dic	DIC	p.waic	WAIC
Statistics	250.0	25327.8	332.3	25437.1

TABLE 10.14: Parameter estimates and model choice criteria values for the spatio-temporal Poisson GLM. Here p.dic and p.waic denote the penalties for the DIC and WAIC, respectively.

The command summary(M2stinla) produces the parameter estimates and model choice statistics reported in Table 10.14. Values of the model choice statistics are comparable to those reported in Table 10.9 for the models fitted using the CARBayesST package with the notable observation that the values of the penalty parameters are much lower for INLA based model fitting. This is expected since the CARBayesST models estimate more random effects than the INLA based models. Perhaps to introduce more required variability the INLA model estimates τ_S^2 to be an order of magnitude higher than the estimate of the spatial variance reported in Table 10.10. However, we note that none of the CARBayesST models is exactly equivalent to the fitted INLA model. Hence a direct comparison between these estimates is not meaningful. The models can be compared using model validation statistics such as those reported in Table 10.11. Here the INLA model produces more accurate predictions according to the RMSE, MAE and CRPS but the coverage value is very low compared to the near perfect value for the AR (2) model, see Table 10.11. Figure 10.4 provides a validation plot which provides further evidence. We do not consider INLA based model fitting any further. However, code lines for INLA based model fitting of binomial and normal distributions are provided in the bmstdr package vignette, see Sahu (2021).

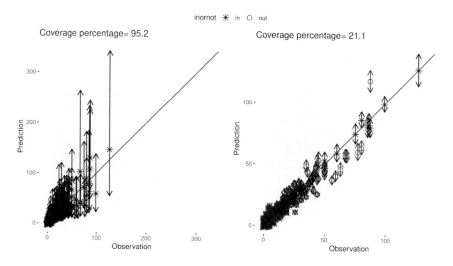

FIGURE 10.4: Predictions with 95% limits against observations for the AR (2) model on the left panel and INLA based model on the right panel.

10.9   Conclusion

This chapter discusses model fitting of areal and areal-temporal data. The generalized linear models, which include the binomial, Poisson and normal distributions, are given an introduction at an elementary level to enable a beginner reader access the modeling topics. These generalized linear models serve as the basic models for observed data. These models are then afforded to have spatial random effects through the use of CAR models for areal unit data and spatio-temporal random effects for spatio-temporal data. A running example on Covid-19 death rates and NO_2 concentration model for England has been used to illustrate the modeling. The example has been intentionally kept same so that the reader is able to appreciate the methodology rather than getting overwhelmed by the novelty of the examples.

All models are fitted using the single command `Bcartime` in the `bmstdr` package. Model fitting is performed using one of the three packages: `CARBayes`, `CARBayesST` and `INLA`. The reader is able to choose which package they want their models to be fitted by. The next chapter illustrates various aspects of model fitting and model based inference methods for a variety of data sets.

10.10 Exercises

1. Reproduce all the results reported in this chapter.

2. Study sensitivity of the prior distributions in a modeling example of your choice from this chapter. How would the reported results change for your example.

3. This chapter has not performed variable selection at all. For the engdeaths data set perform variable selection taking Sahu and Böhning (2021) as a guide.

4. Model the Ohio lung cancer data set as illustrated in the book by Banerjee et al. (2015).

5. Explore the examples provided by different model fitting functions in the CARBayes and CARBayesST packages. Fit the models in those examples using the Bcartime model fitting function. Then perform cross-validation by supplying values for the validrows argument.

6. Fit different spatio-temporal models by using the INLA package. For example, the model argument can be specified by as c("bym", "none") or c("bym", "iid") or c("bym", "ar1"). How do these models differ for modeling the number of Covid-19 deaths in the engdeaths data set.

7. The engdeaths data set also contains the weekly number Covid-19 cases and expected number of cases. Fit suitable models for the number of cases.

11

Further examples of areal data modeling

11.1 Introduction

This chapter showcases and highlights the practical use of areal unit data modeling extending the methodologies where necessary. The spirit of the presentation here is same as that in Chapter 8 for point referenced spatial data. The data sets to be used for modeling illustrations in this chapter have been introduced in Section 1.4 already. We will revisit all the example data sets except for the first one on Covid-19 mortality which has been analyzed in the previous Chapter 10 already.

All the model fitting will be done using the `Bcartime` model fitting function in `bmstdr`. The fitted model object is then explored for making several types of inference required in different practical settings. The full version of the code lines for model fitting and further investigations are not included in the text of this chapter. Instead, those code lines are provided online on github[1]. Such code will allow the reader to reproduce all the results and graphs illustrated here.

11.2 Assessing childhood vaccination coverage in Kenya

The childhood vaccination coverage data set, introduced in Section 1.4.2, contains the number of vaccinated (y_i) and the number of sampled children (m_i) in the age group of 12-23 months for all the 47 $(= n)$ counties in Kenya. The data set also contains three potential county level covariates:

(i) x_1: proportion of women aged 15-49 years who had completed primary education;

(ii) x_2: travel time to the nearest health facility, which are county averages of 5×5 kilometer grid level values;

(iii) x_3: nightlight intensity which are also county averages of the corresponding 5×5 kilometer grid level values.

---

[1]https://github.com/sujit-sahu/bookbmstdr.git

These three covariates and the observed proportion of vaccination are plotted in Figure 11.1. The figure shows high levels of similarity between the observed proportion map and the first two covariates. The third covariate (nightlight) does not seem to correlate highly with the vaccination proportions. For our modeling purposes we standardize these covariates by taking away the mean and then dividing by the standard deviation for their inclusion in our modeling.

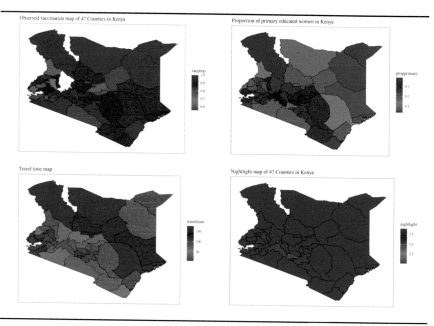

FIGURE 11.1: Observed vaccination proportions in 2014 and three relevant covariates for 47 counties in Kenya.

It is natural to model the vaccination data with the binomial distribution based GLM discussed in Section 10.2. We assume that:

$$Y_i | m_i, p_i \sim \text{Binomial}(m_i, p_i), \quad i = 1, \ldots, n,$$

where p_i is the true proportion of vaccinated children in county i. We model the p_i using the logistic regression model, see Section 10.2.2, having three covariates, x_1, x_2 and x_3 and spatial random effects, denoted by ψ_i, discussed in Section 10.4. The model is written as:

$$\log\left(\frac{p_i}{1 - p_i}\right) = \mathbf{x}_i'\boldsymbol{\beta} + \psi_i, \quad i = 1, \ldots, n.$$

The random effects $\boldsymbol{\psi}$ are assumed to follow the CAR prior distribution $\text{NCAR}(\boldsymbol{\psi} | \rho, \tau^2, W)$ as defined in (10.9). Here W is the $n \times n$ adjacency matrix, which is formed using the map of the counties as described below.

To obtain the W matrix we use the two spdep library functions poly2nb and nb2mat on the map polygon Kmap read before in Section 1.4.2. Here are the code lines:

```
library(spdep)
nbhood <- poly2nb(Kmap)
W <- nb2mat(nbhood, style = "B", zero.policy = T)
```

The help files on these two functions explain the all possible options in the above commands. Here the style option "B" opts for a basic binary adjacency (also called the proximity) matrix, see Section 2.11.

The data for this example is assumed to be read as kdat where the three covariates propprimary, traveltime and nightlight are standardized by taking away the means and dividing by their standard deviations. The bmstdr package function Bcartime is used to fit the logistic regression models with or without random effects. Here are the code lines used to fit these two models:

```
f1 <- yvac ~propprimary + traveltime + nightlight
N <- 220000; burn.in <- 20000; thin <- 100
M1 <- Bcartime(formula=f1, data=kdat, family="binomial", trials=kdat
    $n, N=N, burn.in = burn.in, thin=thin)
# Leroux model
M2 <- Bcartime(formula=f1, data=kdat, scol="id", W=WKenya, family="
    binomial", trials=kdat$n, model="leroux", N=N, burn.in = burn.in,
    thin=thin)
```

Table 11.1 provides the parameter estimates and the model choice criteria values for the independent error regression model, M1 and the Leroux CAR model M2. The Leroux model is seen to be much superior to the independent error regression model as expected. Moreover, the Leroux model does not lose the significance of the first two covariates and the third covariate is not significant in either of the two models. The BYM model, fitted both using the INLA and CARBayes package, produces similar results and hence those results are not included in the discussion here.

After fitting the model we can visualize the spatial random effects and their uncertainties by drawing a pair of maps, as shown in Figure 11.2, as follows in four steps. In the first step we extract the MCMC samples of the random effects. In the second step we obtain the mean and standard deviations of the random effects. In the third step we merge the random effect estimates and their standard deviations with the map data frame adf of the 47 counties, see R Code Notes 1.9. In the final step we draw the map using the ggplot function. All the code lines are detailed in the R Code Notes 11.1. The plotted random effect estimates in Figure 11.2 are seen to capture the spatial variation present in the observed vaccination proportions in the top left panel of Figure 11.1.

	Independent			Leroux		
	Median	2.5%	97.5%	Median	2.5%	97.5%
Intercept	1.905	1.809	2.002	2.031	1.922	2.144
prop primary	0.460	0.347	0.575	0.315	0.049	0.573
travel time	−0.120	−0.221	−0.019	−0.311	−0.597	−0.063
night light	0.099	−0.012	0.227	0.017	−0.198	0.230
τ^2				0.496	0.226	1.041
ρ				0.490	0.067	0.927
	Model choice criteria					
Criteria	Penalty	Value	−	Penalty	Value	
DIC	3.97	326.62		30.71	263.60	
WAIC	8.70	332.20		21.46	261.50	

TABLE 11.1: Parameter estimates and model choice criteria values for the independent error and spatial (Leroux) model fitted to the vaccination data set.

> ♣ **R Code Notes 11.1. Figure 11.2** Here are the code lines to reproduce the map of the random effects. The map of the standard deviations is obtained similarly.
>
> ```
> reffectsamps <- M2$fit$samples$phi
> reffs <- get_parameter_estimates(reffectsamps)
> reffs$id <- 0:46
> udf <- merge(reffs, adf)
> a <- range(udf$mean)
> randomeffectmap <- ggplot(data=udf, aes(x=long, y=lat, group =
>     group, fill=mean)) +
>   scale_fill_gradientn(colours=colpalette, na.value="black",
>       limits=a) +
>   geom_polygon(colour="black",size=0.25) +
>   theme_bw()+theme(text=element_text(family="Times")) +
>   labs(title= "Spatial random effects", x="", y = "", size=2.5) +
>   theme(axis.text.x = element_blank(), axis.text.y =
>       element_blank(),axis.ticks = element_blank())
> randomeffectmap
> ```

We now proceed to estimate the probability of attaining 95% and 80% vaccination coverage for each of the 47 districts as plotted in Figure 4 of the paper by Utazi et al. (2021). These probabilities are easily calculated by using the samples of fitted values provided by the output of the CARBayes package. If M2 is the bmstdr model fitted object using the CARBayes option, then M2$fit $samples$fitted is a two-dimensional array of MCMC samples for the fitted values having 47 columns. The number of rows of this array is the number

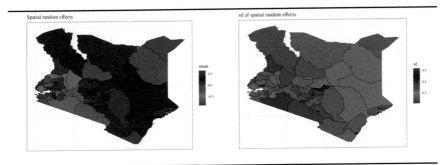

FIGURE 11.2: Spatial random effects from the Leroux model for the 47 counties in Kenya.

of retained MCMC iterations. These fitted values are number of vaccinated children in each county at each MCMC iteration. These are transformed into fitted proportions by dividing the number of sampled children. These proportions are now compared to the threshold values 0.95 and 0.80 and these two indicator variables are averaged over MCMC iterations to estimate the probability of attaining 95% and 80% vaccination coverages. These two coverage maps are plotted in Figure 11.3. These plots look to be almost same as the corresponding ones presented in Figure 4 of Utazi et al. (2021). Only a handful of counties in the Eastern and Central provinces has high probability of achieving 95% coverage. The map for the 80% coverage shows a clear north-south divide in vaccination coverage.

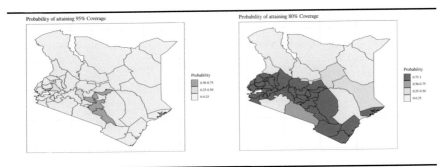

FIGURE 11.3: Probability of attaining 95% (left) and 80% (right) coverage for 47 counties in Kenya.

♣ **R Code Notes 11.2. Figure 11.3** The map and the vaccination data frames used here have been discussed in Figure 1.11. To draw the plots in Figure 11.3 we extract the MCMC samples of the fitted values for each of the 47 counties and then obtain the coverage probability for a given threshold value. These probabilities are then categorized using the cut function and plotted using the `ggplot` function. The essential code lines are given below.

```
# Assume the Kenya map is in the data frame adf and the data is
    kdat
a <- as.matrix(M2$fit$samples$fitted)
ns <- matrix(rep(kdat$n, nrow(a)), byrow = T, nrow=nrow(a))
a <- a/ns

propthresh <- function(b, thresh=0.95) { length(b[b>thresh])/
    length(b) }
pcov95 <- apply(a, 2, propthresh)
pcov80 <- apply(a, 2, propthresh, thresh=0.80)

pcovdat <- data.frame(id=0:46, pcov95=pcov95, pcov80=pcov80)

udf <- merge(pcovdat, adf)
head(udf)
b <- c(-Inf, 0.25, 0.5, 0.75, 1)
labss <- c("0-0.25", "0.25-0.50", "0.50-0.75", "0.75-1")
udf$col195 <- factor(cut(udf$pcov95, breaks=b, labels=labss))
udf$col180 <- factor(cut(udf$pcov80, breaks=b, labels=labss))
com <- c("lightyellow2", "yellow2", "green1", "green4")

cov95map <- ggplot(data=udf, aes(x=long, y=lat, group = group,
    fill=col195)) +
  geom_polygon(colour="black",size=0.25) +
  scale_fill_manual(values =com, guides("Probability"), guide =
      guide_legend(reverse=TRUE))
  plot(cov95map)
```

11.3   Assessing trend in cancer rates in the US

This example models the state-wise annual cancer data for the 48 contiguous states in the USA during 2003 to 2017 previously introduced in Section 1.4.3. The number of annual cancer deaths, denoted by Y_{it} for $i = 1, \ldots, 48$ and

$t = 2003, \ldots, 2017$, varies from 2370 to 176,983 and is clearly dependent on the population size of each individual state. In order to calculate the relative risks we perform internal standardization as detailed in Section 2.12. The expected number of deaths in each year at each state, denoted by E_{it}, is simply obtained as proportional to the corresponding population where the constant of proportionality is the overall cancer death rate. The data set `us48cancer0317` already contains these expected number of deaths.

The observed SMR values are plotted in Figure 11.4. The figure shows a slow gradual increase in the rates from 2003 to 2007. The main purpose of this study is to investigate if there is indeed any significant trend in cancer mortality rates as has been noted in Section 1.4.3. See also Figure 1.13 which provides a time series plot of the SMR values for ten selected states. Our spatio-temporal modeling here includes a linear trend term in the regression part.

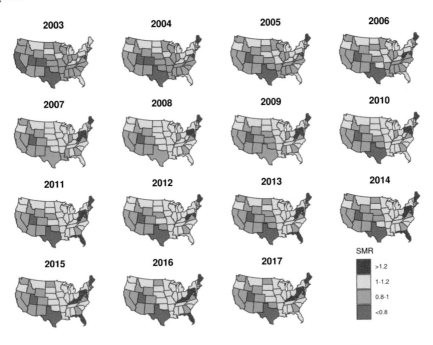

FIGURE 11.4: Observed values of the standardized mortality rates.

♣ **R Code Notes 11.3. Figure 11.4** These maps have been drawn individually for each year and then put together in a single graphsheet using the ggarrange function in the library(ggpubr). To draw the plot for a particular year we proceed as follows. We first look at the range of the observed SMR values and then choose 0.8, 1 and 1.2 as break points where to cut the SMR values using the cut function. Here are the essential code lines.

```
com <- c("green4", "green2", "yellow", "red2")
b <- c(-Inf, 0.8, 1.0, 1.2, Inf)
labss <- c("<0.8", "0.8-1", "1-1.2", ">1.2")
us48cancer0317$catsmrobs <- factor(cut(us48cancer0317$obs_smr,
    breaks=b, labels=labss))
i <- 2003
dp <- mus48cancer0317[mus48cancer0317$Year==i, c("fips", "state
    ", "catsmrobs")]
smr2003 <- plot_usmap(data = dp, values = "catsmrobs", color =
    "red", exclude=c("AK", "HI")) +
  scale_fill_manual(values =com, guides("SMR"), guide =
      guide_legend(reverse=TRUE), na.translate = FALSE) +
  labs(title= i)
  plot(smr2003)
```

The long-term rates can vary because of various factors such as change in socio-economic and demographic variables as well as possible presence of spatio-temporal association. The only predominant socio-economic variable that we use in our analysis is the annual unemployment rate (downloaded from usda.gov) in each of the 48 states. These rates have been plotted in Figure 11.5. Clearly the unemployment rate peaked in 2010 after the economic crisis in 2008.

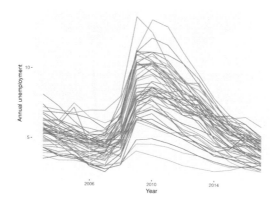

FIGURE 11.5: Annual unemployment rates for the 48 states in the US.

In order to account for the effects of changing demographics our data set includes information regarding the racial composition of each of the states varying annually. There are eight exhaustive broad categories of race and ethnicity combinations which make up the total population. These eight categories are obtained by cross-classifying four levels of race and two levels of ethnicity noted below.

Race	Ethnicity
White (White)	Not Hispanic or Latino (NHL)
American Indian or Alaska Native (AIAN)	Hispanic or Latino (HL)
Asian or Pacific Islander (API)	
Black or African American (BAA)	

Hence the data set `us48cancer0317` contains eight columns containing the percentages of these eight race and ethnicity combinations. The total of these eight percentages for any year and any state is 100% confirming that the eight broad categories are exhaustive. However, this constraint also introduces linear dependencies between the eight columns of data, which is problematic in our regression modeling. Hence, a candidate regression model should not include all eight of these variables in addition to an intercept term. In our investigation below we attempt several combinations of these eight variables and see if those are significant in the overall model. All the covariates including the variable for the linear trend and the unemployment rate have been standardized by subtracting the mean and dividing by the standard deviation.

The number of cancer deaths for each state in each year is modeled as a Poisson random variable with the expected number of deaths as the offset variable. In addition to the fixed effects regression model that includes the linear trend, unemployment percentage and the race-ethnicity proportions we have the option to fit any of the spatio-temporal random effect models described in Section 10.6. Thus, the model is written as

$$
\begin{aligned}
Y_{it} &\sim \text{Poisson}\left(E_{it}\lambda_{it}\right) \\
\log\left(\lambda_{it}\right) &= \mathbf{x}'_{it}\boldsymbol{\beta} + \psi_{it},
\end{aligned}
$$

for $i = 1, \ldots, 48$ and $t = 2003, \ldots, 2017$, where $\mathbf{x}_{it}$ denotes all the covariates, and ψ_{it} is any of the spatio-temporal random effect described in Section 10.6.

Omitting the model selection details for brevity we have decided to adopt the Anova model with interaction, see (10.13), for the random effect. The adopted regression model includes the linear trend term, unemployment rate and only three race-ethnicity proportions AIAN.HL, API.HL and BAA.HL. This model treats the remaining five race-ethnicity proportions: AIAN.NHL, API.NHL, BAA.NHL, White.HL, and White.NHL, as a single base-line category confounded with the intercept term. That is, the model does not estimate an individual intercept term for any of the five left out race-ethnicity proportions. All the commands used for model selection are omitted for brevity.

Table 11.2 provides the parameter estimates of the adopted model. All the included covariates remain significant in the model. Notice that the estimates

	Median	2.5%	97.5%
Intercept	0.017	0.016	0.017
Trend	0.040	0.034	0.051
unemployment	0.008	0.001	0.019
AIAN.HL	-0.064	-0.075	-0.056
API.HL	-0.033	-0.041	-0.022
BAA.HL	0.027	0.016	0.045
τ_S^2	0.042	0.029	0.064
τ_T^2	0.114	0.064	0.232
τ_I^2	0.009	0.008	0.010
ρ_S	0.900	0.679	0.990
ρ_T	0.508	0.062	0.913

TABLE 11.2: Parameter estimates for the spatio-temporal Anova model fitted to the number of cancer deaths data set.

of trend and the coefficient of the unemployment rate are positively significant showing that the cancer rates are increasing over time and also higher unemployment rates are associated with higher rates of cancer deaths. It seems that the Hispanic and Latino members of the two races American Indian or Alaska Native (AIAN) and Asian or Pacific Islander (API) are at significantly lower risk than the other groups. But the Hispanic and Latino members of the Black or African American (BAA) are experiencing significantly higher death rates than the five base line race-ethnicity combinations. The estimates of the variance parameters and the spatial and temporal correlation parameters indicate higher spatial correlation and higher temporal variability. The interaction effect is also important since the model without interaction was not chosen during the model selection investigation noted in the previous paragraph.

Figure 11.6 plots the fitted values of the SMR for all the 15 years (2003-2017) in our data set. This plot shows clear upward trend as many states such as Florida, Maine and Pennsylvania gradually progressed to have higher levels of SMR starting from low levels in 2003. Indeed, this is true for many states in the Eastern USA east of the Mississippi river. Incidentally, we were unable to get a comparable model fit for the same regression model but with a BYM model for space and an AR model for time using the INLA the package. Although it is possible to conduct further investigations and to add more explanatory variables, we end our discussion here.

11.4 Localized modeling of hospitalization data from England

In this section we return to the hospitalization data set from England introduced in Section 1.4.4. In addition to the monthly hospitalization numbers Y_{it}

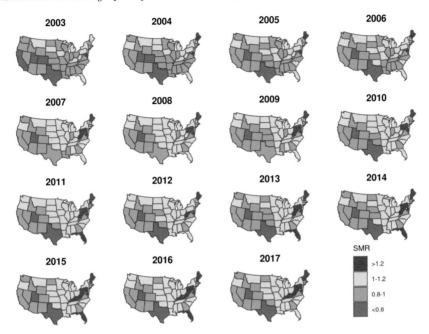

FIGURE 11.6: Fitted values of the standardized mortality rates.

and the expected numbers E_{it}, the paper by Lee et al. (2017) consider two socio-economic covariates: *jsa*, the proportion of the working age population in receipt of Job Seekers Allowance, a benefit paid to those without employment, and *price*), the average property price. These are the key confounders for this analysis because areas that are economically impoverished have worse health on average than more affluent areas. There are two further important covariates. The first one is the mean *temperature* that accounts for the excess seasonality additional to the offsets E_{it}. The last covariate is the average pollution level of NO_2 which has been estimated by a spatio-temporal model as discussed in Mukhopadhyay and Sahu (2018). This data set and the adjacency matrix are downloadable as the objects `Engrespiratory W323` from github[2].

Lee et al. (2017) present a number of summary statistics and graphs to explore the relationships between the response Y_{it}, the number of hospitalization and the covariates. Hence those are not replicated here. But for any new data set we must graphically explore the relationships between the response and the covariayes first before embarking on modeling.

Various spatio-temporal models can be fitted to this data set using the `Bcartime` model fitting routine. In this section we introduce a localized version of the auto-regressive model in Section 10.6.4, due to (Lee et al., 2014).

---

[2]https://github.com/sujit-sahu/bookbmstdr.git

The package vignette Lee (2021) for the `CARBayes` package also provides an accessible discussion of the localized model.

11.4.1   A localized model

As in the previous example we assume the hierarchical models:

$$
\begin{aligned}
Y_{it} &\sim \text{Poisson}\left(E_{it}\lambda_{it}\right)\\
\log\left(\lambda_{it}\right) &= \mathbf{x}'_{it}\boldsymbol{\beta} + \psi_{it},
\end{aligned}
$$

for $i = 1,\ldots,323$ and $t = 1,\ldots,60$, where $\mathbf{x}_{it}$ denotes all the covariates and ψ_{it} is the spatio-temporal random effect parameter. This random effect parameter is assumed to follow the temporal-autoregressive model given in Section 10.6.4 but with the following modifications. In the first stage of the modified specification we assume a new cluster specific intercept $\lambda_{U_{it}}$ in addition to the space-time random effects ϕ_{it} as in Section 10.6.4. The new random variables U_{it} determine cluster membersip of ψ_{it} and hence Y_{it}. Specification for U_{it} is discussed below. As in Section 10.6.4 we write $\phi_t = (\phi_{1t},\ldots,\phi_{nt})$ for $t = 1,\ldots,T$. Hence the full space-time random effect structure is assumed to be:

$$
\begin{aligned}
\psi_{it} &= \lambda_{U_{it}} + \phi_{it}, \quad i = 1,\ldots,n,\\
\phi_t|\phi_{t-1}, W &\sim N\left(\rho_T\phi_{t-1},\, \tau^2 Q(W,1)^-\right), \quad t = 2,\ldots,T\\
\phi_1|\phi_{t-1}, W &\sim N\left(\mathbf{0},\, \tau^2 Q(W,1)^-\right),
\end{aligned}
$$

where ρ_T, τ^2 are parameters already described in Section 10.6.4. Here $Q(W,1)^-$ is a generalized inverse of the matrix $Q(W,1)$. [We say a matrix G is a generalized inverse of a matrix A if $AGA = A$. We use the generalized inverse $Q(W,1)^-$ instead of the inverse since the matrix $Q(W,1)$ is singular, i.e. not invertible.]

The parameters $\lambda_{U_{it}}$ determine the localized structure in the model. The random effect ψ_{it} (corresponding to Y_{it}) has a piece-wise constant clustering or intercept $\lambda_{U_{it}}$ that depends on another random variable U_{it} specified as below. Two data points Y_{it} and Y_{jk} will have similar values if $\lambda_{U_{it}} = \lambda_{U_{jk}}$ but they will show a larger difference if $\lambda_{U_{it}} \neq \lambda_{U_{jk}}$. Suppose that piece-wise constant intercept or clustering process, $\lambda_{U_{it}}$, comprises of at most G distinct levels. The G levels are ordered via the prior specification:

$$
\lambda_j \sim U(\lambda_{j-1}, \lambda_{j+1}), \quad \text{for } j = 1,\ldots,G,
$$

where $\lambda_0 = -\infty$ and $\lambda_{G+1} = \infty$. Here $U_{it} \in \{1,\ldots,G\}$ and this controls the assignment of ψ_{it}, and correspondingly affecting Y_{it}, to have one of the G intercept levels. A penalty based approach is used to model U_{it} and G is chosen larger than necessary and a penalty prior is used to shrink it to the middle intercept level. This middle level is $G^* = (G+1)/2$ if G is odd and $G^* = G/2$ if G is even, and this penalty ensures that Y_{it} is only in the extreme low and high risk classes if supported by the data. Thus, G is the maximum number of distinct intercept terms allowed in the model, and is not the actual number of intercept terms estimated in the model. The allocation prior is independent across areal units but correlated in time, and is given by:

$$f(U_{it}|U_{i,t-1}) = \frac{\exp\left(-\delta[(U_{it} - U_{i,t-1})^2 + (U_{it} - G^*)^2]\right)}{\sum_{r=1}^{G} \exp\left(-\delta[(r - U_{i,t-1})^2 + (r - G^*)^2]\right)}, \quad \text{for } t = 2, \ldots, T$$

$$f(U_{i1}) = \frac{\exp\left(-\delta(U_{i1} - G^*)^2\right)}{\sum_{r=1}^{G} \exp\left(-\delta(r - G^*)^2\right)}$$

$$\delta \sim \text{Uniform}(1, m)$$

where m is a suitable value. Temporal autocorrelation is induced by the term $(U_{it} - U_{i,t-1})^2$ in the penalty, while the term $(U_{it} - G^*)^2$ penalizes class indicators U_{it} towards the middle risk class G^*. The size of this penalty and hence the amount of smoothing that is imparted on **U** is controlled by δ, which is assigned a uniform prior. Further details for this model are given by Lee and Lawson (2016).

11.4.2   Model fitting results

The localized model, described above, can be fitted using the Bcartime function with the model option equal to "localised". This model option calls the ST.CARlocalised function in the CarBayesST (Lee et al., 2018) package. The parameter G should be supplied as an additional argument.

The Bcartime code used to fit the model looks like the following.

```
f0 <- observed ~offset(logExpected) + no2 + price + jsa + temp.mean

mfit <- Bcartime(formula = f0, data =Engrespriratory, W = W323,
    family ="poisson", model="localised", G=2,
scol="spaceid", tcol="timeid", N=N, burn.in = burn.in, thin=thin,
    package="CARBayesST")
```

The resulting parameter estimates are shown in Table 11.3. The covariates *price* and *temperature* are negatively significant showing lesser hospitalization intensity in richer areas and also in the summer months. Higher levels of pollution, as quantified by NO_2, lead to significantly, although moderately, higher level of hospitalization. All these effects are estimated in addition to the presence of the spatio-temporal random effects. These random effects are estimated to have a low level of temporal correlation which may be justified because of the presence of the seasonal covariates, e.g. *temperature*. Estimates of the spatial variance τ^2 and δ are also reasonable.

With the option $G = 2$, the model estimated two localized levels of the intercept in the spatio-temporal random effects denoted by λ_1 and λ_2 in Table 11.3. The interval estimates for these random effects do not overlap and hence these are statistically significantly different. Note that the model has

	Median	2.5%	97.5%
NO_2	0.0082	0.0076	0.0091
price	−0.5073	−0.5316	−0.4838
jsa	0.0465	0.0373	0.0563
temperature	−0.0293	−0.0304	−0.0281
λ_1	−0.7681	−0.7733	−0.7647
λ_2	−0.0810	−0.0861	−0.0762
δ	1.0016	1.0001	1.0089
τ^2	0.1889	0.1834	0.1941
ρ_T	0.1099	0.0897	0.1302

TABLE 11.3: Parameter estimates for the localized model fitted to the England hospitalization data set.

been specified using the parameterisation $\lambda_{U_{it}}$ where U_{it} is binary taking one of the possible values 1 and 2 since G has been assumed to be 2. Hence, at each MCMC iteration j there will be an MCMC iterate $U_{it}^{(j)}$ identifying the λ value for each i and t. These $U_{it}^{(j)}$ are saved as the component `localised.structure` in the `CARBayesST` model fitted object which can be accessed by the `fit` component of the `Bcartime` model fit output, e.g. `mfit$fit$localised.structure` where `mfit` is the above model fitted object.

The MCMC iterates, $U_{it}^{(j)}$, can be summarized to find an estimate of the intercept term for each space-time combination i and t. In this example, we take the maximum a-posteriori estimate. That is, for each i and t we obtain the modal value, $\tilde{U}_{it}$, of $U_{it}^{(j)}$ for $j = 1, \ldots, J$ where J is the total number of saved MCMC iterates. The MCMC iterates, $U_{it}^{(j)}$, also give freedom to summarize over any temporal window of our choice. For example, we may obtain the model value $\tilde{U}_{i_k}$ for the kth year, $k = 2007, \ldots, 2011$. That is, $\tilde{U}_{i_k}$ is the modal value of the iterates $U_{it}^{(j)}$ when time t is within a particular year k. Thus, to find this modal value we use all $U_{it}^{(j)}$ for all the twelve months t in the year k. Similarly, an overall local spatial structure can be obtained by the modal estimates, $\tilde{U}_i$ for each spatial unit i.

Figure 11.7 plots the annual local structures, $\tilde{U}_{i_k}$ for each of the five years $k = 2007, \ldots, 2011$ and also for the overall estimates $\tilde{U}_i$. The two color maps show the two possible levels of the intercepts: λ_1 and λ_2. These maps essentially capture the pre-dominant spatial variation depicted in the exploratory plot of the SMR values in Figure 1.14. The current plot demonstrates that the localized structure can vary temporally as well since the plot does show inter-annual spatial variation.

Of course, it is possible to fit other models to this data set and compare those using the Bayesian model choice criteria such as DIC and WAIC. This can be simply achieved by changing the model option in the fitting function `Bcartime`. Moreover, the BYM model can be fitted using the `INLA` package by changing the package argument. However, we do not undertake such activities

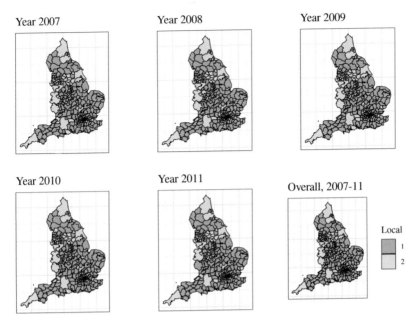

FIGURE 11.7: Fitted local structures for each of the five years and also for the overall.

here for the sake of brevity. Instead, we consider another example illustrating the adaptive CAR model in the CARBayesST package.

♣ **R Code Notes 11.4. Figure 11.7** To draw this figure we first extract the component localised.structure from the fitted CARBayesST model. This component is a vector of the same length as the number of rows in the supplied data frame. This vecor contains one of two unique values 1 and 2 indicating which of the two intercepts λ_1 and λ_2 are there in the model for that row (space-time combination) in the data frame. The modal value of these locality indicators are found for each spatial unit and each year. Thus, there is a unique value, either 1 or 2, for each year and for each of the 323 spatial units. The two color map for each year is then plotted and combined using the ggarrange function as in the other maps. The last map in the plot for the overall level has been obtained by calculating the modal value for each aerial unit over all the years. The full code is provided online.

11.5   Assessing trend in child poverty in London

This example is based on the child poverty data set introduced in Section 1.4.5. Having 33 areal units, see Figure 11.8 for a map, the city of Greater London, provides many opportunities for modeling spatio-temporal data regarding various socio-economic and environmental data sets. Summaries of the child poverty data have been provided in the Figures 1.15 and 1.16. Here we introduce the following relevant covariates and consider an adaptive CAR model as developed by Rushworth et al. (2017).

A map of 32 boroughs and the City of London

FIGURE 11.8: A map of the 32 boroughs and the City of London. 1 = City of London, 2 = Tower Hamlets, 3 = Islington, 4 = Camden, 5 = Westminster, 6 = Kensington and Chelsea, 7 = Hammersmith and Fulham.

There are various determinants of child poverty. Prominent among these are median income and house price and also the percentage of the economically inactive population at the time. There are others but these are the three covariates we consider in our study. The response variable is the proportion of children living in families in receipt of out-of-work (means-tested) benefits or in receipt of tax credits where their reported income is less than 60 per cent of UK median income. We do not have access to a reliable data set on the numbers of families that gave rise to the proportions in our data set. Hence we do not proceed with the binomial distribution based models and instead intend to adopt the Gaussian distribution in our modeling. However, the observed response, being a proportion, is better modeled using a logit transformation in the first place. The logit transformed proportions are then assumed to follow the normal error distribution in the GLM. Since the response is in the logit scale, we also logit transform the covariate assessing the percentage of economically inactive people. The house price covariate is included in the log to the base 10 scale and the median income covariate has been standardized to have mean 0 and sample variance 1.

Figure 11.9 provides pair-wise scatter plots of the response and the covariates in the transformed scale. The diagonal panels in this plot provides kernel density estimates of the variables. Looking at the bottom row of this figure it is clear that child poverty levels increase if the percentage of economically inactive people increases. However, the levels tend to decrease with increasing income and house prices. This is a slight bit worrying since rising house prices may lead to increasing poverty levels. However, this may be confounded by the predominant decreasing trend in poverty levels seen in Figure 1.16. Our regression model will include a linear trend term to account for this large effect.

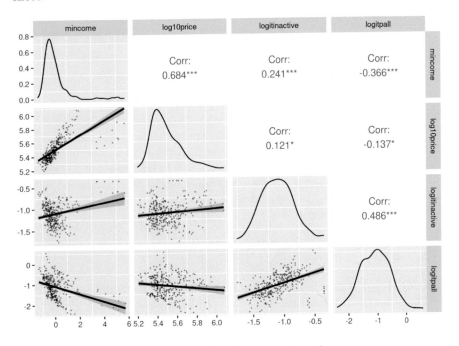

FIGURE 11.9: A pairwise scatter plot.

11.5.1 Adaptive CAR-AR model

To write down the model fully, let Y_{it} for $i = 1, \ldots, n$ and $t = 1, \ldots, T$, denote the logit-transformed estimated proportion of children living in families in receipt of out-of-work (means-tested) benefits or in receipt of tax credits where their reported income is less than 60 per cent of UK median income. Here $n = 33$ is the number of boroughs (including the City of London) and $T = 10$ is the number years for which we have data. Hence

$$Y_{it} = \log\left(\frac{\hat{p}_{it}}{1 - \hat{p}_{it}}\right)$$

for all i and t where $\hat{p}_{it}$ is the estimated proportion in child poverty. Let $\mathbf{x}_{it}$ denote the four dimensional covariate vector at each i and t:

(i) median income – transformed to have zero mean and unit variance, denoted by *income*

(ii) logit transformed proportion of economically inactive population, denoted by *inactive*

(iii) median house price at the log to the base 10 scale, denoted by *price*

(iv) linear trend, t transformed to have zero mean and unit variance, denoted by *trend*.

The top level model is specified as:

$$Y_{it} = \mathbf{x}'_{it}\boldsymbol{\beta} + \psi_{it} + \epsilon_{it}, \quad i = 1, \ldots, n, \ t = 1, \ldots, T,$$

where $\epsilon_{it} \sim N(0, \nu^2)$ independently and ψ_{it} are spatio-temporal random effects which can be any of the models discussed in Section 10.6. The basic linear model is obtained as a special case when $\psi_{it} = 0$ for all values of i and t.

We now recall the temporal auto-regressive model 10.15 discussed in Section 10.6. Let $\boldsymbol{\psi}_t = (\psi_{1t}, \ldots, \psi_{nt})$ denote the vector of random effects at time t. The CAR-AR model specifies:

$$\begin{aligned}
\boldsymbol{\psi}_t | \boldsymbol{\psi}_{t-1}, W &\sim N\left(\rho_T \boldsymbol{\psi}_{t-1}, \tau^2 Q(W, \rho_S)^{-1}\right), \quad t = 2, \ldots, T \\
\boldsymbol{\psi}_1 | W &\sim N\left(\mathbf{0}, \tau^2 Q(W, \rho_S)^{-1}\right),
\end{aligned}$$

where $Q(W, \rho_S)$ is the spatially dependent precision matrix defined in (10.7).

The adaptive CAR-AR model, developed by Rushworth et al. (2017), is a modification of the above CAR-AR model. In the CAR-AR model the random effects have a single level of spatial dependence that is controlled by ρ_S. The adaptive model of this section allows locally varying levels of spatial correlation. As in the localized model in Section 11.4 the adaptive model allows for localized spatial autocorrelation by allowing spatially varying random effects to be correlated (inducing smoothness) or conditionally independent (no smoothing), which is achieved by modeling the non-zero elements of the binary proximity matrix W instead of assuming those to be fixed at 1.

The collection of non-zero W_{ij}'s are denoted by $\mathbf{w}^+$ and those are transformed to the logit scale by writing

$$\mathbf{v}^+ = \log\left(\frac{\mathbf{w}^+}{1 - \mathbf{w}^+}\right).$$

Rushworth et al. (2017) assume a shrinkage prior

$$f(\mathbf{v}^+|\tau_w^2, \mu) \propto \exp\left[-\frac{1}{2\tau_w^2}\left(\sum_{v_{it} \in \mathbf{v}^+}(v_{it} - \mu)^2\right)\right],$$

and then τ_w^2 is assumed to have an inverse gamma prior distribution. The parameter μ is treated to be fixed as in Rushworth et al. (2017) to avoid having numerical problems in MCMC model fitting. The prior distribution for $\mathbf{v}^+$ assumes that the degree of smoothing between pairs of adjacent random effects is not spatially dependent. The reader is referred to the paper Rushworth et al. (2017) for further theoretical discussion regarding this model.

11.5.2 Model fitting results

In order to see if we need the adaptive models at all we first compare four different models including the adaptive model. The first model is the independent error regression model, the second is the Anova model from Section 10.6.2, the third is the AR model without adaptation and the last one is the adaptive model of this section. Table 11.4 provides the DIC and WAIC criteria values for these models. The last column of this table provides the root mean square error between the observed and fitted values of the child poverty response variable. These observed and fitted values are on the original percentage scale, inverse logit-transform of the modeled y_{it}'s.

The table clearly shows that the adaptive model is the chosen one according to both DIC and WAIC. These criteria values are negative because of the small estimates of the variance parameters as seen in Table 11.5. However, it is reassuring to see that the penalty parameters are estimated to be positive. The last column of the table shows that the adaptive model provides best goodness-of-fit since its RMSE is the smallest.

	p.dic	DIC	p.waic	WAIC	RMSE
Independent	5.91	235.97	5.37	235.58	6.32
Anova	39.37	51.94	34.32	50.52	4.26
AR (1)	118.76	−14.26	82.83	−29.91	2.77
Adaptive	125.25	−19.94	85.80	−37.86	2.65

TABLE 11.4: Model choice statistics for the child poverty example.

Table 11.5 provides the parameter estimates of the chosen adaptive model.

As expected, income is negatively significant indicating that higher levels of income reduces child poverty. Higher levels of economic inactivity leads to significantly higher levels of child poverty. The house price covariate is not significant in the model. However, there is significantly strong downward linear trend since the *trend* parameter is significant. There is strong spatial correlation as estimated by ρ_S. There is weak residual temporal correlation, as estimated by ρ_T, after accounting for linear trend. The spatial variance τ^2 is estimated to be higher than the independent error variance ν^2 – reassuring the importance of spatial model fitting. Both of these variance components are estimated to be relatively small which has caused the DIC and WAIC to be negative. The estimate of τ_w^2 is much larger – showing much variability of the adaptation random variables v_{it}'s which have been modeled in the logit scale. Perhaps this would indicate a greater level of uncertainty in the estimation of the positive W_{ij} values in the adaptive model.

Figure 11.10 plots the fitted values of the child poverty response variable. The fitted map agrees very strongly with the observed map seen in Figure 1.15. There are few boroughs where the discretized color categories do not match possibly because of the process of discretization itself. The highlighted borders (see computing notes below) are the borders where, according to the model, there are step changes between the two neighboring boroughs. Most of the borders are highlighted possibly because of the high level of precision of the modeled data – as evidenced by the estimates of τ^2 and ν^2. This requires further investigation which will be pursued elsewhere.

Notes for computing the highlighted borders: The adaptive CARBayesST model fitted object saves a summary of the localized structure in the `localised.structure` component. This component, see the documentation of the CARBayesST function `ST.CARadaptive`, is a list with two $n \times n$ matrices, `Wmedian` and `W99`, which summarizes the estimated adjacency relationships. `Wmedian` contains the posterior median for each positive w_{ij} element estimated in the model for adjacent areal units, while `W99` contains the binary indicator variables for whether $P(w_{ij} < 0.5|\text{data}) > 0.99$. For both matrices, elements corresponding to non-adjacent pairs of areas have missing (NA) values. The **R** Code Notes 11.5 below provides further details for plotting these highlighted borders.

	Median	2.5%	97.5%
Intercept	−1.002	−3.532	1.608
income	−0.269	−0.326	−0.210
inactive	0.618	0.428	0.804
price	0.107	−0.363	0.561
trend	−0.103	−0.148	−0.059
ν^2	0.043	0.034	0.054
τ^2	0.063	0.044	0.089
ρ_S	0.968	0.927	0.988
ρ_T	0.054	0.002	0.215
τ_w^2	71.298	31.361	141.792

TABLE 11.5: Parameter estimates for adaptive CAR model fitted to the child poverty data set.

FIGURE 11.10: A map of 32 boroughs and the City of London. 1 = City of London, 2 = Tower Hamlets, 3 = Islington, 4 = Camden, 5 = Westminster, 6 = Kensington and Chelsea, 7 = Hammersmith and Fulham.

> ♣ **R Code Notes 11.5. Figure 11.10** To draw the highlighted borders we need to use the function `highlight.borders` from the `CARBayes` package. This function takes two arguments `border.locations` and `spdata`. The argument `border.locations` should be set as the `W99` matrix and the `spdata` argument should be set as the map polygon object used to draw the map. The map polygon object is the `sp` object as read from the shape files. The output of the `highlight.borders` is a spatial points data frame which can be plotted on the same plot as the original shape-file object which was passed to the `highlight.borders` function. For further details see the vignette of the `CARBayes` package, Lee (2021).

11.6 Conclusion

Four practical examples of areal unit data modeling have been provided in this chapter. Each of these highlights a particular aspect of data modeling with a different best fit model. The code and data sets are available online from github.

11.7 Exercises

1. Reproduce the results for the Kenya vaccination coverage example. Fit other possible models as indicated in the code file for this example. Perform model choice and compare your results with those reported by Utazi et al. (2021).

2. Perform variable selection and model choice for the US cancer death example.

3. Reproduce all the results for the hospitalization data from England. Use cross-validation for model selection by supplying the `validrows` argument.

4. Perform variable selection for the child poverty example. This example investigated modeling of the child poverty column `pall`. There is another very similar response column called `punder16` for families with children under the age of 16. Repeat the analysis reported in the book but using the `punder16` as the response.

5. Prepare a data set of your interest by downloading data from the web site of the Office for National Statistics. For example, you may download weekly death statistics due to Covid-19. You will also need to download data for some relevant covariates. Perform exploratory data analysis and then use the `Bcartime` function to model the resulting data set.

12

Gaussian processes for data science and other applications

12.1 Introduction

The Gaussian process theory learned and experienced so far has applications in many different application areas including data science, computer experiments and design of experiments. This chapter aims to discuss a few of these applications so that we can further appreciate the enormity and ubiquitousness of the theory of GP. Thus, we expand the overwhelming modeling horizon of this text to other fields of data science and design of experiments.

Looking back at the earlier Chapters 6-11, the overwhelming unifying theme of the presentation has been to model and validate large data sets. Not very surprisingly, this theme also sits in the very heart of data science literature, see e.g. Hastie et al. (2009). In that literature, the primary aim, for example, in machine learning, is to obtain predictive models which validate the best. The main model fitting and validation functions, `Bspatial`, `Bsptime` and `Bcartime`, achieve exactly this by allowing the users to validate rows of data named by them. Numerical illustrations using many data sets have demonstrated that the Gaussian process based models and the Gaussian CAR models perform better than the basic independent error regression models which ignore the spatial and temporal dependencies in the data. The current chapter aims to consolidate this knowledge by introducing the wider context of data science and machine learning following a recent review article by Sambasivan et al. (2020).

Specifically, in the current chapter we aim to present GP based models for machine learning, see e.g. Rasmussen and Williams (2006) showing an immediate connection between the GP based regression models presented in the earlier point referenced data models chapters. In the machine learning literature the GP prior plays an important role in the regression function estimation (and prediction) – which is generically termed as 'learning' in data science. Indeed, there are one-to-one equivalent correspondences between the different terminologies used in the fields of mainstream statistics and data science. Table 12.1, adapted from Wasserman (2004) and Sambasivan et al. (2020), provides a dictionary of similar concepts used in the two fields. In this chapter we further elaborate and use these synergies to build bridges between

DOI: 10.1201/9780429318443-12

the two fields – more importantly between what has been presented so far in the earlier chapters of this book and relevant parts of machine learning.

The main keyword **learning** in machine learning is what is known as *parameter estimation* or estimation of *unknown functions* in Statistics. Learning is an umbrella term in computer science that is also used to mean prediction. There are different ways to perform learning as there are different methods of estimation in statistics. The term **supervised learning** is meant to highlight the learning (or estimation) tasks based on the use of a label, or dependent variable – see Table 12.1, along with independent variables – if there is any available. Supervised learning tasks where the label is a discrete quantity are called *classification* tasks. For example, in the applications involving binary regression models, see e.g. Section 10.2 in Chapter 10, a credit card company may be interested in classifying each card applicant having either good or bad credit based on their past financial behavior. When the label is a continuous variable, the learning task is called *regression*. The goal of a learning task associated with a label is to predict it. Most of what has been presented in the earlier chapters of this book can be seen as regression – or *supervised learning* in data science.

Having understood the concept of supervised learning the reader may be curious to know about unsupervised learning. Indeed, the learning task of grouping together similar data points, known as clustering in the statistics literature, is called **unsupervised learning** in machine learning. These learning tasks do not use the labels explicitly and have many practical applications, see e.g. Hastie et al. (2009)[Chapter 14] for more details. Strictly speaking, unsupervised learning is used much more than clustering. Clustering is only one form of unsupervised learning. However, unsupervised learning is one of the most widely used clustering methods in both research and practice. References to unsupervised learning without further qualification, usually tend to refer to clustering. We, however, do not consider this topic any further in this book. Instead, the model based estimation of the spatial patterns in the data allows us to discover spatial clustering, although the scope of the model based methods is somewhat limited to the setup in Chapter 10.

There are many other modes and concepts of learning, e.g.

- **semi-supervised learning**: learning with a limited amount of labeled data,

- **active learning**: machine learning performed with the help of an activity or experience,

- **deep learning**: supervised learning where the algorithm selects the best predictors as well as the function that maps the feature space to the output space (or the regression function),

- **reinforcement learning**: learning where an agent interacts with its environment to maximize reward,

- **transfer learning**: learning methods which are able to leverage the learning results from a similar problem.

Statistics	Data science	Notation/Explanation
Dependent variable	Label (or Target)	$y_1, \ldots, y_n$
Independent variable (covariates or predictors)	Features, attributes	$\mathbf{x}_1, \ldots, \mathbf{x}_n$
Data	Training or In-sample	$(\mathbf{x}_1, y_1), \ldots, (\mathbf{x}_n, y_n)$ the data to fit/train the model
Validation (or hold out) data	Test or Out of sample	$(\mathbf{x}_{n+1}, y_{n+1}), \ldots, (\mathbf{x}_{n+m}, y_{n+m})$: the data to test the accuracy of prediction from the trained or fitted model
Estimation	Learning	Use data to estimate (or learn) unknown quantity or parameters in the model
Classifier	Hypothesis	Map from covariates to outcome
Classification	Supervised Learning	Predicting a discrete y from $\mathbf{x}$
Clustering	Unsupervised Learning	Putting data into groups
Regression	Supervised Learning	Predicting a continuous y from $\mathbf{x}$
Variable selection	Feature selection	Select which of the predictors ($\mathbf{x}$) are essential for prediction.
Directed acyclic graph	Bayesian net	Multivariate distribution with conditional independent relations
Class of Models	Hypothesis Class	Set of models, e.g. logistic regression for binary classification problem
Type I and Type II errors	False Positive and False Negative	Nature of errors made by a classification rule/model. Sensitivity (true positive rate) and specificity (true negative rate).

TABLE 12.1: Dictionary of similar concepts in statistics and data science.

- **inductive learning**: learning which generalizes previous results/methods to apply on new data sets.

- **transductive learning**: learning where it is not necessary to generalize as in the case of inductive learning but in a setup the user is only interested in prediction performance for a specific data set.

The data science literature is rich with these and other A-Z concepts of learning, such as life-long learning and ensemble learning. Sambasivan et al. (2020) provide a review of these terms and also their adaptation for what are known as 'Big Data' problems. We, however, do not explore these concepts any further in this book.

We end the introduction by having a brief discussion comparing ML and formal statistical methods. The main purpose of statistical modeling is to assess uncertainties in the inference and predictions made using the models. As discussed in Chapter 6 such modeling enables the researchers to associate a precise quantity of uncertainty measure for each individual prediction and individual piece of inference. Machine learning on the other hand is geared more towards finding fast algorithms for making accurate predictions where the accuracy is interpreted as an overall measure rather than individual one. That is, an algorithm may be characterized by its overall error rate rather than the underlying uncertainty for an individual prediction. In ML there is less emphasis on hypothesis testing (to be viewed as an example of statistical inference) because in typical business applications it is only of secondary interest to find association between target variable and features - the primary one being prediction. As a result there is much less occurrence of the use of the P-values for statistical testing of hypothesis in ML. In ML each hypothesis is considered as a model and a Bayesian model selection strategy is a much more comprehensive strategy for large data.

12.2   Learning methods and their Bayesian interpretations

The central problem of statistical learning theory is the estimation of a function from a given set of data. The data for the learning task consist of attributes x with labels Y. We use $\mathcal{Y}$ to denote the output space, i.e., the all possible values y and $\mathcal{X}$ to denote the input space. In a binary classification problem, $\mathcal{Y} = \{1, -1\}$. When the learning task is regression, the output space is infinite, $\mathcal{Y} \subseteq \mathbb{R}$. In the following discussion we assume that both the input and output variables, X and Y, are random and follows a joint probability distribution $P(dx, dy)$.

Important note: So far we have used the notation $f(\cdot)$ to denote a probability (mass or density) function of a random variable but in the chapter we use $P(\cdot)$ to denote the probability function. The notation $f(\cdot)$ in this chapter denotes the unknown function we want to learn about from data. This notational clash has been allowed here so that the notations in this chapter do not deviate too far from the popular notations in the machine learning literature, e.g. Rasmussen and Williams (2006).

The unknown function that we want to learn about is a one-to-one mapping from $\mathcal{X}$ to $\mathcal{Y}$. In a given problem, as before, we use the notation $(x_1, y_1), \ldots, (x_n, y_n)$ to denote the full data set. Following the popular machine learning literature we denote the function by $f(x)$, which is treated to be a random quantity, see e.g. Section 2.4 in Hastie et al. (2009). This function may depend on additional unknown parameters, $\boldsymbol{\beta}$ say, and can be parameterized, for example, $f(\mathbf{x}) = \beta_0 + \beta_1 x$. However, it is not required that the unknown function is either linear or parametric.

The goal of learning is to learn the function $f : \mathcal{X} \mapsto \mathcal{Y}$ that can predict the label y given x. We cannot consider all possible functions. We need some specification of the class of functions we want to consider for the learning task. The class of functions considered for the learning task is called the *hypothesis class or class of models*, $\mathcal{F}$, see Table 12.1. Consider a function $f \in \mathcal{F}$. The hypothesis class $\mathcal{F}$ could be finite or infinite. An algorithm $\mathcal{A}$ is used to pick the best candidate from $\mathcal{F}$ to perform label prediction. To do so, $\mathcal{A}$ measures the *loss* function $L(f(x), y)$, which is a performance indicator for $f(x)$. It measures the loss due to the error in the prediction from $f(x)$ against the true label y. The loss function, $L(f(x), y)$, is a random variable. Therefore, we need the expected value of the loss function to characterize the performance of f. This expected value of the loss is called the *risk* and is defined as:

$$R(f) = E[L] = \int L(f(x), y) P(dx, dy). \tag{12.1}$$

The resolution is to choose a function f^* to be the optimal one if it minimizes the risk function $R(f)$. That is,

$$f^* = \underset{f \in \mathcal{F}}{\operatorname{argmin}} R(f). \tag{12.2}$$

It turns out that for given observed values of x it is sufficient to evaluate and minimize the risk point-wise for each x. That is, we obtain

$$f^*(x) = \underset{f \in \mathcal{F}}{\operatorname{argmin}} E_{y|x}(L(f(x), y)|x). \tag{12.3}$$

Connecting this to the Bayesian decision theoretic estimation in Section 4.8, the solution $f^*(x)$ of (12.3) is the Bayes estimator under the loss function L.

A simple example of the above method is the case when y assumes the values in $\mathcal{Y} = \{1, 2, \cdots, K\}$, i.e., the set of K possible classes. The loss function L can be presented as a $K \times K$ matrix. The $(j, k)^{th}$ elements of the loss matrix L is

$$L(j, k) = \begin{cases} 0 & \text{if } j = k \\ l_{kj} & \text{otherwise,} \end{cases}$$

where $l_{jk} \geq 0$ is the penalty for classifying an observation y_k wrongly to $y_j = f(x)$. A popular choice of L is the 0-1 loss function, where all miss-classification are penalized by a single value, i.e. $l_{jk} = 1$ for all $j \neq k$. We can write the risk as

$$R(f) = \sum_{k=1}^{K} L(f(x), y_k) P(y_k | x).$$

With the $0 - 1$ loss function this simplifies to

$$
\begin{aligned}
R(f) &= \sum_{k \neq f(x)}^{K} P(y_k | x) \\
&= 1 - P\left(y_{k=f(x)} | x\right).
\end{aligned}
$$

Clearly, this is minimized if we take,

$$f(x) = \max_{y \in \mathcal{Y}} P(y | x).$$

This solution is known as *Bayes classifier*, see Berger (1993) and Hastie et al. (2009)[Chapter 2]. This method classifies a point to the most probable class, using the (posterior) probability of $P(y|x)$. The error rate of the Bayes classifier is known as the *Bayes rate* and the decision boundary corresponding to Bayes classifier is known as the Bayes-optimal decision boundary. The term Bayes classifier is also justified from several different viewpoints. For example, when x is treated as random and with the joint distribution $P(dx, dy)$ used to calculate the risk (12.1) then this joint distribution can be treated as a 'likelihood × prior'. Moreover, when x is considered fixed, there is no unknown parameter in the distribution $P(y|x)$, and it is completely known and this is also treated as the posterior predictive distribution. See relevant discussion in Section 4.11.

When y is continuous, as in a regression problem, the approach discussed in (12.1) and (12.2) works, except that we need a suitable loss function for penalizing the error. The most popular loss function is the squared error loss function: $L(f(x), y) = (y - f(x))^2$ and the solution under the loss function is

$$f(x) = E(y|x), \tag{12.4}$$

the conditional expectation also known as the *regression* function. If we replace the squared error loss by the absolute error loss, i.e., $L(f(x), y) = |f(x) - y|$, then the solution is the conditional median,

$$f(x) = \text{median}(y|x).$$

This estimate is more robust than the conditional mean estimate in 12.4.

12.2.1 Learning with empirical risk minimization

The learning of f is performed over a finite set of data often called the *training data set*. To evaluate the expected loss in (12.1), we need to evaluate the expectation over all possible data sets. In practice, the joint distribution of the data is unknown, hence evaluating this expectation is intractable. Instead, a portion of the data set $D_{training} \subset D$, is used to perform the learning and the remainder of the data set $D_{test} = D \backslash D_{training}$ where $D_{training} \cup D_{test} = D$, is used to evaluate the performance of f using the loss function L. The subset of the data used for this evaluation, D_{test}, is called the *test data set*. The expected loss over the training data set is the *empirical risk* and is defined as:

$$\hat{R}(\hat{f}) = \frac{1}{n} \sum_{i=1}^{n} L(f(x_i), y_i). \tag{12.5}$$

Here n represents the number of samples in the training data set, $D_{training}$. The learning algorithm uses the empirical risk, $R(\hat{f})$, as a surrogate for the true risk, $R(f)$, to evaluate the performance of f. The best function in $\mathcal{F}$ for the prediction task is the one associated the lowest empirical risk. This principle is called *Empirical Risk Minimization*, see e.g. Vapnik (1998), and is defined as

$$f^* = \inf_{f \in \mathcal{F}} \hat{R}(f). \tag{12.6}$$

The task of applying algorithm $\mathcal{A}$ to determine f^* is called the *learning task*. Implementation of algorithm $\mathcal{A}$ is called the *learner*. For example, the maximum likelihood procedure is an example of the *learner*. The lowest possible risk for the learning problem, R^* associated with the function f^*, is called the *Bayes risk* and hence finding the Bayesian learner should be the main purpose in scientific problems.

If a Bayesian learner cannot be found then the user may need to compare performances of different learning algorithms using what is known as *bias-variance trade-off*, see e.g. Hastie et al. (2009), to determine the best hypothesis class for a problem. This is similar in spirit to the model choice criteria PMCC describe in Section 4.16.3. Recall that the PMCC has been split up in two parts: goodness-of-fit and penalty. In a similar vein the risk associated with an arbitrary learner f can be split up in two parts: square of a bias term in estimation and variance of predictions.

Choosing hypothesis classes that are more complex than what is optimal can lead to a phenomenon called **over-fitting**. Often, over-fitting implies very good performance of the class on the training data set but very poor performance on the test data. The capability of the function determined by algorithm $\mathcal{A}$ to maintain the same level of precision on both the training and test data sets is called **generalization**. If the algorithm $\mathcal{A}$ generalizes well, then the new insight learned from modeled data is likely to be reproducible in the new data set, provided the training data set is true representation of the population.

12.2.2   Learning by complexity penalization

Data sets with complex structure occur in many applications. Using a complex hypothesis class on a simple learning problem and simple hypothesis class on a complex problem, both result in poor performance. Hence we need methods that are sophisticated to handle the complexity of the problem. The method should consider a set of hypothesis, and pick an optimal hypothesis class based on an assessment of the training data. The method must also achieve good generalization and must avoid over-fitting. A class of methods that can achieve good generalization is known as *Complexity Penalization Method*. Such a method includes a penalty for the complexity of the function while evaluating the risk associated with it, see Hastie et al. (2009) for details. The general template to determine the solution f^* of complexity penalization method is:

$$f^* = \operatorname*{argmin}_{f \in \mathcal{F}} \left\{ \hat{R}(f) + C(f) \right\}, \tag{12.7}$$

where $C(f)$ is the term associated with the complexity of the function f. The solution f^* in (12.7) is the solution of constrained optimization of the risk $\hat{R}(f)$, where $C(f)$ is the cost or constrained on $\hat{R}(f)$.

We have not yet discussed ways of specifying the complexity $C(f)$ of the hypothesis class. There are many available methods and the right choice depends on the learning problem and hypothesis class $\mathcal{F}$. The intent here is to point out that methods to specify the complexity of the hypothesis class exists. Examples of choices used to specify the complexity of the hypothesis class include *VC dimension (Vapnik Chevronenkis dimension)*, *Covering Number* and *Radamacher Complexity*, see Bousquet et al. (2004) for details.

The Bayesian approach to making inference has one-to-one correspondence with the complexity penalization method. In the Bayesian approach, we consider a prior probability distribution, denoted by $P(f)$, of the unknown f over $\mathcal{F}$ so that $\int_{\mathcal{F}} P(f)df = 1$. The Bayes rule obtains the posterior distribution,

$$P(f|y) = \frac{P(y|f)P(f)}{P(y)}, \tag{12.8}$$

where y represents the data or the labels. The denominator in Equation (12.8) is the normalizing constant to ensure that probabilities associated with the functions in $\mathcal{F}$ integrate to 1 and the denominator is free from f. So after taking the log transformation on both sides, the Equation (12.8) can be expressed as

$$\log(P(f|y)) \propto \log(P(y|f)) + \log(P(f)).$$

Consider the right hand side of Equation (12.7). The first term, $\hat{R}(f)$ called the risk, is proportional to the negative log-likelihood of the function f, i.e., $-\log(P(y|f))$. The second term of Equation (12.7), $C(f)$ can be interpreted as the negative log-prior distribution, i.e., $-\log(P(f))$ for the problem under consideration, see Hastie et al. (2009). The $C(f)$ can also be viewed as a cost function. The cost $C(f)$, is large when the function f is less likely and is small when f is more likely. The solution f^* in Equation (12.7) is the mode of the posterior distribution of f, i.e.,

$$f^* = \underset{f \in \mathcal{F}}{\operatorname{argmin}} \left\{ - \log(P(f|x)) \right\}. \tag{12.9}$$

The posterior mode is the Bayes estimator under Kullback-Libeler type loss function, see Das and Dey (2010). This shows the theoretical equivalence of Bayesian and complexity penalization methods. The prior distribution in Bayesian inference basically acts as a penalty function in estimation or equivalently learning in data science.

12.2.3   Supervised learning and generalized linear models

As noted in Table 12.1 the data science community sees supervised learning as solution to two different problems, namely (i) regression and (ii) classification. However, in model based statistics these two are special cases of one set of umbrella methods termed as generalized linear models (GLM) in Chapter 10. The approach in statistics is to model the dependent variable y using the natural exponential family (10.1) and then use a suitable link function to model the mean of the data as a function of the covariates. Thus, what matters most is the distribution of the data – if the distribution is discrete then we fit and predict using a discrete distribution model as one of the members of the exponential family (10.1). As seen in Chapters 10 and 11 these models classify future data using the Bayesian posterior predictive distribution (4.8). MCMC based methods reconstruct such a predictive distribution using sampling as discussed in Chapter 5. Thus, the predictive classification categories are always the ones which are present in the data in the first place. We, however, note that the GLMs in Chapter 10 have been presented for areal data only but those can also be assumed for point referenced data, although this book does not discuss those models.

12.2.4   Ridge regression, LASSO and elastic net

These three are oft-quoted keywords in statistical learning theory and hence these require special attention. These methods are most meaningful in the context of regression or supervised learning with many explanatory variables. These are often used to avoid the multicollinearity problem in regression where the covariates are highly correlated so that the least squares regression technique becomes problematic because of the near singularity of the $X'X$ matrix discussed below. Following the spirit of this chapter and the section we now introduce these methods and discuss their connections with the Bayesian methods.

The three methods assume the function f is linear

$$f(X) = X\beta, \tag{12.10}$$

where $X = [x_{ij}]_{n \times p}$ is the collection of p covariates or features and $\boldsymbol{\beta} = (\beta_1, \cdots, \beta_p)$ are regression coefficients. Thus, in this set we parameterize the unknown function f we aim to learn about using p unknown parameters and a strict linear functional form (12.10). The ordinary least square solutions of β is

$$\hat{\boldsymbol{\beta}}_{OLS} = \underset{\boldsymbol{\beta}}{\operatorname{argmin}}\{(\mathbf{y} - X\boldsymbol{\beta})'(\mathbf{y} - X\boldsymbol{\beta})\},$$

can be obtained by solving the normal equations

$$X'X\boldsymbol{\beta} = X'\mathbf{y}. \tag{12.11}$$

If two (or more) predictors are highly correlated, that makes the system of equations (12.11) "near singular". It makes the solution unreliable. The "near singular" undesirable property of many problems are known as *multicollinearity*, and the L_2 penalty on $\boldsymbol{\beta}$ can fix the problem. The approach is known as the Ridge solution of the multicollinearity, see Hoerl and Kennard (1970),

$$\hat{\boldsymbol{\beta}}_{Ridge} = \underset{\boldsymbol{\beta}}{\operatorname{argmin}}\{(\mathbf{y} - X\boldsymbol{\beta})'(\mathbf{y} - X\boldsymbol{\beta}) + \lambda \boldsymbol{\beta}'\boldsymbol{\beta}\}, \tag{12.12}$$

where $\lambda > 0$ is a tuning parameter discussed below. If we compare the Ridge solution in (12.12) with (12.7), the first term

$$R(\boldsymbol{\beta}) = (\mathbf{y} - X\boldsymbol{\beta})'(\mathbf{y} - X\boldsymbol{\beta}),$$

is the residual sum of squares and $C(\boldsymbol{\beta}) = \boldsymbol{\beta}'\boldsymbol{\beta}$ is the L_2 penalty on $\boldsymbol{\beta}$. The objective function in Equation (12.12) can be presented as

$$p(\boldsymbol{\beta}|\mathbf{y}, X, \sigma^2) \propto \exp\left\{-\frac{1}{2\sigma^2}(\mathbf{y} - X\boldsymbol{\beta})'(\mathbf{y} - X\boldsymbol{\beta})\right\} \cdot \exp\left\{-\frac{\lambda}{2\sigma^2}\boldsymbol{\beta}'\boldsymbol{\beta}\right\},$$

where $p(\boldsymbol{\beta}|\mathbf{y}, X, \sigma^2)$ is the posterior distribution of $\boldsymbol{\beta}$, the L_2 penalty is proportional to the Gaussian prior distribution on $\boldsymbol{\beta}$, where $(\boldsymbol{\beta}|\sigma^2, \lambda) \sim N(0, \sigma^2/\lambda)$, and $y \sim N(X\boldsymbol{\beta}, \sigma^2 I)$ yields the likelihood function. In this case, the Ridge solution is the *posterior mode*, and it has a mathematically closed form solution:

$$\hat{\boldsymbol{\beta}}_{Ridge} = (X'X + \lambda I)^{-1}X'\mathbf{y}.$$

This result implies that the Ridge learning method is the Bayesian solution, which is also known as the shrinkage estimator, see e.g., Hastie et al. (2009). One more point we must note is that, if two predictors are highly correlated, i.e., both the predictors inherently contained similar kind of information, then they are naturally expected to have a similar functional relationship with y. Hence we need an algorithm, which keeps the predictors which are most relevant in predicting y and drop the less crucial features and come up with a parsimonious model, see Tibshirani (1996). Managing the complexity of the hypothesis class involves reducing the number of features in f and the task is known as *feature selection*, see Tibshirani (1996). In Bayesian statistics, the same task is known as the *model selection*, see e.g. Gelfand and Dey (1994).

Hence, the learning algorithm $\mathcal{A}$ should figure out the best subset of p features from X, for which a performance metric like the RMSE and others detailed in Section 6.8. One can apply the *best subset selection*, see Hastie et al. (2009) [Chapter 3], but the best model has to search through 2^p many models. So the complexity of model space makes it impossible to implement even for a data set with even a moderate number of features, say 20.

The shrinkage methods are a popular technique to manage complexity for linear hyper-planes hypothesis class, see Tibshirani (1996). The Least Absolute Shrinkage and Selection Operator (LASSO) can be a particularly effective technique for feature selection, see Tibshirani (1996). If the values of coefficients are estimated to be zero, then effectively the solution is to drop that feature from the model. Such solutions are called *sparse* solutions. The LASSO yields the desired *sparse* solutions with L_1 penalty on $\boldsymbol{\beta}$, defined as

$$C(\boldsymbol{\beta}) = \lambda \sum_{j=1}^{p} \left| \boldsymbol{\beta}_j \right|.$$

Although the Ridge solution handles the multicollinearity issue, it, however, fails to yield the *sparse* solutions. The LASSO estimate is defined as:

$$\hat{\beta}_{lasso} = \underset{\boldsymbol{\beta}}{\mathrm{argmin}} \left\{ (\mathbf{y} - X\boldsymbol{\beta})'(\mathbf{y} - X\boldsymbol{\beta}) + \lambda \sum_{j=1}^{p} \left| \boldsymbol{\beta}_j \right| \right\}, \qquad (12.13)$$

where λ is a parameter that affects the *sparsity* of the solution. The L_1 penalty on $\boldsymbol{\beta}$ is equivalent to the Laplace or double exponential prior distribution, see e.g., Park and Casella (2008). The least angle regression algorithm (LARS) (Efron et al., 2004) for LASSO solution is a popular algorithm which makes the LASSO solution highly scalable for large data sets.

Note that λ is a parameter that must be provided to the learning algorithm $\mathcal{A}$. There are several approaches to learn (estimate) λ. In one approach, λ is learned using a grid search with k-fold cross-validation technique, see Section 6.8.2. In another approach, full Bayesian methodology elicits a prior on λ, known as the Bayesian LASSO presented in Park and Casella (2008). The Bayesian LASSO focuses on estimating the posterior mean of $\boldsymbol{\beta}$ using the Gibbs sampler. The slow implementation of the Gibbs sampler makes the full Bayesian implementation of the LASSO less attractive for practitioners. On the contrary, the fast, scalable implementation of the LARS makes it very attractive with partial Bayes solution for the practitioner.

The convex combination of the L_1 and L_2 penalty yields a new kind of penalty, known as the *elastic net*,

$$\hat{\beta}_{EN} = \underset{\boldsymbol{\beta}}{\mathrm{argmin}} \left\{ (\mathbf{y} - X\boldsymbol{\beta})'(\mathbf{y} - X\boldsymbol{\beta}) + \lambda \sum_{j=1}^{p} \left(\alpha \left| \boldsymbol{\beta}_j \right| + (1-\alpha)\beta_j^2 \right) \right\}, \quad (12.14)$$

where $0 \leq \alpha \leq 1$, see Zou and Hastie (2005). Like LASSO and Ridge, we can similarly argue that the Elastic Net solution is a Bayesian solution and fully Bayesian Elastic Net implementation is also available, see Li and Lin (2010). One of the advantages of the Elastic Net is that it can address the multicollinearity problem and feature selection together. The *copula prior* proposed in a recent paper showed that the Ridge, LASSO, elastic net etc. are special cases of the copula prior solution, see Sharma and Das (2017).

12.2.5 Regression trees and random forests

A learning method using regression trees partition the input space $\mathcal{X}$ into regions with a constant response for each region, see Breiman et al. (1984). If there are M regions in this partition then

$$\mathcal{X}_1 \cup \mathcal{X}_2 \cdots \cup \mathcal{X}_M = \mathcal{X}$$

where $\mathcal{X}_i \cap \mathcal{X}_j = \emptyset$ if $i \neq j$. Here M represents the number of terminal nodes and is an important parameter for the tree hypothesis class discussed below. The hypothesis class for trees takes the following form:

$$f(x) = \sum_{m=1}^{M} c_m I(x \in \mathcal{X}_m), \tag{12.15}$$

where c_m represents the constant response for region $\mathcal{X}_m$ and $I(x \in \mathcal{X}_i)$ is the indicator function that is defined as

$$I(x) = \begin{cases} 1 & \text{if } x \in \mathcal{X}_m \\ 0 & \text{otherwise.} \end{cases}$$

If we use the square error loss function then the optimal choice for c_m is the average of the response values y_i in the region $\mathcal{X}_m$. The input space is partitioned into regions $\mathcal{X}_1, \cdots, \mathcal{X}_m$ using a greedy algorithm as detailed in Hastie et al. (2009)[Chapter 9, Section 9.2.2]. This algorithm recursively partitions $\mathcal{X}$ as follows. Using the squared error loss function, it is justified to estimate each c_m by

$$\hat{c}_m = \frac{1}{N} \sum_{x_i \in \mathcal{X}_m} y_i$$

where N is the number of instances of the response y_i's that belong to region $\mathcal{X}_m$. The greedy algorithm starts with all the data in one group so that $M = 1$ and splits into two as follows. We choose a splitting variable j out of the p covariates and a split point s, and define the pair of regions

$$\mathcal{X}_1(j, s) = \{X | X_j \leq s\} \text{ and } \mathcal{X}_2(j, s) = \{X | X_j \geq s\}.$$

We find the optimal values of j and s by minimizing

$$\min_{j,s} \left[\min_{c_1} \sum_{x_i \in \mathcal{X}_1(j,s)} (y_i - c_1)^2 + \min_{c_2} \sum_{x_i \in \mathcal{X}_2(j,s)} (y_i - c_2)^2 \right].$$

The minimizer for c_1 and c_2 are the averages $\hat{c}_1$ and $\hat{c}_2$ given above. The outer minimization is then performed by inspecting each of the p covariates. The optimal values of j and s, $\hat{j}$ and $\hat{s}$ say, have allowed us to partition $\mathcal{X}$ into two $\mathcal{X}_1(\hat{j}, \hat{s})$ and $\mathcal{X}_2(\hat{j}, \hat{s})$. These are also called nodes of the tree. The splitting algorithm is then applied separately on each of these two regions.

The algorithm needs to be stopped at an optimal number of final partition size M. This parameter M represents the height of the tree. It determines the complexity of the solution and the complexity management strategy must monitor the parameter. A strategy that works well is to partition the input space until there is a minimum (threshold) number of instances in each region. This tree is then shortened using pruning. Pruning is facilitated by minimization of a cost function, which is defined as follows. Let T be the tree that is subject to pruning having $|T|$ nodes. The cost function to be minimized for pruning is defined as:

$$C_\alpha(T) = \sum_{m=1}^{|T|} \sum_{y_i \in \mathcal{X}_m} (y_i - c_m)^2 + \alpha |T|, \qquad (12.16)$$

where α is a parameter that controls the complexity associated with T. Note that $C_\alpha(T)$ is the penalized sum of square of errors. As with the linear model, for each value of α, we obtain a hypothesis f_α by applying the empirical risk minimization technique of Section 12.2.1 where (12.16) is minimized.

Many variations of tree models have been developed. One of the most popular is the *Random Forest* where multiple number of trees, often called an ensemble, are generated at the training stage and then prediction is performed by averaging from the output of the individual trees. That is why the random forest technique is called an ensemble learning method, see e.g. Breiman (2001).

12.3   Gaussian Process (GP) prior-based machine learning

The final hypothesis class we consider is the Gaussian Process priors. For this, we consider a full Bayesian approach to learning. We still use the template defined by (12.7); however we now use the Bayesian approach, explained in the Equation (12.8), to pick the best model from the hypothesis class $\mathcal{F}$.

A GP prior distribution is assumed over a space of functions with zero mean. Hence the GP prior is not assumed for the data per se but it is assumed for the error function. The following development assumes that data y_i's have zero mean. A mean function can be added to model data with non-zero means as we shall illustrate. The GP prior distribution is described as follows.

As in Bayesian hierarchical modeling we first assume,

$$y_i = f_i(\mathbf{x}_i) + \epsilon_i,$$

where $\epsilon_i \sim N(0, \sigma_\epsilon^2)$. This means $\mathbf{y} \sim N(\mathbf{f}, \sigma_\epsilon^2 I)$, where $\mathbf{f} = (f_1(\mathbf{x}_1), \ldots, f_n(\mathbf{x}_n))$. The functional values $f_1(\mathbf{x}_1), \ldots, f_n(\mathbf{x}_n)$ are assumed to be realization of a GP with zero mean and a covariance function, which is discussed below, following Rasmussen and Williams (2006). This assumption implies that

$$\mathbf{f} \sim N(\mathbf{0}, K)$$

where K is the $n \times n$ matrix containing the covariances between the functionals $f_i(\mathbf{x}_i)$ and $f_j(\mathbf{x}_j)$ for all i and j. Now the problem is how do we define the elements of $K_{ij} = K(\mathbf{x}_i, \mathbf{x}_j)$. The function $K(\mathbf{x}_i, \mathbf{x}_j)$ is also known as the kernel function in machine learning.

There are different choices for the Kernel K. For example,

- Linear: $K(\mathbf{x}_i, \mathbf{x}_j) = \sigma_0^2 + \mathbf{x}_i \mathbf{x}_j'$

- Periodic: $K(\mathbf{x}_i, \mathbf{x}_j) = \exp\left(-\frac{2\sin^2\left(\frac{\mathbf{x}_i - \mathbf{x}_j}{2}\right)}{\lambda^2}\right).$

These define non-stationary covariance functions which we do not discuss any further in this book. Instead, we discuss the GP prior based on the development in Section 2.7, which is the reader is familiar with. Recall the definition of GP presented in Section 2.7. The covariates $\mathbf{x}$ here serve as the spatial locations $\mathbf{s}$ used there. Also note the conflict in the notation f as commented in the second paragraph of Section 12.2.

The covariance function used to define the GP prior on f can still be assumed to be a member of the Matèrn family of covariance functions (2.1). The distance, which is in the argument of the Matèrn correlation function (2.2), is now calculated as the distance between the two covariate vectors $\mathbf{x}_i$ and $\mathbf{x}_j$ given by $d_{ij} = ||\mathbf{x}_i - \mathbf{x}_j||$. The intuition in choosing this correlation function, characterized by the distance measure, is same as before – function variables close in input space are highly correlated, whilst those far away are uncorrelated.

The assumption of Matèrn family of covariance function implies that marginally the data $\mathbf{y}$ has the distribution

$$\mathbf{y} \sim N(\mathbf{0}, K + \sigma_\epsilon^2 I)$$

where K is the matrix having elements $K_{ij} = \sigma^2 \rho(d_{ij}|\boldsymbol{\psi})$, see (2.2). The reader may now recall that this model is the marginal model with nugget effect as in Section 6.5.1. Hence, the GP prior based machine learning has already been extensively analyzed previously in Chapter 6. Moreover, note that Chapter 7 discusses these marginal models for spatio-temporal data.

One very strong point in favor of the GP prior model is that the estimates, $\hat{f}(\mathbf{x})$ say, approximates $f(\mathbf{x})$ well, i.e.,

$$P\left(\sup_{x} |\hat{f}(\mathbf{x}) - f(\mathbf{x})| < \epsilon\right) > 0 \quad \forall \epsilon > 0,$$

see Ghoshal and Vaart (2017) for further details. However, the solution $\hat{f}$ involves the inversion of the covariance matrix of order n as we have seen in previous chapters.

12.3.1   Example: predicting house prices

Taken from Sambasivan et al. (2020) this example illustrates the finer accuracy of GP based learning methods when compared with LASSO and regression trees discussed above. The data set, originally created by Harrison and Rubinfeld (1978) and can be downloaded from the web site cited by Lichman (2016), provides the median house prices in each of the census tracts in 1970 in the Boston region along with covariates described in Table 12.2. For our illustration, we consider a regression task where the objective is to predict the value of median house price when we know other information about the census tract.

	Attribute	Description
1	CRIM	per capita crime rate by town
2	ZN	proportion of residential land zoned for lots over 25,000 sq.ft
3	INDUS	proportion of non-retail business acres per town
4	CHAS	Charles River dummy variable ($= 1$ if tract is adjacent to river; 0 otherwise)
5	NOX	nitric oxides concentration (parts per 10 million)
6	RM	average number of rooms per dwelling
7	AGE	proportion of owner-occupied units built prior to 1940
8	DIS	weighted distances to five Boston employment centres
9	RAD	index of accessibility to radial highways
10	TAX	full-value property-tax rate per USD 10,000
11	PTRATIO	pupil-teacher ratio by town
12	B	Proportion of Black Minority
13	LSTAT	percentage of lower status of the population
14	MEDV	median value of owner-occupied homes in USD 1000's

TABLE 12.2: Description of the Boston house price data set.

The quantity to be predicted is the median house value (attribute *MEDV*) for a census tract. This data set does not contain the coordinates of the locations of houses as the median price refers to the median house price in a census tract. Hence, this is not an example of a typical point referenced spatial data modeled in Chapter 6. Hence the spatial modeling functions in the `bmstdr` package are not suitable for analyzing this data set. An areal data model of Chapter 10 may perhaps be more appropriate. However, that is also not feasible here since the boundary information of the census tracts is not included in the data set. Hence, we have to use alternative software packages for fitting three ML methods, (i) LASSO, (ii) regression trees see e.g. Breiman et al. (1984) using the `rpart` package (Therneau et al., 2017), and (iii) GP regression. The data and code for doing these have been provided by Sambasivan et al. (2020) on the website [1]. There are 505 samples in the data set. We use 70% of the data for training and 30% of the data for testing. We use the root mean square error as a metric to evaluate the performance of the model produced by the hypothesis class.

Sambasivan et al. (2020) attempt a range of α values in (12.16) and they have picked the solution that had the lowest penalized sum of squared errors. For each α, (12.16), i.e., $C_\alpha(T)$ provides the optimal tree size. This is illustrated in Table 12.3. A review of Table 12.3 shows that a tree with 6 splits produces the best result.

α	Num Splits $|T|$	$C_\alpha(T)$
0.51	0	1.01
0.17	1	0.52
0.06	2	0.36
0.04	3	0.30
0.03	4	0.29
0.01	5	0.27
0.01	6	0.25
0.01	7	0.26

TABLE 12.3: Selection of α for regression trees, based on $C_\alpha(T)$. Lower the $C_\alpha(T)$ better it is. The best choice corresponds to $|T| = 6$ and $\alpha = 0.01$.

Finally, Sambasivan et al. (2020) consider the Gaussian process prior models. The kernel used for this problem is a sum of a linear kernel and squared exponential kernel. Table 12.4 presents the RMSE in the test set, for each of the hypothesis classes. Predictions and observations in the test data set for each of the three methods are shown in three panels are shown in Figure 12.1. Clearly, the GP based ML method is doing a lot better than the other two. The discrete nature of predictions seen in the middle plot for regression tree method is due to the discrete nature of predictions by node averaging as noted in 12.15. The plots also show that it is particularly challenging to predict the very large and very small house prices.

---

[1]https://github.com/cmimlg/SMLReview

Hypothesis Class	RMSE
LASSO	4.97
Regression Tree	5.06
Gaussian Process	3.21

TABLE 12.4: RMSE for the test data set in the Boston house price example.

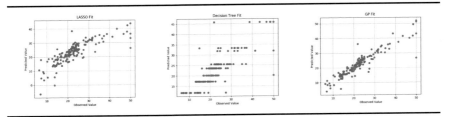

FIGURE 12.1: Predictions versus observations using three methods.

12.4 Use of GP in Bayesian calibration of computer codes

This section discusses the popular use of the GP in computer experiments. The discussion paper by Kennedy and O'Hagan (2001) on this topic sparked a huge growth in the literature in this area. The primary problem, as stated in Section 2.3 of Kennedy and O'Hagan (2001), is concerned with interpolation of the output of computer code at new input configuration given observed output data at a sample of input configurations. It is assumed that it is very expensive to run the code so that interpolation is preferred to predict the output of the code at the new input configurations. This is also a very important problem in engineering design where emulation (a surrogate) is used to predict the behavior when real time observation is expensive at new input values. A related reference in hydrological models has been provided by Kennedy and O'Hagan (2001).

A more recent comprehensive reference in this area is the book *Surrogates: Gaussian Process Modeling, Design and Optimization for the Applied Sciences* by Gramacy (2020). Gramacy notes, "A surrogate is a substitute for the real thing. ... Gathering data is expensive, and sometimes getting exactly the data you want is impossible or unethical. A surrogate could represent a much cheaper way to explore relationships, and entertain "what ifs?"." Chapter 5 of Gramacy (2020) describes the Gaussian Process regression much akin to what we have discussed in Chapter 6 of this book. In Chapter 6 of his book Gramacy discusses model based designs for GPs. Much of the GP based methodologies presented in this book can be applied to solve such problems especially when it is important to consider temporal variation as well.

In the remainder of this section we briefly discuss the GP setup of Kennedy and O'Hagan (2001). In their presentation Section 4.2, they distinguish between two groups of inputs to the computer model. The first group, referred to as *calibration inputs*, denoted by $\boldsymbol{\theta} = (\theta_1, \theta_2, \theta_{q_2})$, are context specific and these are objects of inferential interests. This is the vector of calibration parameters.

The second group of inputs comprises of all the other model inputs whose values may change in the calibrated model. These inputs are Called as *variable inputs*, and denoted by $\mathbf{x} = (x_1, \ldots, x_{q_1})$, these are assumed to be known for the observations defined below. However, these inputs may be assumed to be known or subject to parametric variability in any later use of the model. The true values of these are denoted $\zeta(\mathbf{x})$.

The calibration data are made up of n observations $\mathbf{z} = (z_1, \ldots, z_n)$, where z_i is an observation of $\zeta(\mathbf{x}_i)$ for known values of $\mathbf{x}_i$. In addition, let $\mathbf{y} = (y_1, \ldots, y_N)$ denote the output from $N(\gg n)$ runs of the computer code, where

$$y_j = \eta(\mathbf{x}_j^*, \mathbf{t}_j)$$

where both the variable inputs $\mathbf{x}_j^*$ and calibration inputs $\mathbf{t}_j$ are known for each run j. The full data set obtained is the collection of $\mathbf{y}$ and $\mathbf{z}$.

The model linking these data and inputs as formulated by Kennedy and O'Hagan (2001) is that

$$z_i = \zeta(x_i) + e_i = \rho\eta(\mathbf{x}_i, \boldsymbol{\theta}) + \delta(\mathbf{x}_i) + e_i, \qquad (12.17)$$

for $i = 1, \ldots, n$, where ρ is the regression coefficient. The observation errors e_i are given independent normal distribution with mean zero and variance σ_ϵ^2 – much like the nugget effect in the model (6.12). The processes $\eta(\mathbf{x}_i, \boldsymbol{\theta})$ and $\delta(\mathbf{x}_i)$ are assumed to be independent and also independent of the nugget effect. Kennedy and O'Hagan (2001) provide justification to assume that both of these unknown processes are given GP prior distributions with non-zero means respectively given by

$$m_1(\mathbf{x}, \mathbf{t}) = \mathbf{h}_1(\mathbf{x}, \mathbf{t})'\boldsymbol{\beta}_1, \quad m_2(\mathbf{x}) = \mathbf{h}_2(\mathbf{x})'\boldsymbol{\beta}_2,$$

where $\mathbf{h}_1(\cdot, \cdot)$ and $\mathbf{h}_2(\cdot)$ are known functions. The GP specifications use the correlation function

$$r(\mathbf{x} - \mathbf{x}') = \exp\left\{-\sum_{j=1}^{q}\phi_j(x_j - x_j')^2\right\}$$

where ϕ_j's are parameters to be estimated. The full Bayesian model is completed by assuming prior distributions for all the unknown parameters. The full estimation procedure has been detailed and illustrated by Kennedy and O'Hagan (2001) in their paper and we do not reproduce that here. We, instead, conclude the ubiquitousness of the GP based modeling in the scientific literature. We also note that it would be desirable to have **R** software packages that can perform these sorts of model fitting and predictions.

12.5 Conclusion

This chapter has introduced many key concepts in the vast area of machine learning and data science. Many keywords in machine learning such as supervised learning, LASSO, ridge regression are explained using a Bayesian perspective. The ubiquitous use of Gaussian processes and the superiority of machine learning methods using Gaussian processes have been discussed and illustrated. This chapter also discusses the use of Gaussian process in the Bayesian calibration of computer codes.

12.6 Exercises

1. Apply machine learning methods such as ridge regression and LASSO to the `piemonte` data set. Perform model validation and compare your results with those from GP based spatio-temporal modeling.

2. Solve the previous exercse for the `nysptime` data set. Obtain and compare the methods using the four validation criteria discussed in Section 6.8.

Appendix A: Statistical densities used in the book

A.1 Continuous

1. **Uniform** A random variable X follows the uniform distribution $U(a, b)$ if it has the pdf

$$f(x|a, b) = \frac{1}{b - a}, \quad \text{when } a < x < b$$

   It can be shown that $E(X) = \frac{1}{2}(a + b)$ and $\text{Var}(X) = (b - a)^2/12$.

2. **Cauchy** A random variable X follows the Cauchy distribution $C(a, b)$ if it has the pdf

$$f(x|a, b) = \frac{b}{\pi \left[b^2 + (x - a)^2 \right]}, \quad \text{for } -\infty < x < \infty, \ b > 0.$$

   The mean and variance do not exist for this distribution. The standard Cauchy distribution is the special case of this distribution when $a = 0$ and $b = 1$.

3. **Half-Cauchy** This distribution is a special case of the Cauchy distribution when the range is restricted to one side only. A random variable X follows the half-Cauchy distribution $C(a, b)$ if it has the pdf

$$f(x|a, b) = \frac{b}{2\pi \left[b^2 + (x - a)^2 \right]}, \quad \text{for } a < x < \infty, \ b > 0.$$

   Notice the factor 2 in the denominator and the range of x starting at a. The standard Cauchy distribution is the special case of this distribution when $a = 0$ and $b = 1$.

4. **Gamma** A random variable X follows the gamma distribution with parameters $a > 0$ and $b > 0$, denoted by $G(a, b)$, if it has the pdf

$$f(x|a, b) = \frac{b^a}{\Gamma(a)} x^{a-1} e^{-bx}, x > 0. \tag{A.18}$$

DOI: 10.1201/9780429318443-A

The fact that the above is a density function implies that

$$\int_0^\infty x^{a-1} e^{-bx} dx = \frac{\Gamma(a)}{b^a}, \quad \text{for } x > 0. \tag{A.19}$$

Thus, $\Gamma(a)$ is the definite integral above when $b = 1$. Using the gamma integral (A.19) it can be shown that:

$$E(X) = \frac{a}{b}, \text{ and } \text{Var}(X) = \frac{a}{b^2}.$$

We also can prove the results:

$$E\left(\frac{1}{X}\right) = \frac{b}{a-1}, \text{ if } a > 1 \text{ and } \text{Var}\left(\frac{1}{X}\right) = \frac{b^2}{(a-1)^2(a-2)}, \text{ if } a > 2.$$

In fact, the distribution of the random variable $\frac{1}{X}$ is known as the inverse gamma distribution with parameters a and b, denoted by IG(a, b). We shall not require the inverse gamma distribution in this book but for the sake of completeness we formally define the inverse gamma distribution below.

The Gamma distribution has two important special cases:

(a) **Exponential** When $a = 1$ the gamma distribution reduces to the exponential distribution which has pdf

$$f(x|b) = be^{-bx}, x > 0, \tag{A.20}$$

for $b > 0$.

(b) χ^2 When $a = \frac{n}{2}$ and $b = \frac{1}{2}$ the gamma distribution reduces to the χ^2-distribution with n degrees of freedom.

5. **Inverse Gamma** A random variable X follows the inverse gamma distribution with parameters $a > 0$ and $b > 0$ if it has the pdf

$$f(x|a, b) = \frac{b^a}{\Gamma(a)} \frac{1}{x^{a+1}} e^{-\frac{b}{x}}, x > 0. \tag{A.21}$$

As stated above, it can be shown that $E(X) = \frac{b}{a-1}$ if $a > 1$ and $\text{Var}(X) = \frac{b^2}{(a-1)^2(a-2)}$ if $a > 2$.

6. **Beta** A random variable X follows the beta distribution with parameters $a > 0$ and $b > 0$ if it has the pdf

$$f(x|a, b) = \frac{1}{B(a, b)} x^{a-1} (1-x)^{b-1}, 0 < x < 1, \tag{A.22}$$

where

$$B(a, b) = \int_0^1 x^{a-1} (1-x)^{-1} dx.$$

It can be shown that $E(X) = \frac{a}{a+b}$ and $\text{Var}(X) = \frac{ab}{(a+b)^2(a+b+1)}$ and

$$B(a,b) = \frac{\Gamma(a)\Gamma(b)}{\Gamma(a+b)}, \quad a > 0, b > 0.$$

7. **Univariate normal**: A random variable X has the normal distribution, denoted by $N(\mu, \sigma^2)$, if it has the probability density function

$$f(x|\mu, \sigma^2) = \frac{1}{\sqrt{2\pi\sigma^2}} e^{-\frac{1}{2\sigma^2}(x-\mu)^2}, \quad -\infty < x < \infty, \qquad (A.23)$$

where $\sigma^2 > 0$ and μ is unrestricted. It can be shown that $E(X) = \mu$ and $\text{Var}(X) = \sigma^2$.

8. **Multivariate normal**: A p dimensional random variable $\mathbf{X}$ has the multivariate normal distribution, denoted by $N(\boldsymbol{\mu}, \Sigma)$, if it has the probability density function

$$f(\mathbf{x}|\boldsymbol{\mu}, \Sigma) = \left(\frac{1}{2\pi}\right)^{\frac{p}{2}} |\Sigma|^{-\frac{1}{2}} e^{-\frac{1}{2}(\mathbf{x}-\boldsymbol{\mu})'\Sigma^{-1}(\mathbf{x}-\boldsymbol{\mu})}, \qquad (A.24)$$

where $-\infty < x_i < \infty$, $i = 1, \ldots, p$ and Σ is a $p \times p$ positive semi-definite matrix, $|\Sigma|$ is the determinant of the matrix Σ. It can be shown that $E(\mathbf{X}) = \boldsymbol{\mu}$ and $\text{Var}(X) = \Sigma$. The matrix Σ is also called the covariance matrix of X and the inverse matrix Σ^{-1} is called the inverse covariance matrix. This distribution is a generalization of the univariate normal distribution.

Note that when Σ^{-1} is singular, i.e. $|\Sigma^{-1}| = 0$, the covariance matrix Σ, which is the inverse matrix of Σ^{-1} does not exist. However, the above density (A.24) is written as

$$f(\mathbf{x}|\boldsymbol{\mu}, \Sigma) \propto e^{-\frac{1}{2}(\mathbf{x}-\boldsymbol{\mu})'\Sigma^{-1}(\mathbf{x}-\boldsymbol{\mu})}, \qquad (A.25)$$

without the up front normalizing constant. This distribution is called a *singular normal distribution* and in Bayesian modeling this is often used as a prior distribution, see e.g. Section 2.14.

We now state the conditional distribution of a subset of the random variables $\mathbf{X}$ given the other random variables. Suppose that we partition the p-dimensional vector X into one p_1 and another $p_2 = p - p_1$ dimensional random variable $\mathbf{X}_1$ and $\mathbf{X}_2$. Similarly partition $\boldsymbol{\mu}$ into two parts $\boldsymbol{\mu}_1$ and $\boldsymbol{\mu}_2$ so that we have:

$$\mathbf{X} = \begin{pmatrix} \mathbf{X}_1 \\ \mathbf{X}_2 \end{pmatrix}, \quad \boldsymbol{\mu} = \begin{pmatrix} \boldsymbol{\mu}_1 \\ \boldsymbol{\mu}_2 \end{pmatrix}.$$

Partition the $p \times p$ matrix Σ into four matrices: Σ_{11} having dimension $p_1 \times p_1$, Σ_{12} having dimension $p_1 \times p_2$, $\Sigma_{21} = \Sigma'_{12}$ having dimension $p_2 \times p_1$, and Σ_{22} having dimension $p_2 \times p_2$ so that we can write

$$\Sigma = \begin{pmatrix} \Sigma_{11} & \Sigma_{12} \\ \Sigma_{21} & \Sigma_{22} \end{pmatrix}.$$

The conditional distribution of $\mathbf{X}_1|\mathbf{X}_2 = \mathbf{x}_2$ is the following normal distribution:

$$N\left(\mu_1 + \Sigma_{12}\Sigma_{22}^{-1}(\mathbf{x}_2 - \mu_2),\ \Sigma_{11} - \Sigma_{12}\Sigma_{22}^{-1}\Sigma_{21}\right).$$

The marginal distribution of $\mathbf{X}_i$ is $N(\mu_i, \Sigma_{ii})$ for $i = 1, 2$.

A key distribution theory result that we use in the book is the distribution of a linear function of the multivariate random variable $\mathbf{X}$. Let

$$\mathbf{Y} = \mathbf{a} + B\mathbf{X}$$

where $\mathbf{a}$ is a $m\ (< p)$ dimensional vector of constants and B is an $m \times p$ matrix. The first part of the result is that

$$E(\mathbf{Y}) = \mathbf{a} + B\mu \quad \text{and} \quad \text{Var}(\mathbf{Y}) = B\Sigma B'$$

if $E(\mathbf{X}) = \mu$ and $\text{Var}(\mathbf{X}) = \Sigma$. If in addition we assume that $X \sim N(\mu, \Sigma)$ then

$$\mathbf{Y} \sim N(\mathbf{a} + B\mu, B\Sigma B').$$

9. **Univariate t**: A random variable X has the t-distribution, $t(\mu, \sigma^2, \nu)$ if it has the density function:

$$f(x|\mu, \sigma^2, \nu) = \left(1 + \frac{(x-\mu)^2}{\nu\sigma^2}\right)^{-\frac{\nu+1}{2}}, \quad -\infty < x < \infty, \quad \text{(A.26)}$$

when $\nu > 0$. It can be shown that

$$E(X) = \mu \text{ if } \nu > 1 \text{ and } \text{Var}(X) = \frac{\nu}{\nu - 2}\sigma^2 \text{ if } \nu > 2.$$

Also,

$$E(X^2) = \mu^2 + \sigma^2 \frac{\nu}{\nu - 2}, \text{Var}(X^2) = \frac{2\sigma^4\nu^2(\nu - 1)}{(\nu - 4)(\nu - 2)^2} + 8\sigma^2\mu^2\frac{\nu}{\nu - 2},$$
$$\text{(A.27)}$$

when $\nu > 4$.

10. **Multivariate t**: A p dimensional random variable $\mathbf{X} \sim t_p(\mu, \Sigma, \nu)$ has the probability density function

$$f(\mathbf{x}|\mu, \Sigma, \nu) = \frac{\Gamma\left(\frac{\nu+p}{2}\right)}{\Gamma\left(\frac{\nu}{2}\right)(\nu\pi)^{p/2}}|\Sigma|^{-1/2}\left\{1 + \frac{(\mathbf{x} - \mu)'\Sigma^{-1}(\mathbf{x} - \mu)}{\nu}\right\}^{-(\nu+p)/2}$$
$$\text{(A.28)}$$

for $-\infty < x_i < \infty$, $i = 1, \ldots, p$ where $\nu > 0$ and Σ is a positive definite matrix. Like the univariate t-distribution, it can be shown that $E(\mathbf{X}) = \mu$ and $\text{var}(X) = \frac{\nu}{\nu-2}\Sigma$ if $\nu > 2$.

A.2   Discrete

1. **Binomial** A random variable X is said to follow the binomial distribution, denoted by $B(n, p)$, if it has the probability mass function:

$$f(x|n, p) = \binom{n}{x} p^x (1 - p)^{n-x}, \quad x = 0, 1, \ldots, n, \qquad (A.29)$$

where n is a positive integer and $0 < p < 1$ and

$$\binom{n}{x} = \frac{n!}{x!(n - x)!}.$$

It can be shown that $E(X) = np$ and $\text{Var}(X) = np(1 - p)$. The Bernoulli distribution is a special case when $n = 1$.

2. **Poisson** A random variable X is said to follow the Poisson distribution, denoted by $P(\lambda)$, if it has the probability mass function:

$$f(x|\lambda) = e^{-\lambda} \frac{\lambda^x}{x!}, \quad x = 0, 1, \ldots, \qquad (A.30)$$

where $\lambda > 0$. It can be shown that $E(X) = \lambda$ and $\text{Var}(X) = \lambda$. This is a limiting case of the Binomial distribution when $n \to \infty$, $p \to 0$ but $\lambda = np$ remains finite in the limit.

3. **Negative binomial** A random variable X is said to follow the negative binomial distribution, denoted by $NB(r, p)$, if it has the probability mass function:

$$f(x|r, p) = \frac{\Gamma(r + x)}{\Gamma(x + 1)\Gamma(r)} p^r (1 - p)^x, \quad x = 0, 1, \ldots, \qquad (A.31)$$

for a positive integer $r > 0$ and $0 < p < 1$. It can be shown that $E(X) = \frac{r(1-p)}{p}$ and $\text{Var}(X) = \frac{r(1-p)}{p^2}$. Here X can be interpreted as the number of failures in a sequence of independent Bernoulli trials before the occurrence of the rth success, where the success probability is p in each trial.

The **geometric** distribution is a special case when $r = 1$.

Appendix B: Answers to selected exercises

B.1   Solutions to Exercises in Chapter 4

1. Let D denote the event that a randomly selected person has the disease. We are given $P(D) = 0.001$.

   Let +ve denote the event that the diagnostic test is positive for a randomly selected person. We are given

   $$P(+\text{ve}|D) = 0.95, \quad P(+\text{ve}|\bar{D}) = 0.002.$$

   We have to find $P(D|+\text{ve})$. By the Bayes theorem,

   $$
   \begin{aligned}
   P(D|+\text{ve}) &= \frac{P(D \cap +\text{ve})}{P(+\text{ve})} \\
   &= \frac{P(D)P(+\text{ve}|D)}{P(D)P(+\text{ve}|D) + P(\bar{D})P(+\text{ve}|\bar{D})} \\
   &= \frac{0.001 \times 0.95}{0.001 \times 0.95 + 0.999 \times 0.002} \\
   &= 0.322
   \end{aligned}
   $$

2. Here we have the probability table:

$P(J) = 0.5$	$P(F|J) = 0.01$
$P(E) = 0.2$	$P(F|E) = 0.04$
$P(G) = 0.3$	$P(F|G) = 0.02$

   where J stands for Japan, E stands for England and G stands for Germany and F stands for the event that the Professor is faulty.

   $$
   \begin{aligned}
   P(F) &= P(F \cap J) + P(F \cap E) + P(F \cap G) \\
   &= P(J)P(F|J) + P(E)P(F|E) + P(G)P(F|G) \\
   &= 0.5 \times 0.01 + 0.2 \times 0.04 + 0.3 \times 0.02 \\
   &= 0.019
   \end{aligned}
   $$

   Now

   $$
   \begin{aligned}
   P(E|F) &= \frac{P(E)P(F|E)}{P(F)} \\
   &= \frac{0.2 \times 0.04}{0.019} \\
   &= \frac{9}{19}
   \end{aligned}
   $$

DOI: 10.1201/9780429318443-B

3. By following the Poisson distribution example in the text, here the posterior distribution is given by

$$\pi(\theta|\mathbf{y}) \propto e^{-(n+\beta)\theta}\theta^{\sum_{i=1}^{n} y_i + \alpha},$$

which is the gamma distribution with parameters $\sum_{i=1}^{n} y_i + \alpha$ and $n + \beta$. The Bayes estimator is the posterior mean,

$$E(\theta|\mathbf{y}) = \frac{\sum_{i=1}^{n} y_i + \alpha}{n + \beta} = \frac{13 + 3}{5 + 1} = \frac{8}{3}.$$

4.

$$\text{Likelihood: } f(\mathbf{y}|\theta) = \theta^n (y_1 \cdots y_n)^{\theta - 1}$$

$$\text{Prior: } \pi(\theta) = \frac{\beta^\alpha}{\Gamma(\alpha)}\theta^{\alpha - 1}e^{-\beta\theta}$$

$$\begin{aligned}\pi(\theta|\mathbf{y}) &\propto \theta^{n+\alpha-1}e^{-\beta\theta + \theta\sum_{i=1}^{n} \log y_i}\\ &= \theta^{n+\alpha-1}e^{-\theta(\beta - \sum_{i=1}^{n} \log y_i)}\end{aligned}$$

Hence $\theta|\mathbf{y} \sim G(n + \alpha, \beta - \sum_{i=1}^{n} \log y_i)$.

Therefore, the Bayes estimator under squared error loss is

$$E(\theta|\mathbf{y}) = \frac{n + \alpha}{\beta - \sum_{i=1}^{n} \log y_i}.$$

5. (i) Here

$$\begin{aligned}\pi(\mu|\bar{y}) &= \frac{\sqrt{n}}{\sqrt{2\pi\sigma^2}}\frac{\sqrt{n_0}}{\sqrt{2\pi\sigma^2}}e^{-\frac{n}{2\sigma^2}(\bar{y}-\mu)^2 - \frac{n_0}{2\sigma^2}(\mu-\gamma)^2}\\ &\propto e^{-\frac{n+n_0}{2\sigma^2}\mu^2 - \frac{1}{\sigma^2}\mu(n\bar{y}+n_0\gamma)}\\ &\propto e^{-\frac{n+n_0}{2\sigma^2}\left[\mu - \frac{n\bar{y}+n_0\gamma}{n_0+n}\right]^2}.\end{aligned}$$

Hence the posterior distribution for μ is normal with

$$\text{mean} = E(\mu|\bar{y}) = \frac{n_0\gamma + n\bar{y}}{n_0 + n}, \quad \text{variance} = \text{var}(\mu|\bar{y}) = \frac{\sigma^2}{n_0 + n}.$$

(ii) The posterior mean is a convex combination of the data and prior means since

$$E(\mu|\bar{y}) = \frac{n_0}{n_0 + n}\gamma + \frac{n}{n_0 + n}\bar{y}.$$

where the weights are proportional to the sample sizes.

The posterior variance is interpreted as the variance of the sample mean of the total number $n_0 + n$ observations.

(iii) Since ϵ and $\mu|\bar{y}$ are independently normally distributed, their sum $\tilde{Y} = \mu + \epsilon$ must be normally distributed and

$$E(\tilde{Y}|\bar{y}) = E(\mu|\bar{y}) + E(\epsilon) = \frac{n_0\gamma + n\bar{y}}{n_0 + n}.$$

Also,

$$\text{var}(\tilde{Y}|\bar{y}) = \text{var}(\mu|\bar{y}) + \text{var}(\epsilon) = \frac{\sigma^2}{n_0 + n} + \sigma^2$$

(iv) Now

$$E(\mu|\bar{y}) = \frac{0.25 \times 120 + 2 \times 130}{0.25 + 2} = 128.9.$$

$$\text{var}(\mu|\bar{y}) = \frac{25}{0.25 + 2} = 11.11$$

Hence a 95% credible interval for μ is $(128.9 \pm 1.96\sqrt{11.11}) = (122.4, 135.4)$. Also,

$$E(\tilde{Y}|\bar{y}) = E(\mu|\bar{y}) = 128.9.$$

and

$$\text{var}(\tilde{Y}|\bar{y}) = \frac{\sigma^2}{n_0 + n} + \sigma^2 = 11.11 + 25 = 36.11$$

Thus, a 95% prediction interval is given by: $(128.9 \pm 1.96\sqrt{36.11}) = (117.1, 140.7)$.

6. (i) We have $Y_1, \ldots Y_n \overset{iid}{\sim} N(0, \sigma^2)$. Therefore,

$$
\begin{aligned}
f(\mathbf{y}|\sigma^2) &= \prod_{i=1}^{n} \frac{1}{\sqrt{2\pi\sigma^2}} e^{-\frac{1}{2\sigma^2} y_i^2} \\
&= \frac{1}{(2\pi)^{n/2}(\sigma^2)^{n/2}} e^{-\frac{1}{2\sigma^2} \sum_{i=1}^{n} y_i^2}
\end{aligned}
$$

From Section A.1, the prior density of σ^2 is

$$\pi(\sigma^2) = \frac{\beta^m}{\Gamma(m)} \frac{1}{(\sigma^2)^{m+1}} e^{-\beta/\sigma^2}.$$

Therefore, the posterior density is:

$$
\begin{aligned}
\pi(\sigma^2|\mathbf{x}) &\propto \frac{1}{(\sigma^2)^{n/2}} e^{-\frac{1}{2\sigma^2} \sum_{i=1}^{n} y_i^2} \frac{1}{(\sigma^2)^{m+1}} e^{-\beta/\sigma^2}, \\
&= \frac{1}{(\sigma^2)^{n/2+m+1}} e^{-\frac{1}{\sigma^2}(\beta + \frac{1}{2}\sum_{i=1}^{n} y_i^2)}, \quad \sigma^2 > 0.
\end{aligned}
$$

Clearly this is the density of the inverse gamma distribution with parameters $m^* = n/2 + m$ and $\beta^* = \beta + \frac{1}{2}\sum_{i=1}^{n} y_i^2$.

(ii) We proceed as follows for the posterior predictive distribution.

$$
\begin{aligned}
f(x_{n+1}|\mathbf{x}) &= \int_{-\infty}^{\infty} f(x_{n+1}|\theta)\pi(\theta|\mathbf{x})d\theta \\
&= \int_{0}^{\infty} \frac{1}{\sqrt{2\pi\sigma^2}} e^{-\frac{1}{2\sigma^2}x_{n+1}^2} \frac{(\beta^*)^{m^*}}{\Gamma(m^*)} \frac{1}{(\sigma^2)^{m^*+1}} e^{-\beta^*/\sigma^2} d\sigma^2, \\
&\propto \int_{0}^{\infty} \frac{1}{\sqrt{2\pi\sigma^2}} \frac{1}{(\sigma^2)^{m^*+\frac{1}{2}+1}} e^{-\frac{1}{\sigma^2}(\beta^*+\frac{1}{2}x_{n+1}^2)} d\sigma^2.
\end{aligned}
$$

Now the integrand looks like the inverse gamma density with $\tilde{m} = m^* + \frac{1}{2}$ and $\tilde{\beta} = \beta^* + \frac{1}{2}x_{n+1}^2$

$$
\begin{aligned}
f(x_{n+1}|\mathbf{x}) &\propto \frac{\Gamma(\tilde{m})}{(\tilde{\beta})^{\tilde{m}}} \\
&= \frac{\Gamma(m^*+\frac{1}{2})}{(\beta^*+\frac{1}{2}x_{n+1}^2)^{m^*+\frac{1}{2}}} \\
&\propto \left(\beta^* + \frac{1}{2}x_{n+1}^2\right)^{-m^*-\frac{1}{2}} \\
&\propto \left(1 + \frac{x_{n+1}^2}{2\beta^*}\right)^{-m^*-\frac{1}{2}}.
\end{aligned}
$$

7. (i) We have $Y_1, \ldots, Y_n \overset{iid}{\sim} N(\beta x_i, \sigma^2)$.
Therefore,

$$
\begin{aligned}
f(y_1, \ldots, y_n|\beta) &= \prod_{i=1}^{n} \frac{1}{\sqrt{2\pi\sigma^2}} e^{-\frac{1}{2\sigma^2}(y_i-\beta x_i)^2} \\
&\propto e^{-\frac{1}{2\sigma^2}\sum_{i=1}^{n}(y_i-\beta x_i)^2}.
\end{aligned}
$$

Prior is

$$
\pi(\beta) = \frac{1}{\sqrt{2\pi\tau^2}} e^{-\frac{1}{2\tau^2}(\beta-\beta_0)^2}
$$

Therefore, posterior is:

$$
\begin{aligned}
\pi(\beta|\mathbf{y}) &\propto e^{-\frac{1}{2\sigma^2}\sum_{i=1}^{n}(y_i-\beta x_i)^2 - \frac{1}{2\tau^2}(\beta-\beta_0)^2} \\
&= e^{-\frac{1}{2}\left\{\frac{1}{\sigma^2}\sum_{i=1}^{n}(y_i-\beta x_i)^2 + \frac{1}{\tau^2}(\beta-\beta_0)^2\right\}} \\
&= e^{-\frac{1}{2}M}, \quad \text{say.}
\end{aligned}
$$

Now

$$
\begin{aligned}
M &= \frac{\sum_{i=1}^{n}y_i^2}{\sigma^2} - 2\beta\frac{\sum_{i=1}^{n}y_i x_i}{\sigma^2} + \beta^2\frac{\sum_{i=1}^{n}x_i^2}{\sigma^2} + \beta^2\frac{1}{\tau^2} - 2\beta\frac{\beta_0}{\sigma^2} + \frac{\beta_0^2}{\tau^2} \\
&= \beta^2\left(\frac{\sum_{i=1}^{n}x_i^2}{\sigma^2} + \frac{1}{\tau^2}\right) - 2\beta\left(\frac{\sum_{i=1}^{n}y_i x_i}{\sigma^2} + \frac{\beta_0}{\tau^2}\right) + \frac{\sum_{i=1}^{n}y_i^2}{\sigma^2} + \frac{\beta_0^2}{\tau^2} \\
&= \beta^2\left(\frac{1}{\sigma_1^2}\right) - 2\beta\frac{\beta_1}{\sigma_1^2} + \frac{\sum_{i=1}^{n}y_i^2}{\sigma^2} + \frac{\beta_0^2}{\tau^2} \\
&= \frac{(\beta-\beta_1)^2}{\sigma_1^2} - \frac{1}{\sigma_1^2}\left(\frac{\sum_{i=1}^{n}y_i x_i}{\sigma^2} + \frac{\beta_0}{\tau^2}\right)^2 + \frac{\sum_{i=1}^{n}y_i^2}{\sigma^2} + \frac{\beta_0^2}{\tau^2}
\end{aligned}
$$

where

$$
\sigma_1^2 = \frac{1}{\frac{\sum_{i=1}^{n}x_i^2}{\sigma^2} + \frac{1}{\tau^2}} \quad \text{and} \quad \beta_1 = \sigma_1^2\left(\frac{\sum_{i=1}^{n}y_i x_i}{\sigma^2} + \frac{\beta_0}{\tau^2}\right).
$$

Clearly,

$$
\beta|\mathbf{y} \sim N(\beta_1, \sigma_1^2).
$$

(ii) Define

$$\hat{\beta} = \frac{\sum_{i=1}^{n} y_i x_i}{\sum_{i=1}^{n} x_i^2}$$

which is the maximum likelihood estimate of β. We have

$$
\begin{aligned}
\beta_1 &= \sigma_1^2 \left(\frac{\sum_{i=1}^{n} y_i x_i}{\sigma^2} + \frac{\beta_0}{\tau^2} \right) \\
&= \frac{\frac{\sum_{i=1}^{n} y_i x_i}{\sigma^2} + \frac{\beta_0}{\tau^2}}{\frac{\sum_{i=1}^{n} x_i^2}{\sigma^2} + \frac{1}{\tau^2}} \\
&= \frac{\tau^2 \sum_{i=1}^{n} y_i x_i + \sigma^2 \beta_0}{\tau^2 \sum_{i=1}^{n} x_i^2 + \sigma^2} \\
&= \frac{\tau^2 \frac{\sum_{i=1}^{n} y_i x_i}{\sum_{i=1}^{n} x_i^2} + \frac{\sigma^2}{\sum_{i=1}^{n} x_i^2} \beta_0}{\tau^2 + \frac{\sigma^2}{\sum_{i=1}^{n} x_i^2}} \\
&= \frac{w_1 \hat{\beta} + w_2 \beta_0}{w_1 + w_2}
\end{aligned}
$$

where

$$w_1 = \tau^2 \quad \text{and} \quad w_2 = \frac{\sigma^2}{\sum_{i=1}^{n} x_i^2}.$$

(iii) As $\tau^2 \to \infty$, $\sigma_1^2 \to \frac{\sigma^2}{\sum_{i=1}^{n} x_i^2}$. That is

$$
\begin{aligned}
\beta | \mathbf{y} &\sim N \left(\hat{\beta}, \frac{\sigma^2}{\sum_{i=1}^{n} x_i^2} \right) \\
\text{i.e.} \quad \beta | \mathbf{y} &\sim N \left(\hat{\beta}, \; \text{var}(\hat{\beta}) \right).
\end{aligned}
$$

Hence inference for β using the posterior will be same as that based on the maximum likelihood estimate.

(iv) We want

$$f(y_{n+1} | \mathbf{y}) = \int_{-\infty}^{\infty} f(y_{n+1} | \beta) \pi(\beta | \mathbf{y}) d\beta.$$

Although this can be derived from the first principles, we take a different approach to solve this.

We use two results on conditional expectation:

$$E(X) = EE(X|Y), \quad \text{var}(X) = E\text{var}(X|Y) + \text{var} E(X|Y).$$

We take $X = Y_{n+1}$ and $Y = \beta$. We also have $Y_{n+1}|\beta \sim N(\beta x_{n+1}, \sigma^2)$ and $\beta | \mathbf{y} \sim N(\beta_1, \sigma_1^2)$. Now

$$E(Y_{n+1} | \mathbf{y}) = E(\beta x_{n+1}) = \beta_1 x_{n+1}.$$

$$
\begin{aligned}
\text{var}(Y_{n+1} | \mathbf{y}) &= E[\text{var}(Y_{n+1} | \beta)] + \text{var}[E(Y_{n+1} | \beta)] \\
&= E[\sigma^2] + \text{var}[\beta x_{n+1}] \\
&= \sigma^2 + x_{n+1}^2 \sigma_1^2
\end{aligned}
$$

Also we can write

$$Y_{n+1}|\mathbf{y} = x_{n+1}\beta|\mathbf{y} + \epsilon$$

where $\beta|\mathbf{y}$ follows $N(\beta_1, \sigma_1^2)$ and ϵ follows $N(0, \sigma^2)$ independently. Hence $Y_{n+1}|\mathbf{y}$ follows a normal distribution. Therefore,

$$Y_{n+1}|\mathbf{y} \sim N(\beta_1 x_{n+1}, \sigma^2 + x_{n+1}^2 \sigma_1^2).$$

8. (i) Here

$$f(\mathbf{y}|\lambda) = \lambda^n e^{-\lambda \sum_{i=1}^n y_i}$$

and

$$\pi(\lambda) = \frac{\beta^m}{\Gamma(m)} \lambda^{m-1} e^{-\beta\lambda}$$

The posterior is

$$\begin{aligned}
\pi(\lambda|\mathbf{y}) &\propto f(\mathbf{y}|\lambda) \times \pi(\lambda) \\
&\propto \lambda^{m+n-1} e^{-\lambda(\beta + \sum_{i=1}^n y_i)}.
\end{aligned}$$

Clearly $\lambda|\mathbf{y} \sim G(m+n, \beta + \sum_{i=1}^n y_i)$.

(ii) We have $f(y_{n+1}|\lambda) = \lambda e^{-\lambda y_{n+1}}$. Now

$$\begin{aligned}
f(y_{n+1}|\mathbf{y}) &= \int_0^\infty f(y_{n+1}|\lambda) \pi(\lambda|\mathbf{y}) d\lambda \\
&= \int_0^\infty \lambda e^{-\lambda y_{n+1}} \frac{(\beta+t)^{m+n}}{\Gamma(m+n)} \lambda^{m+n-1} e^{-(\beta+t)\lambda} \\
&= \frac{(\beta+t)^{m+n}}{\Gamma(m+n)} \int_0^\infty \lambda^{m+n+1-1} e^{-(\beta+t+y_{n+1})\lambda} \\
&= \frac{(\beta+t)^{m+n}}{\Gamma(m+n)} \frac{\Gamma(m+n+1)}{(\beta+t+y_{n+1})^{m+n}} \\
&= \frac{(n+m)(\beta+t)^{n+m}}{(y_{n+1}+\beta+t)^{n+m+1}},
\end{aligned}$$

where $y_{n+1} > 0$.

(iii) We have $f(y_{n+2}, y_{n+1}|\lambda) = \lambda^2 e^{-\lambda(y_{n+2}+y_{n+1})}$, since Y_{n+2} and Y_{n+1} are conditionally independent given λ. Now

$$\begin{aligned}
f(y_{n+2}, y_{n+1}|\mathbf{y}) &= \int_0^\infty f(y_{n+2}, y_{n+1}|\lambda) \pi(\lambda|\mathbf{y}) d\lambda \\
&= \int_0^\infty \lambda^2 e^{-\lambda(y_{n+2}+y_{n+1})} \frac{(\beta+t)^{m+n}}{\Gamma(m+n)} \lambda^{m+n-1} e^{-(\beta+t)\lambda} \\
&= \frac{(n+m+1)(n+m)(\beta+t)^{n+m}}{(y_{n+2}+y_{n+1}+\beta+t)^{n+m+2}},
\end{aligned}$$

where $y_{n+2} > 0$ and $y_{n+1} > 0$.

9. Recall that

$$P(M_i|\mathbf{y}) = \frac{P(M_i)f(\mathbf{y}|M_i)}{P(M_0)f(\mathbf{y}|M_0) + P(M_1)f(\mathbf{y}|M_0)}$$

where

$$f(\mathbf{y}|M_i) = \int f(\mathbf{y}|\theta, M_i) \times \pi_i(\theta) d\theta$$

and $P(M_i)$ is the prior probability of model i. We have

Model 0	Model 1
$\theta = \frac{1}{2}$	$\frac{1}{2} < \theta < 1$
$f(\mathbf{y}|\theta = \frac{1}{2}) = (\frac{1}{2})^6$	$f(\mathbf{y}|\theta) = \theta^5(1 - \theta)$
$f(\mathbf{y}|M_0) = (\frac{1}{2})^6$	$f(\mathbf{y}|M_1) = \int_{\frac{1}{2}}^1 \theta^5(1 - \theta)\,\pi_1(\theta)\,d\theta$

Now we calculate the model probabilities.

Model 0	Model 1	$P(M_0|\mathbf{y})$
$P(M_0) = \frac{1}{2}$	$P(M_1) = \frac{1}{2}$	
$f(\mathbf{y}|M_0) = \frac{1}{64}$	$f(\mathbf{y}|M_1) = \int_{\frac{1}{2}}^1 \theta^5(1 - \theta)\,2\,d\theta = \frac{5}{112}$	$P(M_0|\mathbf{y}) = 0.26$
$P(M_0) = 0.8$	$P(M_1) = 0.2$	
$f(\mathbf{y}|M_0) = \frac{1}{64}$	$f(\mathbf{y}|M_1) = \int_{\frac{1}{2}}^1 \theta^5(1 - \theta)\,8(1 - \theta)\,d\theta = \frac{73}{1792}$	$P(M_0|\mathbf{y}) = 0.60$
$P(M_0) = 0.2$	$P(M_1) = 0.8$	
$f(\mathbf{y}|M_0) = \frac{1}{64}$	$f(\mathbf{y}|M_1) = \int_{\frac{1}{2}}^1 \theta^5(1 - \theta)\,48(\theta - \frac{1}{2})(1 - \theta)\,d\theta = 0.051$	$P(M_0|\mathbf{y}) = 0.07$

10. Here

$$E(Y_i|\theta_0) = \frac{1 - \theta_0}{\theta_0}, \quad E(Y_i|\theta_1) = \theta_1.$$

We have

$$\theta_0 \sim Beta(\alpha_0, \beta_0), \quad \theta_1 \sim G(\alpha_1, \beta_1)$$

Therefore,

$$
\begin{aligned}
E(Y_i|M_0) &= \int_0^1 \frac{1-\theta_0}{\theta_0} \frac{1}{B(\alpha_0,\beta_0)} \theta_0^{\alpha_0-1}(1 - \theta_0)^{\beta_0-1} d\theta_0 \\
&= \frac{1}{B(\alpha_0,\beta_0)} \int_0^1 \theta_0^{\alpha_0-1-1}(1 - \theta_0)^{\beta_0+1-1} d\theta_0 \\
&= \frac{B(\alpha_0-1,\beta_0+1)}{B(\alpha_0,\beta_0)} \\
&= \frac{\beta_0}{\alpha_0-1}.
\end{aligned}
$$

Now $E(Y_i|M_1) = E(\theta_1)$ where $\theta_1 \sim G(\alpha_1, \beta_1)$. Therefore, $E(Y_i|M_1) = \frac{\alpha_1}{\beta_1}$.

Two predictive means are equal if

$$\frac{\beta_0}{\alpha_0 - 1} = \frac{\alpha_1}{\beta_1}.$$

The Bayes factor for Model 0 is

$$B_{01}(\mathbf{y}) = \frac{f(\mathbf{y}|M_0)}{f(\mathbf{y}|M_1)}$$

where $f(\mathbf{y}|M_i)$ is the marginal likelihood under Model i.
Let $t = \sum y_i$. Here

$$
\begin{aligned}
f(\mathbf{y}|M_0) &= \int_0^1 \theta_0^n (1-\theta_0)^t \frac{1}{B(\alpha_0,\,\beta_0)} \theta_0^{\alpha_0-1}(1-\theta_0)^{\beta_0-1} d\theta_0 \\
&= \frac{1}{B(\alpha_0,\,\beta_0)} \int_0^1 \theta_0^{n+\alpha_0-1}(1-\theta_0)^{t+\beta_0-1} d\theta_0 \\
&= \frac{B(n+\alpha_0,\,t+\beta_0)}{B(\alpha_0,\,\beta_0)}
\end{aligned}
$$

For the Poisson model

$$
\begin{aligned}
f(\mathbf{y}|M_1) &= \int_0^\infty \frac{e^{-n\theta_1}\theta_1^t}{\prod_{i=1}^n y_i!} \frac{\beta_1^{\alpha_1}}{\Gamma(\alpha_1)} \theta_1^{\alpha_1-1} e^{-\beta_1\theta_1} d\theta_1 \\
&= \frac{1}{\prod_{i=1}^n y_i!} \frac{\beta_1^{\alpha_1}}{\Gamma(\alpha_1)} \int_0^\infty \theta_1^{t+\alpha_1-1} e^{-\theta(n+\beta_1)} d\theta_1 \\
&= \frac{1}{\prod_{i=1}^n y_i!} \frac{\beta_1^{\alpha_1}}{\Gamma(\alpha_1)} \frac{\Gamma(t+\alpha_1)}{(n+\beta_1)^{t+\alpha_1}}
\end{aligned}
$$

We now calculate numerical values of the Bayes factor.

	$\alpha_0 = 1, \beta_0 = 2$ $\alpha_1 = 2, \beta_1 = 1$	$\alpha_0 = 30, \beta_0 = 60$ $\alpha_0 = 60, \beta_0 = 30$
$y_1 = y_2 = 0$	1.5	2.7
$y_1 = y_2 = 2$	0.29	0.38

B.2   Solutions to Exercises in Chapter 5

1. Let $\phi = \log \frac{\theta}{1-\theta}$. It is given that $\phi \sim N(0,1)$. Therefore,

$$
\pi(\phi) = \frac{1}{\sqrt{2\pi}} \exp\left(-\frac{1}{2}\phi^2\right)
$$

The question asks us to find the pdf of θ. We calculate the Jacobian

$$
\frac{d\,\text{old}}{d\,\text{new}} = \frac{d\phi}{d\theta} = \frac{1}{\theta(1-\theta)}.
$$

Therefore,

$$
\pi(\theta) = \frac{1}{\sqrt{2\pi}\theta(1-\theta)} \exp\left(-\frac{1}{2}\left(\log\frac{\theta}{1-\theta}\right)^2\right),
$$

if $0 < \theta < 1$.

Since $Y_1, \ldots, Y_n \sim Bernoulli(\theta)$ we have the likelihood

$$
L(\theta) = \theta^t (1-\theta)^{n-t}
$$

where $t = \sum y_i$.

Since the proposal distribution is the prior distribution the Metropolis-Hastings acceptance ratio is the ratio of the likelihood function, i.e.

$$\alpha(x,y) = \min\left\{1, \frac{y^t(1-y)^{n-t}}{x^t(1-x)^{n-t}}\right\}.$$

2. Here

$$\begin{aligned}L(\theta) &= \frac{1}{(2\pi)^{n/2}}\exp\left(-\tfrac{1}{2}\sum(y_i-\theta)^2\right)\\ &\propto \exp\left(-\tfrac{n}{2}(\theta-\bar{y})^2\right),\end{aligned}$$

and the prior is

$$\pi(\theta) = \frac{1}{\pi}\frac{1}{1+\theta^2}.$$

Let us write $a = \bar{y}$ and $x = \theta$ then we have the posterior

$$\pi(x) \propto \exp\left(-\frac{n}{2}(x-a)^2\right) \times \frac{1}{\pi}\frac{1}{1+x^2}.$$

In the rejection method $g(x) = \frac{1}{\pi}\frac{1}{1+x^2}$. Therefore,

$$\begin{aligned}M &= \sup_{-\infty<x<\infty}\frac{\pi(x)}{g(x)}\\ &= \sup_{-\infty<x<\infty}\exp\left(-\tfrac{n}{2}(x-a)^2\right)\\ &= 1,\end{aligned}$$

since the supremum is achieved at $x = a$. The acceptance probability of the rejection method is

$$\frac{1}{M}\frac{\pi(x)}{g(x)} = \exp\left[-\frac{n}{2}(x-a)^2\right].$$

Now we consider the Metropolis-Hastings algorithm. Since the proposal distribution is the prior distribution, the Metropolis-Hastings acceptance ratio is the ratio of the likelihood function, i.e.

$$\alpha(x,y) = \min\left\{1, \frac{\exp\left[-\frac{n}{2}(y-a)^2\right]}{\exp\left[-\frac{n}{2}(x-a)^2\right]}\right\}.$$

3. (a) Here

$$L(\theta,\sigma^2) \propto \frac{1}{(\sigma^2)^{n/2}}\exp\left[-\frac{1}{2\sigma^2}\sum_{i=1}^{n}(y_i-\theta)^2\right]$$

Therefore,

$$\begin{aligned}\pi(\theta,\sigma^2|\mathbf{y}) &\propto \frac{1}{(\sigma^2)^{n/2}}\exp\left[-\frac{1}{2\sigma^2}\sum_{i=1}^{n}(y_i-\theta)^2\right]\frac{1}{\sigma^2}\\ &= \frac{1}{(\sigma^2)^{n/2+1}}\exp\left[-\frac{1}{2\sigma^2}\sum_{i=1}^{n}(y_i-\theta)^2\right]\end{aligned}$$

(b) Since
$$\sum_{i=1}^{n}(y_i - \theta)^2 = \sum_{i=1}^{n}(y_i - \bar{y})^2 + n(\theta - \bar{y})^2$$

we have
$$\theta|\sigma^2, \mathbf{y} \sim N(\bar{y}, \sigma^2/n).$$

Also
$$\sigma^2|\theta, \mathbf{y} \sim IG\left(m = n/2, \beta = \sum_{i=1}^{n}(y_i - \theta)^2/2\right),$$

IG denote the inverse gamma distribution.

(c)
$$
\begin{aligned}
\pi(\theta|\mathbf{y}) &= \int_0^\infty \pi(\theta, \sigma^2|\mathbf{y})d\sigma^2 \\
&\propto \int_0^\infty \frac{1}{(\sigma^2)^{n/2+1}} \exp\left[-\frac{1}{2\sigma^2}\sum_{i=1}^{n}(y_i - \theta)^2\right] d\sigma^2 \\
&= \frac{\Gamma(n/2)}{\left[\sum_{i=1}^{n}(y_i - \theta)^2/2\right]^{n/2}} \\
&\propto \left[\sum_{i=1}^{n}(y_i - \theta)^2\right]^{-n/2} \\
&= \left[\sum_{i=1}^{n}(y_i - \bar{y})^2 + n(\theta - \bar{y})^2\right]^{-n/2} \\
&\propto \left[1 + \frac{n(\theta - \bar{y})^2}{\sum_{i=1}^{n}(y_i - \bar{y})^2}\right]^{-n/2} = \left[1 + \frac{t^2}{\alpha}\right]^{-\frac{\alpha+1}{2}},
\end{aligned}
$$

where $\alpha = n - 1$ and
$$t^2 = \frac{n(n-1)(\theta - \bar{y})^2}{\sum_{i=1}^{n}(y_i - \bar{y})^2} = \frac{n(\theta - \bar{y})^2}{s^2},$$

and now
$$s^2 = \frac{1}{n-1}\sum_{i=1}^{n}(y_i - \bar{y})^2.$$

Clearly we see that
$$t = \frac{\theta - \bar{y}}{s/\sqrt{n}}$$

follows the Student t-distribution with $n - 1$ df.
Therefore, $\theta|\mathbf{y} \sim t$-distribution with $n - 1$ df and
$$E(\theta|\mathbf{y}) = \bar{y}$$

and
$$\text{var}(\theta|\mathbf{y}) = \frac{s^2}{n}\text{var}(t_{n-1}) = \frac{s^2}{n}\frac{n-1}{n-3}, \quad \text{if } n > 3.$$

4. (a) Here

$$
\begin{aligned}
E\left[L(\tilde{\mathbf{Y}}, \mathbf{y})|\mathbf{y}\right] &= E\left[\sum_{i=1}^{n}(\tilde{Y}_i - y_i)^2|\mathbf{y}\right] \\
&= \sum_{i=1}^{n} E\left[\tilde{Y}_i - E\left(\tilde{Y}_i|\mathbf{y}\right) - y_i + E\left(\tilde{Y}_i|\mathbf{y}\right)|\mathbf{y}\right]^2 \\
&= \sum_{i=1}^{n} E\left[\tilde{Y}_i - E(\tilde{Y}_i|\mathbf{y})\right]^2 + \sum_{i=1}^{n}\left[E\left(\tilde{Y}_i|\mathbf{y}_i\right) - y_i\right]^2 \\
&\quad -2\sum_{i=1}^{n} E\left[\left\{E\left(\tilde{Y}_i|\mathbf{y}\right) - y_i\right\}\left\{\tilde{Y}_i - E(\tilde{Y}_i|\mathbf{y})\right\}\right] \\
&= \sum_{i=1}^{n} E\left[\tilde{Y}_i - E(\tilde{Y}_i|\mathbf{y})\right]^2 + \sum_{i=1}^{n}\left[E\left(\tilde{Y}_i|\mathbf{y}_i\right) - y_i\right]^2 \\
&\quad -2\sum_{i=1}^{n}\left[\left\{E\left(\tilde{Y}_i|\mathbf{y}\right) - y_i\right\}\left\{E\left(\tilde{Y}_i|\mathbf{y}\right) - E(\tilde{Y}_i|\mathbf{y})\right\}\right] \\
&= \sum_{i=1}^{n} E\left[\tilde{Y}_i - E(\tilde{Y}_i|\mathbf{y})\right]^2 + \sum_{i=1}^{n}\left[E\left(\tilde{Y}_i|\mathbf{y}_i\right) - y_i\right]^2 \\
&= \sum_{i=1}^{n} \text{Var}(\tilde{Y}_i|\mathbf{y}) + \sum_{i=1}^{n}\left[E\left(\tilde{Y}_i|\mathbf{y}_i\right) - y_i\right]^2
\end{aligned}
$$

The first term is a penalty term for prediction and the second term is a goodness of fit term.

If N Monte Carlo samples are available for each $\tilde{Y}_i|\mathbf{y}$, then we form the ergodic average:

$$
\frac{1}{N}\sum_{j=1}^{N}\left[\sum_{i=1}^{n}\left(\tilde{Y}_i^{(j)} - y_i\right)^2\right]
$$

to estimate the expected loss function.

We choose the model for which the expected loss is minimum.

(b) Define the deviance statistic:

$$
D(\boldsymbol{\theta}) = 2\log f(\mathbf{y}|\boldsymbol{\theta}) + 2\log h(\mathbf{y})
$$

where $f(\mathbf{y}|\boldsymbol{\theta})$ is the likelihood and $h(\mathbf{y})$ is the standardizing function which depends on $\mathbf{y}$ only. The goodness-of fit of a model is given by

$$
\bar{D} = E_{\boldsymbol{\theta}|\mathbf{y}}\left[D(\boldsymbol{\theta})\right]
$$

Ther penalty is given by

$$
p_D = E_{\boldsymbol{\theta}|\mathbf{y}}\left[D(\boldsymbol{\theta})\right] - D\left[E\left(\boldsymbol{\theta}|\mathbf{y}\right)\right] = \bar{D} - D(\bar{\boldsymbol{\theta}}).
$$

The DIC is given by

$$
\bar{D} + p_D = 2\bar{D} + D(\bar{\boldsymbol{\theta}}).
$$

We choose the model for which the DIC is minimum.

Bibliography

Akaike, H. (1973). Information theory and an extension of the maximum likelihood principle. In B. N. Petrov and F. Csáki (Eds.), *2nd International Symposium on Information Theory*, pp. 267–281. Budapest: Akadémiai Kiadó.

Akima, R. (2013). *akima: Interpolation of Irregularly Space Data*. R package version 0.5-11.

Baddeley, A., E. Rubak, and R. Turner (2015). *Spatial Point Patterns: Methodology and Applications with R* (1st ed.). Boca Raton: Chapman & Hall/CRC.

Bakar, K. S. (2012). *Bayesian Analysis of Daily Maximum Ozone Levels*. Southampton, United Kingdom: PhD Thesis, University of Southampton.

Bakar, K. S. (2020). Interpolation of daily rainfall data using censored bayesian spatially varying model. *Computational Statistics 35*, 135–152.

Bakar, K. S., P. Kokic, and H. Jin (2016). Hierarchical spatially varying coefficient and temporal dynamic process models using sptdyn. *Journal of Statistical Computation and Simulation 86*, 820–840.

Bakar, K. S., P. Kokic, and W. Jin (2015). A spatio-dynamic model for assessing frost risk in south-eastern australia. *Journal of the Royal Statistical Society, Series C 64*, 755–778.

Bakar, K. S. and S. K. Sahu (2015). sptimer: Spatio-temporal bayesian modeling using r. *Journal of Statistical Software, Articles 63*(15), 1–32.

Banerjee, S., B. P. Carlin, and A. E. Gelfand (2015). *Hierarchical Modeling and Analysis for Spatial Data* (2nd ed.). Boca Raton: CRC Press.

Banerjee, S., A. E. Gelfand, A. O. Finley, and H. Sang (2008). Gaussian predictive process models for large spatial data sets. *Journal of Royal Statistical Society, Series B 70*, 825–848.

Bass, M. R. and S. K. Sahu (2017). A comparison of centering parameterisations of gaussian process based models for bayesian computation using mcmc. *Statistics and Computing 27*, 1491–1512.

Bauer, G., M. Deistler, and W. Scherrer (2001). Time series models for short term forecasting of ozone in the eastern part of austria. *Environmetrics 12*, 117–130.

Berger, J. O. (1985). *Statistical Decision Theory and Bayesian Analysis* (2nd ed.). New York: Springer-Verlag.

Berger, J. O. (1993). *Statistical Decision Theory and Bayesian Analysis* (2nd ed.). Springer Series in Statistics.

Berger, J. O. (2006). The case for objective bayesian analysis. *Bayesian Analysis 1*(3), 385–402.

Bernardo, J. M. and A. Smith (1994). *Bayesian Theory*. Chichester: John Wiley and Sons, Ltd.

Berrocal, V., A. Gelfand, and D. Holland (2010). A spatio-temporal down-scaler for output from numerical models. *Journal of Agricultural, Biological and Environmental Statistics 15*, 176–197.

Besag, J., J. York, and A. Mollié (1991). Bayesian image restoration with two applications in spatial statistics. *Annals of the Institute of Statistics and Mathematics 43*, 1–59.

Bivand, R. (2020). Creating neighbours.

Bland, M. (2000). *An Instruction to Medical Statistics* (3rd ed.). Oxford University Press: Oxford.

Blangiardo, M. and M. Cameletti (2015). *Spatial and Spatio-temporal Bayesian Models with R - INLA*. Chichester: John Wiley and Sons.

Bousquet, O., S. Boucheron, and G. Lugosi (2004). Introduction to statistical learning theory. In *Advanced lectures on machine learning*, pp. 169–207. Springer.

Breiman, L. (2001). Random forests. *Machine learning 45*(1), 5–32.

Breiman, L., J. Friedman, C. J. Stone, and R. A. Olshen (1984). *Classification and regression trees*. CRC Press.

Cameletti, M., R. Ignaccolo, and S. Bande (2011). Comparing spatio-temporal models for particulate matter in piemonte. *Environmetrics 22*, 985–996.

Chatfield, C. (2003). *The Analysis of Time Series: An Introduction*. Chapman & Hall.

Clayton, D. and M. Hills (1993). *Statistical Models in Epidemiology*. Oxford: Oxford University Press.

Cressie, N. A. C. and G. Johannesson (2008). Fixed rank kriging for very large spatial data sets. *Journal of the Royal Statistical Society, Series: B 70*, 209–226.

Cressie, N. A. C. and C. K. Wikle (2011). *Statistics for Spatio-Temporal Data*. New York: John Wiley & Sons.

Damon, J. and S. Guillas (2002). The inclusion of exogenous variables in functional autoregressive ozone forecasting. *Environmetrics 13*, 759–774.

Das, S. and D. Dey (2010). On bayesian inference for generalized multivariate gamma distribution. *Statistics and Probability Letters 80*, 1492–1499.

Diggle, P. (2014). *Statistical analysis of spatial point patterns* (3rd ed.). Boca Raton: Chapman & Hall/CRC.

Diggle, P. and P. J. Ribeiro (2007). *Model-based Geostatistics*. New York: Springer-Verlag.

Efron, B., T. Hastie, I. Johnstone, and R. Tibshirani (2004). Least angle regression. *The Annals of Statistics 32*(2), 407–451.

Feister, U. and K. Balzer (1991). Surface ozone and meteorological predictors on a subregional scale. *Atmospheric Environment 25*, 1781–1790.

Finley, A. O., S. Banerjee, and A. E. Gelfand (2015). spBayes for large univariate and multivariate point-referenced spatio-temporal data models. *Journal of Statistical Software 63*(13), 1–28.

Fuglstad, G.-A., D. Simpson, F. Lindgren, and H. Rue (2018). Constructing priors that penalize the complexity of gaussian random fields. *Journal of the American Statistical Association 114*, 445–452.

Geary, R. C. (1954). The contiguity ratio and statistical mapping. *The Incorporated Statistician 5*(3), 115–146.

Gelfand, A. E. and D. K. Dey (1994). Bayesian model choice: Asymptotics and exact calculations. *Journal of the Royal Statistical Society. Series B (Methodological) 56*(3), 501–514.

Gelfand, A. E., D. K. Dey, and H. Chang (1992). Model determination using predictive distributions with implementation via sampling-based methods. In J. M. Bernardo, J. O. Berger, A. P. Dawid, and A. F. M. Smith (Eds.), *Bayesian Statistics 4*, pp. 147–167. Oxford: Oxford University Press.

Gelfand, A. E. and S. K. Ghosh (1998). Model choice: A minimum posterior predictive loss approach. *Biometrika 85*, 1–11.

Gelfand, A. E. and S. K. Sahu (1999). Identifiability, improper priors, and gibbs sampling for generalized linear models. *Journal of the American Statistical Association 94*(445), 247–253.

Gelfand, A. E. and S. K. Sahu (2009). Monitoring data and computer model output in assessing environmental exposure. In A. O'Hagan and M. West (Eds.), *Handbook of Applied Bayesian Analysis*, pp. 482–510. Oxford University Press.

Gelfand, A. E., S. K. Sahu, and B. P. Carlin (1995a). Efficient parameterisations for normal linear mixed models. *Biometrika 82*(3), 479–488.

Gelfand, A. E., S. K. Sahu, and B. P. Carlin (1995b). Efficient parametrisations for normal linear mixed models. *Biometrika 82*(3), 479–488.

Gelfand, A. E., S. K. Sahu, and B. P. Carlin (1996). Efficient parametrizations for generalized linear mixed models, (with discussion). In J. Bernardo, J. Berger, and A. Dawid (Eds.), *Bayesian Statistics 5*, pp. 165–180. Oxford: Clarendon Press.

Gelman, A., J. Hwang, and A. Vehtari (2014). Understanding predictive information criteria for bayesian models. *Statistics and Computing 24*(6), 997–1016.

Gelman, A., G. Roberts, and W. Gilks (1996). Efficient Metropolis jumping rules. *Bayesian Statistics 5*, 599–608.

Gelman, A. and D. B. Rubin (1992). Inference from iterative simulation using multiple sequences. *Statistical Science 7*(4), 457–472.

Geweke, J. (1992). Evaluating the accuracy of sampling-based approaches to calculating posterior moments. In J. Bernardo, J. Berger, and A. Dawid (Eds.), *Bayesian Statistics 4*, pp. 169–193. Oxford: Clarendon Press.

Geyer, C. J. (1992). Practical markov chain monte carlo. *Statistical Science 7*, 473–483.

Ghoshal, S. and A. V. d. Vaart (2017). *Fundamentals of Nonparametric Bayesian Inference*. Cambridge University Press.

Gilks, W. and P. Wild (1992). Adaptive rejection sampling for gibbs sampling. *Applied Statistics 41*, 337–348.

Gilks, W. R., S. Richardson, and D. J. Spiegelhalter (1996). *Markov Chain Monte Carlo in Practice*. CRC Press.

Gneiting, T. (2002). Nonseparable, stationary covariance functions for space-time data. *Journal of the American Statistical Association 97*, 590–600.

Gneiting, T., F. Balabdaoui, and A. Raftery (2007). Probabilistic forecasts, calibration and sharpness. *Journal of the Royal Statistical Society, Series B 69*, 243–268.

Gómez-Rubio, V. (2020). *Bayesian inference with INLA*. Boca Raton: Chapman and Hall/CRC.

Gramacy, R. B. (2020). *Surrogates: Gaussian Process Modeling, Design and Optimization for the Applied Sciences.* Boca Raton, Florida: Chapman Hall/CRC. https://bobby.gramacy.com/surrogates/.

Green, M. B., J. L. Campbell, R. D. Yanai, S. W. Bailey, A. S. Bailey, N. Grant, I. Halm, E. P. Kelsey, and L. E. Rustad (2018). Downsizing a long-term precipitation network: Using a quantitative approach to inform difficult decisions. *Plos One.*

Greven, S., F. Dominici, and S. Zeger (2011). An Approach to the Estimation of Chronic Air Pollution Effects Using Spatio-Temporal Information. *Journal of the American Statistical Association 106*, 396–406.

Hammond, M. L., C. Beaulieu, S. K. Sahu, and S. A. Henson (2017). Assessing trends and uncertainties in satellite-era ocean chlorophyll using space-time modeling. *Global Biogeochemical Cycles 31*, 1103–1117.

Harrison, D. and D. L. Rubinfeld (1978). Hedonic prices and the demand for clean air. *Journal of Environmental Economics & Management 5*, 81–102.

Hastie, T., R. Tibshirani, and J. Friedman (2009). *The Elements of Statistical Learning* (2nd ed.). New York: Springer.

Hastings, W. K. (1970). Monte Carlo sampling methods using Markov chains and their applications. *Biometrika 57*(1), 97–109.

Higdon, D. (1998, JUN). A process-convolution approach to modelling temperatures in the North Atlantic Ocean. *Environmental and Ecological Statistics 5*(2), 173–190.

Hinde, A. (1998). *Demographic Methods.* Arnold: London.

Hoerl, A. E. and R. W. Kennard (1970). Ridge regression: Biased estimation for nonorthogonal problems. *Technometrics 12*(1), 55–67.

Hoffman, M. D. and A. Gelman (2014). The no-u-turn sampler: Adaptively setting path lengths in hamiltonian monte carlo. *Journal of Machine Learning Research 15*, 1593–1623.

Huerta, G., B. Sanso, and J. R. Stroud (2004). A spatio-temporal model for maxico city ozone levels. *Journal of the Royal Statistical Society, Series C 53*, 231–248.

Jeffreys, H. (1961). *Theory of Probability* (3rd ed.). Oxford: Clarendon Press.

Jona Lasinio, G., S. K. Sahu, and K. V. Mardia (2007). Modeling rainfall data using a bayesian kriged-kalman model. In U. S. S. K. Upadhyay and D. K. Dey (Eds.), *Bayesian Statistics and its Applocations.* London: Anshan Ltd.

Kanevski, M. and M. Maignan (2004). *Analysis and Modelling of Spatial Environmental Data.* Boca Raton: Chapman & Hall/CRC.

Kang, D., R. Mathur, S. T. Rao, and S. Yu (2008). Bias adjustment techniques for improving ozone air quality forecasts. *Journal of Geophysical Research 113*, 10.1029/2008JD010151.

Kaufman, C. G., M. J. Schervish, and D. W. Nychka (2008). Covariance tapering for likelihood-based estimation in large spatial data sets. *Journal of the American Statistical Association 103*(484), 1545–1555.

Keiding, N. (1987). The method of expected number of deaths, 1786-1886-1986. *International Statistical Review 55*, 1–20.

Kennedy, M. C. and A. O'Hagan (2001). Bayesian calibration of computer models. *Journal of the Royal Statistical Society, Series B 63*, 425–464.

Knorr-Held, L. (2000). Bayesian modelling of inseparable space-time variation in disease risk. *Statistics in Medicine 19*(17–18), 2555–2567.

Kumar, U. and K. D. Ridder (2010). Garch modelling in association with fft-arima to forecast ozone episodes. *Atmospheric Environment 44*, 4252–4265.

Lambert, B. (2018). *A student's Guide to Bayesian Statistics*. Los Angeles: Sage.

Last, J. M. (2000). *A Dictionary of Epidemiology* (3rd ed.). Oxford: Oxford University Press.

Lee, D. (2021). Carbayes version 5.2.3: An r package for spatial areal unit modelling with conditional autoregressive priors. Technical report, University of Glasgow.

Lee, D., C. Ferguson, and R. Mitchell (2009). Air pollution and health in Scotland: a multicity study. *Biostatistics 10*, 409–423.

Lee, D. and A. B. Lawson (2016). Quantifying the spatial inequality and temporal trends in maternal smoking rates in glasgow. *Annals of Applied Statistics 10*, 1427–1446.

Lee, D., S. Mukhopadhyay, A. Rushworth, and S. K. Sahu (2017). A rigorous statistical framework for spatio-temporal pollution prediction and estimation of its long-term impact on health. *Biostatistics 18*(2), 370–385.

Lee, D., A. Rushworth, and G. Napier (2018). Spatio-temporal areal unit modeling in r with conditional autoregressive priors using the carbayesst package. *Journal of Statistical Software 84*(9), 10.18637/jss.v084.i09.

Lee, D., A. Rushworth, and S. K. Sahu (2014). A bayesian localised conditional auto-regressive model for estimating the health effects of air pollution. *Biometrics 70*, 419–429.

Leroux, B. G., X. Lei, and N. Breslow (2000). Estimation of disease rates in small areas: A new mixed model for spatial dependence. In M. E. Halloran and D. Berry (Eds.), *Statistical Models in Epidemiology, the Environment, and Clinical Trials*, pp. 179–191. New York: Springer-Verlag.

Li, Q. and N. Lin (2010). The bayesian elastic net. *Bayesian Analysis 5*(1), 151–170.

Lichman, M. (2016). UCI machine learning repository. https://archive.ics.uci.edu/ml/machine-learning-databases/housing/.

Lilienfeld, A. M. and D. E. Lilienfeld (1980). *Foundations of Epidemiology* (2nd ed.). Oxford: Oxford University Press.

Longhurst, A. (1995). Seasonal cycles of pelagic production and consumption. *Prog. Oceanogr. 36*(2), 77–167.

Longhurst, A. (1998). *Ecological Geography of the Sea*. San Diego, California: Academic Press.

Lunn, D., C. Jackson, N. Best, A. Thomas, and D. Spiegelhalter (2013). *The BUGS Book: A Practical Introduction to Bayesian Analysis*. Boca Raton: Chapman & Hall.

Lunn, D. J., A. Thomas, N. Best, and D. Spiegelhalter (2013). *The BUGS Book: A Practical Introduction to Bayesian Analysis*. Chapman & Hall.

MacEachern, S. N. and L. M. Berliner (1994). Subsampling the Gibbs sampler. *The American Statistician 48*(3), 188–190.

Mardia, K. V. and C. Goodall (1993). Spatial-temporal analysis of multivariate environmental monitoring data. In G. P. Patil and C. R. Rao (Eds.), *Multivariate Environmental Statistics*, pp. 347–386. Amsterdam: Elsevier.

Mardia, K. V., C. Goodall, E. J. Redfern, and F. Alonso (1998). The kriged kalman filter (with discussion). *Test 7*, 217–252.

Matérn, B. (1986). *Spatial Variation* (2nd ed.). Berlin: Springer-Verlag.

McCullagh, P. and J. A. Nelder (1989). *Generalized Linear Models* (2nd ed.). Boca Raton: Chapman and Hall.

McMillan, N., S. M. Bortnick, M. E. Irwin, and M. Berliner (2005). A hierarchical bayesian model to estimate and forecast ozone through space and time. *Atmospheric Environment 39*, 1373–1382.

Metropolis, N., A. W. Rosenbluth, M. N. Rosenbluth, A. H. Teller, and E. Teller (1953). Equation of state calculations by fast computing machines. *The Journal of Chemical Physics 21*, 1087.

Miettinen, O. S. (1985). *Theoretical Epidemiology. Principles of Occurrence Research in Medicine.* New York: Wiley.

Møller, J. and R. P. Waagepetersen (2003). *Statistical Inference and Simulation for Spatial Point Processes.* Taylor & Francis.

Moore, G. (1975). Progress in digital integrated electronics. In *Technical Digest 1975. International Electron Devices Meeting*, pp. 11–13. IEEE.

Moran, P. A. P. (1950). Notes on continuous stochastic phenomena. *Biometrika 37*(1/2), 17–23.

Mukhopadhyay, S. and S. K. Sahu (2018). A bayesian spatio-temporal model to estimate long term exposure to outdoor air pollution at coarser administrative geographies in England and Wales. *Journal of the Royal Statistical Society, Series A 181*, 465–486.

Neal, R. M. (2011). Mcmc using hamiltonian dynamics. In S. Brooks, A. Gelman, G. L. Jones, and X.-L. Meng (Eds.), *Handbook of Markov Chain Monte Carlo*, pp. 113–162. Boca Raton: Chapman and Hall/CRC.

Park, T. and G. Casella (2008). The bayesian lasso. *Journal of the American Statistical Association 103*(482), 681–686.

Pebesma, E., B. Bivand, B. Rowlingson, and V. G. Rubio (2012). sp*: Classes and Methods for Spatial Data.* R package version 1.0-5.

Petris, G., S. Petrone, and C. Patrizia (2010). *Dynamic Linear Models with R.* Dordrecht: Springer.

Plummer, M., N. Best, C. K., and K. Vines (2006, March). Coda: Convergence diagnosis and output analysis for MCMC. *R News 6*(1), 7–11.

Raftery, A. E. (1996). Hypothesis testing and model selection. In W. R. Gilks, S. Richardson, and D. J. Spiegelhalter (Eds.), *Markov Chain Monte Carlo in Practice*, pp. 163–187. CRC Press.

Raftery, A. E. and S. M. Lewis (1992). One long run with diagnostics: Implementation strategies for markov chain monte carlo. *Statistical Science 7*, 493–497.

Rasmussen, C. E. and C. Williams (2006). *Gaussian Processes for Machine Learning.* MIT Press.

Rimm, A. A., A. J. Hartz, J. H. Kalbfleisch, A. J. Anderson, and R. G. Hoffmann (1980). *Basic Biostatistics in Medicine and Epidemiology.* New York: Appleton-Century-Crofts.

Ripley, B. D. (1987). *Stochastic Simulation.* New York: Wiley.

Robert, C. O. and G. Casella (2004). *Monte Carlo Statistical Methods* (2nd. ed.). New York: Springer-Verlag.

Roberts, G. O. (1996). Markov chain concepts related to sampling algorithms. In W. R. Gilks, S. Richardson, and D. J. Spiegelhalter (Eds.), *Markov Chain Monte Carlo in Practice*, pp. 45–57. CRC Press.

Roberts, G. O., A. Gelman, and W. R. Gilks (1997). Weak convergence and optimal scaling of random walk Metropolis algorithms. *The Annals of Applied Probability 7*(1), 110–120.

Roberts, G. O. and S. K. Sahu (1997). Updating schemes, correlation structure, blocking and parameterization for the Gibbs sampler. *Journal of the Royal Statistical Society: Series B (Statistical Methodology) 59*(2), 291–317.

Rue, H., S. Martino, and N. Chopin (2009). Approximate Bayesian inference for latent Gaussian models by using integrated nested laplace approximations. *Journal of the royal statistical society: Series B (statistical methodology) 71*(2), 319–392.

Rushworth, A., D. Lee, and C. Sarran (2017). An adaptive spatio-temporal smoothing model for estimating trends and step changes in disease risk. *Journal of the Royal Statistical Society, Series C 66*, 141–157.

Sahu, S. (2012). Hierarchical bayesian models for space-time air pollution data. In S. S. R. T Subba Rao and C. R. Rao (Eds.), *Handbook of Statistics-Vol 30. Time Series Analysis: Methods and Applications*, pp. 477–495. Elsevier.

Sahu, S. K. (2015). Bayesian spatio-temporal modelling to deliver more accurate and instantaneous air pollution forecasts. In P. Aston, T. Mulholland, and K. Tant. (Eds.), *UK Success Stories in Industrial Mathematics*, pp. 67–74. Springer International.

Sahu, S. K. (2021). *bmstdr: Bayesian Modeling of Spatio-Temporal Data with R.* https://www.soton.ac.uk/~sks/bmbook/bmstdr-vignette.html.

Sahu, S. K. and K. S. Bakar (2012a). A comparison of bayesian models for daily ozone concentration levels. *Statistical Methodology 9*(1), 144–157.

Sahu, S. K. and K. S. Bakar (2012b). Hierarchical bayesian auto-regressive models for large space time data with applications to ozone concentration modelling. *Applied Stochastic Models in Business and Industry 28*, 395–415.

Sahu, S. K., K. S. Bakar, and N. Awang (2015). Bayesian forecasting using hierarchical spatio-temporal models with applications to ozone levels in the eastern united states. In I. L. Dryden and J. Kent (Eds.), *Geometry Driven Statistics*, pp. 260–281. Chichester: John Wiley and Sons.

Sahu, S. K., K. S. Bakar, J. Zhan, J. L. Campbell, and R. D. Yanai (2020). Spatio-temporal bayesian modeling of precipitation using rain gauge data from the hubbard brook experimental forest, new hampshire, usa. In *Proceedings of the Joint Statistical Meetings*, pp. 77–92. American Statistical Association.

Sahu, S. K. and D. Böhning (2021). Bayesian spatio-temporal joint disease mapping of Covid-19 cases and deaths in local authorities of england. *Spatial Statistics*.

Sahu, S. K. and P. Challenor (2008). A space-time model for joint modeling of ocean temperature and salinity levels as measured by argo floats. *Environmetrics 19*, 509–528.

Sahu, S. K., D. K. Dey, and M. Branco (2003). A new class of multivariate skew distributions with applications to bayesian regression models. *Canadian Journal of Statistics 31*, 129–150.

Sahu, S. K., A. E. Gelfand, and D. M. Holland (2006). Spatio-temporal modeling of fine particulate matter. *Journal of Agricultural, Biological, and Environmental Statistics 11*, 61–86.

Sahu, S. K., A. E. Gelfand, and D. M. Holland (2007). High-resolution space-time ozone modeling for assessing trends. *Journal of the American Statistical Association 102*, 1221–1234.

Sahu, S. K. and K. V. Mardia (2005a). A bayesian kriged-kalman model for short-term forecasting of air pollution levels. *Journal of the Royal Statistical Society, Series C 54*, 223–244.

Sahu, S. K. and K. V. Mardia (2005b, September). Recent trends in modeling spatio-temporal data. In *Proceedings of the Special meeting on Statistics and Environment*, pp. 69–83. Università Di Messina.

Sahu, S. K., S. Yip, and D. M. Holland (2009). Improved space-time forecasting of next day ozone concentrations in the eastern u.s. *Atmospheric Environment 43*, 494–501.

Sahu, S. K., S. Yip, and D. M. Holland (2011). A fast bayesian method for updating and forecasting hourly ozone levels. *Environmental and Ecological Statistics 18*, 185–207.

Sambasivan, R., S. Das, and S. K. Sahu (2020). A bayesian perspective of statistical machine learning for big data. *Computational Statistics*, https://doi.org/10.1007/s00180–020–00970–8.

Sansó, B. and L. Guenni (1999). Venezuelan rainfall data analysed by using a bayesian space-time model. *Journal of the Royal Statistical Society, Series C, 48*, 345–362.

Sansó, B. and L. Guenni (2000). A nonstationary multisite model for rainfall. *Journal of the American Statistical Association 95*, 1089–1100.

Savage, N. H., P. Agnew, L. S. Davis, C. Ordóñez, R. Thorpe, C. E. Johnson, F. M. O'Connor, and M. Dalvi (2013). Air quality modelling using the met office unified model (aqum os24-26): model description and initial evaluation. *Geoscientific Model Development 6*(2), 353–372.

Schmidt, A. and A. O'Hagan (2003). Bayesian inference for non-stationary spatial covariance structure via spatial deformations. *Journal of the Royal Statistical Society, Series B 65*(3), 743–758.

Schwartz, G. E. (1978). Estimating the dimension of a model. *Annals of Statistics 6*, 461–464.

Shaddick, G. and J. V. Zidek (2015). *Spatio-Temporal Methods in Environmental Epidemiology* (1st ed.). Boca Raton: Chapman & Hall/CRC.

Sharma, R. and S. Das (2017). Regularization and variable selection with copula prior. *In Corespondence abs/1709.05514.*

Simpson, D., H. Rue, A. Riebler, T. G. Martins, and S. H. Sørbye (2017). Penalising model component complexity: A principled, practical approach to constructing priors. *Statistical Science 32*(1), 1–28.

Smith, A. F. and G. O. Roberts (1993). Bayesian computation via the Gibbs sampler and related Markov chain Monte Carlo methods. *Journal of the Royal Statistical Society. Series B (Methodological) 55*, 3–23.

Sousa, S. I. V., J. C. M. Pires, F. Martins, M. C. Pereira, and M. C. M. Alvim-Ferraz (2009). Potentialities of quantile regression to predict ozone concentrations. *Environmetrics 20*, 147–158.

Spiegelhalter, S. D., N. G. Best, B. P. Carlin, and A. V. D. Linde (2002). Bayesian measures of model complexity and fit. *Journal of the Royal Statistical Society B 64*(4), 583–639.

Stan Development Team (2015). *Stan modeling language: Users guide and reference manual.* Columbia, New York: Columbia University.

Stan Development Team (2020). RStan: the R interface to Stan. R package version 2.21.2.

Stein, M. L. (1999). *Statistical Interpolation of Spatial Data: Some Theory for Kriging.* New York: Springer-Verlag.

Stein, M. L. (2008). A modelling approach for large spatial datasets. *Journal of the Korian Statistical Society 37*, 3–10.

Stroud, J. R., P. Muller, and B. Sanso (2001). Dynamic models for spatio-temporal data. *Journal of the Royal Statistical Society, Series B 63*, 673–689.

Therneau, T., B. Atkinson, and B. Ripley (2017). *rpart: Recursive Partitioning and Regression Trees*. R package version 4.1-11.

Tibshirani, R. (1996). Regression shrinkage and selection via the lasso. *Journal of the Royal Statistical Society, Series B 58*, 267–288.

Tierney, L. (1996). Introduction to general state-space markov chain theory. In W. R. Gilks, S. Richardson, and D. J. Spiegelhalter (Eds.), *Markov Chain Monte Carlo in Practice*, pp. 59–74. CRC Press.

Tobler, W. (1970). A computer movie simulating urban growth in the detroit region. *Economic Geography 46*, 234–240.

Utazi, C. E., K. Nilsen, O. Pannell, W. Dotse-Gborgbortsi, and A. J. Tatem (2021). District-level estimation of vaccination coverage: Discrete vs continuous spatial models. *Statistics in Medicine*, doi: 10.1002/sim.8897.

Vapnik, V. (1998). *Statistical learning theory*. New York: Wiley.

Wasserman, L. (2004). *All of statistics: a concise course in statistical inference*. Springer Texts in Statistics.

Watanabe, S. (2010). Asymptotic equivalence of bayes cross validation and widely applicable information criterion in singular learning theory. *Journal of Machine Learning Research 11*, 3571–3594.

West, M. and J. Harrison (1997). *Bayesian Forecasting and Dynamic Models* (2nd ed.). New York: Springer.

Wikle, C. K., A. Zammit-Mangion, and N. Cressie (2019). *Spatio-Temporal Statistics with R*. Boca Raton: Chapman & Hall/CRC.

Zhang, H. (2004). Inconsistent estimation and asymptotically equal interpolations in model-based geostatistics. *Journal of the American Statistical Association 99*, 250–261.

Zidek, J. V., N. D. Le, and Z. Liu (2012). Combining data and simulated data for space-time fields: application to ozone. *Environmental and Ecological Statistics 19*, 37–56.

Zou, H. and T. Hastie (2005). Regularization and variable selection via the elastic net. *Journal of the Royal Statistical Society, Series B 67*, 301–320.

Glossary

geom_abline Adds a line with given slope and intercept. 9

geom_bar Draws a bar plot. 64

geom_circle Adds circles to the plot. 248

geom_contour Adds contours to the plot. 275

geom_dl Adds labels to the plot. 18

geom_histogram Draws a histogram. 50

geom_hline Adds a horizontal line. 229

geom_line Adds a line. 64

geom_path Adds paths to the plot. 13

geom_point Adds the points. 275

geom_polygon Draws the polygons. 11, 275

geom_raster Draws a raster map. 275

geom_rect Adds a rectangular box to the plot. 188

geom_ribbon Draws a ribbon. 211

geom_smooth Adds a smooth line. 56

geom_text Adds text to the plot. 11, 19

geom_vline Adds a vertical line. 253

Index